CONTEMPORARY READINGS IN BIOLOGY

Edited by

Gideon E. Nelson and James D. Ray, Jr.
University of South Florida

John Wiley & Sons, Inc., New York ● London ● Sydney ● Toronto

Library of Congress Cataloging in Publication Data

Nelson, Gideon E   comp.
  Contemporary readings in biology.

    1.   Biology–Addresses, essays, lectures.
I. Ray, James Davis, 1918–   joint comp. II. Title
[DNLM: 1. Biology–Collected works. QH302 n426c
1974]
QH311.N37   574'.08   73-14853
ISBN 0–471–63156–6

Printed in the United States of America

10 9 8 7 6 5 4 3 2 1

PREFACE

Once again we offer a selection of readings for use in introductory biology courses. As previously, we have selected these articles on the basis of readability, relevance to today's problems, presentation of new advances in biology, and unusually clear explanations of complex biological concepts.

Selections are from widely circulated journals such as *Science, American Scientist, Bioscience, Bulletin of the Atomic Scientists, Science Teacher, Journal of the American Medical Association, Advancement of Science, Endeavour,* and *Science Journal* as well as *Human Development, Engineering and Science, Discovery, Natural History,* and *Atlantic.* Their arrangement follows subject matter organization of *Fundamental Concepts of Biology, 3rd Edition* by Nelson, Robinson, and Boolootian.

This selection of readings, when used as an integral part of a course, should extend not only the text but also the experiences of the instructor. The issues and problems examined go beyond the classroom into the everyday affairs of the student. We hope this book will function significantly in the development of an informed and active community.

December, 1973

Gideon E. Nelson
James D. Ray, Jr.

CONTENTS

PART   XI   *EPILOGUE*

PART I

INTRODUCTION

The fairy tern, a tropical species, is shown here hovering as it examines the environment below.

1. *The Significance of Science*

VICTOR F. WEISSKOPF

"All phenomena and all human experiences are supposed to fit into the context of natural laws and have found or will probably find a scientific description or explanation. . . . Scientific interpretation of human experiences does not always shed light on . . . feelings and value judgments . . . decisive in the realm of human decision making . . . but scientific reasoning can and should provide information about predictable consequences. The actual decision however, remains outside of science, it represents a kind of reasoning which necessarily is complementary to scientific thought." — Dr. Victor F. Weisskopf, a native of Vienna, received his doctorate from Göttingen. He is Institute Professor (physics) at Massachusetts Institute of Technology and a member of the National Academy of Sciences. His interests include quantum mechanics, electron theory, and nuclear physics.

"And I gave my heart to seek and search out by wisdom concerning all things that are done under heaven. This sore task hath God given to the sons of man to be exercised herewith."

Ecclesiastes 1:13

"For in much wisdom is much grief and he that increaseth knowledge increases sorrow."

Ecclesiastes 1:18

The development of science and technology during the last centuries has been very fast and overwhelming. All aspects of human society have been deeply influenced by it, the quality of life has been changed and often gravely disturbed. Today we have become very sensitive to the problems raised by this fast development, and we are faced with important questions regarding the role of science in society.

Science is under severe attack from some quarters; it is considered a panacea for the cure of all ills by others. I will sketch here three positions in regard to science that characterize some of the common attitudes toward this problem.

Position I. Many branches of science have grown excessively during the recent decades; too large amounts of public support and too much scientific manpower are devoted to esoteric research in fields that have little to do with practical problems. Only

Source: Copyright 1972 American Association for the Advancement of Science. Reprinted by permission of the publisher and author from *Science, 76:* 138-145, 14 April 1972. This article was reprinted in *Physics in the Twentieth Century, Selected Essays* by V. F. Weisskopf (The MIT Press).

such scientific research should be supported as that promising reasonable payoff in terms of practical applications for industry, public welfare, medicine, or national defense. Science as a study of nature for its own sake is appreciated by only a few people and has very limited public value. Its support should be reduced to a much more modest scale.

Position 2. Most of today's scientific research is detrimental to society because the source of industrial innovations, most of which have led and will lead to further deterioration of our environment, to an inhuman computerized way of life destroying the social fabric of our society, to more dangerous and destructive applications in weaponry leading to wars of annihilation, and to further development of our society toward Orwell's world of 1984. At best, science is a waste of resources that should be devoted to some immediate, socially useful purpose.

Position 3. The methods and approaches used in the natural sciences and in technology—the so-called scientific method—has proved overwhelmingly successful in resolving problems, in elucidating situations, in explaining phenomena of the natural world, and in attaining well-defined aims. It should be extended to all problems confronting humanity because it promises to be as successful in any area of human endeavor and human interest as it has been in the realm of natural science and technology.

These three positions are, to a large extent, mutually exclusive. They point in three almost orthogonal directions. In this essay I contend that each of these positions takes a narrow and one-sided view of the role of science in human society. Society is involved in man's thought and action in many different and often contradictory ways. Science must coexist with other forms of human urges, feelings, and self-realizations. Science is based on a very fundamental human urge: man's innate desire to know and understand the universe in which he lives and to gain insight into the driving forces that govern the world around us. This urge is paired with another one: the desire to improve the precarious conditions of human existence in a hostile world, in a hostile natural environment, and in hostile societies. Man desires to influence and to change the material and social conditions of life with the help of acquired knowledge and experience, which, in modern times, are mainly derived from science. As in all human situations, the urges and desires do not always lead to actions that serve the intended purposes, and the intended purposes are not always such that real benefits accrue for the people involved. These are the basic elements for our discussion of the role of science in human affairs.

Basic Science and Practical Applications

Let us return to position 1, the excessive cost of basic science. It is based on the supposition that most of research is unimportant and irrelevant if it is carried out without regard to practical applications. It is commonplace that technology and medicine owe an enormous debt to the study of nature for its own sake, that is to basic science. It is hardly necessary to mention here the many instances which prove that modern industry and modern care for the sick are based on past results of basic science. Nor is basic science such an expensive luxury when its cost is compared with its services. The total cost of all basic science from Archimedes to the present day is probably near $30 billion (1), less than 12 days' worth of production of the United States whose gadgets and machines are to a large extent the product of earlier scientific achievement. The practical value of those parts of pure science which seemingly have no immediate connections with applications has been clearly brought out by H. B. G. Casimir, who collected a number of interesting examples of how decisive technical progress was made by scientists who did not work at all for a well-defined practical aim (2):

I have heard statements that the role of academic research in innovation is slight. It is about the most blatant piece of nonsense it has been my fortune to stumble upon.

Certainly, one might speculate idly whether transistors might have been discovered by people who had not been trained in and had not contributed to wave mechanics or the theory of electrons in solids. It so happened that inventors of transistors were versed in and contributed to the quantum theory of solids.

One might ask whether basic circuits in computers might have been found by people who wanted to build computers. As it happens, they were discovered in the thirties by physicists dealing with the counting of nuclear particles because they were interested in nuclear physics.

One might ask whether there would be nuclear power because people wanted new power sources or whether the urge to have new power would have led to the discovery of the nucleus. Perhaps—only it didn't happen that way, and there were the Curies and Rutherford and Fermi and a few others.

One might ask whether an electronic industry could exist without the previous discovery of electrons by people like Thomson and H. A. Lorentz. Again, it didn't happen that way.

One might ask whether induction coils in motor cars might have been made by enterprises which wanted to make motor transport and whether then they would have stumbled on the laws of induction. But the laws of induction had been found by Faraday many decades before that.

Or whether, in an urge to provide better communication, one might have found electromagnetic waves. They weren't found that way. They were found by Hertz who emphasized the beauty of physics and who based his work on the theoretical considerations of Maxwell. I think there is hardly any example of twentieth century innovation which is not indebted in this way to the basic scientific thought.

Some of these examples are evidences of the fact that experimentation and observation at the frontier of science require technical means beyond the capabilities of ordinary technology. Therefore, the scientist in his search for new insights is forced and often succeeds to extend the technological frontier. This is why a large number of technologically important inventions had their origin not in the desire to fulfill a certain practical aim but in the attempts to sharpen the tools for the penetration of the unknown.

The examples quoted are taken from past developments and it is frequently asserted that some branches of modern fundamental science are so far removed from the human environment that practical applications are most improbable. In particular, the physics of elementary particles and astronomy are considered to be in this category. These sciences deal with far-off objects; elementary particles in the modern sense are also "far off," because mesons and baryons appear only when matter is subject to extremely high energy which is commonly not available on Earth but probably occurs only at a few distant spots in the universe. The "far-off" feature of these sciences is also what makes them expensive. It costs much money to create in our laboratory conditions that may be realized only in some exploding galaxy. It costs much to build instruments for the study of the limits of the universe. The argument against these sciences is that they are dealing with subjects far removed from our human environment and that therefore they are of minor relevance.

Let us consider the question of what constitutes the human environment. Ten thousand years ago there were no metals in the human environment. Metals rarely are found in pure form in nature. But after man had found out how to create them from ores, metals played an important role in our environment. The first piece of copper must have looked very esoteric and useless. In fact, man used it for a long time only for decorative purposes. Later, the introduction of this new material into man's ken gave rise to interesting possibilities that ultimately led to the dominant role of metals in his environment. In short, we have created a metallic environment. To choose another example, electricity appears rarely in nature in observable forms; for example, only in lightning discharges and in frictional electrostatics, which is not an important part of the human environment. After long years of research into minute effects, it was possible to recognize the nature of electric phenomena and then to find out what dominant role they play in the atom. The introduction of these new phenomena into the human world created a completely new electric environment in which we live today with 120-volt outlets in every wall. The most recent example is in nuclear physics. In the early days, prying into the problems of nuclear structure was considered a purely academic pursuit, directed only toward the advancement of knowledge concerning the innermost structure of matter. Rutherford said in 1933, "Anyone who expects a source of power from transformation of these atoms is talking moonshine." His conclusion was based on the same reasoning: The nuclear phenomena are too far removed from our human environment. True enough, apart from the rare cases of natural radioactivity, nuclear reactions must be artificially created at high cost with energetic particle beams. Most nuclear phenomena on Earth are man-made; they occur naturally only in the center of stars. Here again, the introduction of these man-made phenomena into our human world has led to a large number of interactions. Artificial radioactivity has revolutionized many branches of medicine, biology, chemistry, and metallurgy; the process of fission is an ever-increasing source of energy, for the better or the worse. Nuclear phenomena are now an important part of new human environment.

These examples show the weakness of the argument that certain natural phenomena are too far removed to be relevant to the human environment. Natural laws are universal; in principle, any natural process can be generated on Earth under suitable conditions. Modern instruments did create a cosmic environment in our laboratories when they produced processes that do not ordinarily take place in a terrestrial environment. Astronomy and particle physics deal with previously unknown and mostly unexplained phenomena. There is every possibility that some of them one day could also be reproduced on Earth in some form or another and be applied in a reasonable way for some useful purpose. Today already some special medical effects have been found for pion beams, effects that cannot be brought about by any other means. Purcell (3) once said about the applicability of frontier fields such as particle physics: "In our ignorance, it would be presumptuous to dismiss the possibility of useful application as it would be irresponsible to guarantee it."

One cannot divide the different branches of science into those that are important for practical applications and others that are not. The primary aim of science is not application, it is gaining insights into the causes and laws which govern natural processes. But a better understanding of a natural process almost always leads to possibilities of influencing it, or of influencing other processes related to the one that was investigated. The further science developed, the more relations between seemingly unrelated processes were discovered. The study of the solar corona, a phenomenon far off the earth, may lead to a better understanding of the behavior of highly ionized gases in magnetic fields, a topic of great technological importance. These relations between pure and applied science are part of the many-sided involvement of science in all aspects of human endeavor, from the urge to know more about the environment to the desire to improve and to dominate it.

Basic Science and Today's Problems

Position 2 is the expression of an attitude that makes science bear the brunt of public reaction against the mounting difficulties of modern life. This is not the place to analyze the predicaments of modern civilization whose difficulties are related to the increased rate of technological expansion, a rate that today has seemingly reached a critical value in both time and space. With regard to time, the changes in our way of life are now so rapid that marked differences are observable within one generation. This is a new and unsettling phenomenon for mankind; the experiences of the older generation are no longer as useful as they once were in coping with the problems of today. With regard to space, the effects of technology on our environment are no longer small; the parts of the earth's surface, of the water, of the air, which are changed by man or could be destroyed by man are no longer negligible compared to those left untouched. These are unexpected and disturbing consequences with which we do not yet know how to deal.

Since technology, particularly the increasing rate of technological change, is based largely upon science, it is not surprising that science is blamed for its difficulties. An obvious reaction to this situation would be to declare a moratorium on science; this would supposedly stop technological innovation and give us time to settle the problems that are already with us, instead of creating new ones. Recent cuts in scientific support reflect this attitude to some extent. We intend to refute, not the facts on which position 2 is based, but the conclusions drawn from that position.

The call for a moratorium in science is based on its inexorable way of progressing; one discovery leads to many others, and it seems impossible to prevent the application of new discoveries to unintentionally destructive purposes and socially detrimental technologies. Must we conclude, therefore, that it is harmful to continue the search for further knowledge and understanding of the world in which we live? It would seem that this search should be valuable under any circumstances, since knowing less about the world should hardly be better than knowing more.

Ignorance is of no value in itself; cruelty of man against his fellow man or thoughtless exploitation of man and nature existed before the industrial revolution. To stop the growth of scientific knowledge would not prevent its abuses, but it would deprive us of finding new means for avoiding them and deprive us also of an important source of philosophical insight. New scientific knowledge is neither good nor bad. New knowledge usually leads to a better way of predicting consequences and sometimes also to an ability to do something that one could not do before. It will be applied for good or for bad purposes, depending on the decision-making structure of the society, just as in the case of any social and political measure. In this respect science and technology are not different from other human activities.

Today it is fashionable to emphasize the negative aspects of technological progress and to take the positive aspects for granted. One should remember, however, that medical science has doubled the average life span of man, has eliminated many diseases, and has abolished pain in many forms. It has provided the means of effective birth control. The so-called "green revolution" created the potential to eliminate starvation among all presently living people. This is a scientific-technical achievement of momentous significance, even though the actual situation is a far cry from what could be achieved. One should also remember the developments in transportation, construction, and power supply provided by modern technology and their great potentialities for improving the quality of life.

The trouble comes from the fact that, in too many instances, technology has not achieved that purpose. On the contrary, it has contributed to a definite deterioration of life. Medicine may have abolished pain, but modern weapons are producing wholesale pain and suffering. Medical progress has achieved a great measure of death-control which has caused a population explosion; the available means of birth control are far from being effectively used. The blessings of modern medicine are unevenly distributed; lack of adequate medical care for the poor in some important countries causes mounting social tensions. The green revolution produces ten times more food than before, but the distribution is so uneven that starvation still prevails in many parts of the globe; furthermore, the massive use of fertilizers causes eutrophication of many waters. Power production and the internal combustion engine as a means of transportation have polluted the atmosphere. It is really impossible to avoid harmful effects when we apply our knowledge of natural processes for practical purposes? It should not be so.

There are two distinct sides to these problems; the social and political aspect and the technical aspect. In some instances the technical aspects do not pose any serious problems. The most important example is the use of technology for war or suppression. The only way to prevent the application of scientific results to the development of weapons is to reduce and prevent armed conflicts; certainly, this is a sociopolitical problem in which scientists and nonscientists should be equally interested, but it is not per se a problem of natural science. Other more benign examples are the problems of congested transportation, of city construction, and of some, but not all, of the problems of pollution. In these cases we know what causes the trouble and we know what measures can be taken to avoid it. But we don't know how to convince people to accept these measures. The problems are political and social. The natural scientists cannot help except by pointing out as clearly as possible what the consequences of certain actions or inactions will be. It is beyond the scope of this article to discuss whether it is possible to resolve these political and social problems. We take the only possible attitude in this dilemma: We assume that there will be a solution at some time, in some form, to some of these problems.

However, there are also many detrimental effects of technology of which the physical causes or the remedies are not known to a sufficient degree. Many detrimental effects of industrialization upon the environment belong in this category; among these are carbon dioxide production, long-range influences on atmospheric currents and on climatic conditions, the influence of urbanization on health, the problem of better means of birth control, and many more. Here science has enormous tasks to do in discovering, observing, and explaining unexplored phenomena, relations, and effects. The problems deal with our natural environment and therefore necessarily pose prime questions pertaining to natural science.

What role does basic science play in these efforts? One could conclude that the tasks are for applied science only and that research for its own sake, research that is not directed toward one of the specific problems, is not necessary. It may even be harmful since it takes away talented manpower and resources. This is not so. The spirit of basic research is composed of the following elements: an interest in understanding nature; an urge to observe, to classify, and to follow up observed phenomena for the sake of the phenomena themselves; a drive to probe deeper into a subject by experimenting with nature, by using ingenuity to study phenomena under special and unusual conditions—all in order to find connections and dependencies, causes and effects, laws and principles.

This attitude of basic research is necessary for the solution of today's pressing problems because it leads people to search for causes and effects in a systematic way, regardless of any ulterior aim. Many of today's troubles are caused by unforeseen consequences of human action on the environment, by interference with the natural cycle of events. The effects of accumulated technological developments are about to cover the entire surface of the earth. We face a complicated network of physical, chemical, and biological causes and effects, many of them only partially understood. Much painstaking basic research will be required before these problems can be tackled efficiently. If technical solutions are introduced before the conditions are thoroughly understood, one may well worsen the situation in the attempt to improve it.

Why is basic research needed for this kind of training? Why couldn't one train people directly by putting them to work on socially pressing problems? Those who ask these questions compare the situation with teaching Greek and Latin to youngsters in order to give them experience in learning foreign languages. The comparison is fallacious. Polanyi (4) has expressed the reason most lucidly:

"The scientific method was devised precisely for the purpose of elucidating the nature of things under more carefully controlled conditions and by more rigorous criteria than are present in situations created by practical problems. These conditions and criteria can be discovered only by taking a purely scientific interest in the matter, which again can exist only in minds educated in the appreciation of scientific value. Such sensibility cannot be switched on at will for purposes alien to its inherent passion."

There are two sides to the argument. One concerns the analysis of a situation, and the other concerns the search for ways to improve it. The attitude engendered by pure science is most conducive to getting a clearer picture of the facts and the problems that may have to be faced in coping with air pollution, the population explosion, or the effects of technical innovations upon our environment. In basic science the search is for phenomena and connections in all possible directions, whereas in applied science the search is directed toward a specific goal.

Furthermore, when new technical ideas are needed—and they will be needed—the attitude of basic science is that of looking more toward innovative ideas and less toward the application of known devices because the problems at the frontier are exactly those that cannot be solved with established methods. In basic research a pool of young men and women is formed who are accustomed to tackle unexplained phenomena and who are ready to find new ways to deal with them. They are trained to work under the most exacting conditions in open competition with the scientific world community. Instead of "environmentalists" we should train physicists, chemists, geologists, and biologists capable of dealing with the problems of environment.

Whenever large practical projects have been carried out under emergency conditions—projects that were apparently immensely difficult or impossible—scientists from basic fields have played a decisive role. In the past most of the examples have come from war related projects, such as the development of radar or the atomic bomb. There is no doubt, however, that this kind of development can be transferred to more constructive problems. In fact, many basic scientists have made important contributions toward a solution of the arms control problem. Their activities have initiated the discussions that led to the halt of bomb tests. Today they are deeply involved in environmental problems.

Two qualifications are in order. Today's problems certainly will require the methods and results of natural science, but they cannot be solved by these methods alone. As was mentioned earlier, the problems are to a great extent social and political, dealing with the behavior of man in complicated and rapidly evolving situations.

These are aspects of human experience to which today's methods of natural science are not applicable. Seen within the framework of that science, these phenomena exhibit a degree of instability, a multidimensionality for which our present scientific thinking is inadequate and to which such thinking must be applied with circumspection. There is great temptation to transfer the methods that were so successful in natural science directly to social or political problems. This is not possible in most cases. Different methods may be developed in the future. The social sciences are working hard at the task.

The second qualification concerns the need for scientists trained in basic science. We do not argue that only those trained in basic science can solve our problems to the exclusion of others. Far from it; a collaboration between all kinds of people is needed—basic and applied scientists, engineers, physicians, social scientists, psychologists, lawyers, and politicians. The argument submits that people trained in basic science will play an important and irreplaceable role. They are necessary, but not sufficient. But their necessity emphasizes the importance of keeping basic science activities alive.

To keep basic science vigorous is today much harder than it was in the past; it would be harder even if the financial support were as generous as before. The reason is quite natural; the world situation has become so serious that many scientists or potential scientists find it difficult to worry about some unexplained natural phenomena or undiscovered laws of nature when there are more immediate things to worry about. Some scientists feel that we are in an emergency situation and that we should stop basic science for the duration as we did during World War II. But the war lasted only 4 years for the United States, while the present crisis will endure for at least two decades. If we cripple basic science today, it will not be long before there will be no new generations of devoted young scientists for the tests that mankind must face in the future.

Limitations of Science

Another motivation for the antiscience attitude expressed by position 2 is connected with widespread critical view of science and the ways of thinking it fosters. In this view, science is considered as materialistic and inhuman, as an instrument of defining everything in terms of numbers and thus excluding and denying the irrational and emotive approach to human experience. Value judgments, the distinction between good and evil, and personal feelings supposedly have no place in science. Therefore, it is said, the one-sided development of the scientific approach has suppressed some most important and valuable parts of human experience in that it has produced an alienated individual in a world dominated by science and technology in which everything is reduced to impersonal data.

The foregoing arguments are diametrically opposite to the views expressed in position 3, which contends that the supposedly rational, inemotive approach of science is the only successful way to deal with human problems of all sorts. Many of today's trends against science are based on the feeling that the scientific view neglects or is unable to take into account some of the most important experiences in human life.

This widely held belief seems to be in contradiction to the claim of "completeness" of science, which is the basis of position 3. It is the claim that every experience— whether caused by a natural phenomenon or by a social or psychic circumstance—is potentially amenable to scientific analysis and to scientific understanding. Of course, many experiences, in particular in the social and psychic realm, are far from being understood today by science, but it is claimed that there is no limit in principle to such scientific insights.

I believe that both the defenders and the attackers of this view could be correct, because we are facing here a typical "complementary" situation (5). A system of description can be complete in the sense that there is no experience that does not have a logical place in it, but it still could leave out important aspects which, in principle, have no place within the system. The most famous example in physics is the complementarity between the classical description and the quantum properties of a mechanical system. The classical view of an atom is a little planetary system of electrons running around the nucleus in well-defined orbits. This view cannot be disproved by experiment; any attempt to observe accurately the position of an electron in the atom with suitable light beams or other devices would find the electron there is a real particle, but the attempt to observe it would have destroyed the subtle individuality of the quantum state which is so essential for the atomic properties. Classical physics is "complete" in the sense that it never could be proven false within its own framework of concepts, but it does not encompass the all-important quantum effects. There is a difference between "complete" and something we may call "all-encompassing."

The well-known claim of science for universal validity of its insights may also have its complementary aspects. There is a scientific way to understand every phenomenon, but this does not exclude the existence of human experiences that remain outside science. Let us illustrate the situation by a simple example: How is a Beethoven sonata described in the realm of science? From the point of view of physics, it is a complicated quasi-periodic oscillation of air pressure; from the point of view of physiology it is a complicated sequence of nerve impulses. This is a complete description in scientific terms, but it does not contain the elements of the phenomenon that we consider most relevant. Even a psychological study in depth of what makes the listening to these tone-sequences so exciting cannot do justice to the immediate and direct experience of the music.

Such complementary aspects are found in every human situation. There exist human experiences in the realms of emotion, art, ethics, and personal relations that as "real" as any measurable experience of our five senses; surely the impact of these experiences is amenable to scientific analysis, but their significance and immediate relevance may get lost in such analysis, just as the quantum nature of the atom is lost when it is subject to observation.

Today one is rather unaccustomed to think in those terms because of the rapid rise of science and the increasing success of the application of scientific ideas to the manipulation of our natural environment in order to make the process of living less strenuous.

Whenever in the history of human thought one way of thinking has developed with force, other ways of thinking become unduly neglected and subjugated to an overriding philosophy claiming to encompass all human experience. The preponderance of religious thought in medieval Europe is an obvious example; the preponderance of scientific thought today is another. This situation has its root in a strong human desire for clear-cut universally valid principles containing the answers to every question. However, the nature of most human problems is such that universally valid answers do not exist, because there is more than one aspect to each of these problems. In either of the two examples, great creative forces were released, and great human suffering resulted from abuses, exaggerations and from the neglect of complementary ways of thinking.

These complementary aspects of human experience play an important role when science is applied to practical aims. Science and technology can provide the means and methods to ease the strain of physical labor, to prolong life, to grow more food, to reach the moon, or to move with supersonic velocity from one place to another. Science and technology are needed to predict what would be the effects of such actions on the total environment. However, the decisions to act or not to act are based on judgments that are outside the realm of science. They are mainly derived from two strong human motives: the desire to improve the conditions of life, and the drive for power and influence over other people. These urges can perhaps be scientifically explained by the evolution of the human race, but they must be regarded as a reality of human experience outside the scientific realm. Science cannot tell us which of the urges is good or bad. Referring to the first rather than to the second urge, Archibald MacLeish has put this idea into verse: "No equation can divine the quality of life, no instrument record, no computer conceive it/only bit by bit can feeling man lovingly retrieve it."

The true significance of science would become clearer if scientists and nonscientists were more aware of the existence of these aspects that are outside the realm of science. If this situation were better appreciated, the prejudice against science would lose much of its basis and the intrinsic value of our growing knowledge of natural phenomena would be much better recognized.

Intrinsic Value of Science

Since the beginnings of culture, man was curious about the world in which he lives and eager to explain it. The explanations have taken different forms—mythologic, religious, or magic—and they usually encompass all and everything from the beginning to the end. About 500 years ago man's curiosity took a special turn toward detailed experimentation with nature. It was the beginning of science as we know it today. Instead of reaching directly at the whole truth in an explanation for the entire universe—its creation and present form—it tried to acquire partial truths in a small measure about some definable and reasonably separable group of phenomena. Science developed only when men began to refrain from asking general questions such as: What is matter made of? How was the universe created? What is the essence of life? Instead, they asked limited questions, such as: How does an object fall? How does water flow in a tube? Thus, in place of asking general questions and receiving limited answers, they asked limited questions and found general answers. It remains a great miracle, that this process succeeded, and that the answerable questions became gradually more and more universal. As Einstein said: "The most incomprehensible fact is that nature is comprehensible."

Indeed, today one is able to give a reasonably definite answer to the question of what matter is made of. One begins to understand the essence of life and the origin of the universe. Only a renunciation of immediate contact with the "one and absolute truth," only endless detours through the diversity of experience could allow the methods of science to become more penetrating, and their insights to become more fundamental. It resulted in the recognition of universal principles such as gravitation, the wave nature of light, the conservation of energy, heat as a form of motion, the electric and magnetic fields, the existence of fundamental units of matter, the living cell, the Darwinian evolution. It reached its culmination in the 20th century with the discovery of the connections between space and time by Einstein, the recognition of the electric nature of matter and of the principles of quantum mechanics; the 20th century yielded some answers of how nature manages to produce specific materials, qualities, shapes, colors, and structures, and gave rise to new insights into the nature of life as a result of the development of molecular biology. A framework has been created for a unified description and understanding of the natural world on a cosmic and microcosmic level, and its evolution from a disordered hydrogen cloud to the existence of life on our planet. This framework allows us to see fundamental connections between the properties of nuclei, atoms, molecules, living cells, and stars; it tells us in terms of a few constants of nature why matter in its different forms exhibits the qualities we observe. The scientific insight is not complete, it is still being developed; but its universal character and its success in disclosing the essential features of our natural world makes it one of the great cultural creations of our era.

As part of our culture it has much in common with the arts; new forms and ideas are created in order to express the relations of man to his environment. However, the influence of science on society and on our lives and our thinking is much greater today in the positive and negative sense. There were times in the past in which the arts had a similar influence. Science is a unique product of our period.

Science differs from contemporary artistic creations by its collective character. A scientific achievement may be the result of the work of one individual, but its significance rests solely on its role as a part of a single edifice erected by the collective effort of past and present generations of scientists. This effort was and is made by scientists all over the world; the character of the contributions does not reflect their national, racial, or geographic origin. Science is a truly universal human enterprise; the same questions are asked by all men involved in science, the same joy of insight is experienced when a new aspect of deeper coherence is found in the fabric of nature. The choices of problems and the directions of research at the frontier of fundamental science depend much less on the economic, social, and political needs than most people assume; they are determined mainly by the instrumental possibilities of observation and by the internal logics of fundamental science itself. This is not so in applied science and technology which obviously are much more—though not completely—subject to societal demands of all kinds. The rapid developments of applied electronics and acoustics during World War II were certainly determined by military needs. But there are exceptions on both sides. The progress in nuclear physics and in plasma physics—these are to a large extent fundamental branches—was certainly much accelerated by the possibilities of practical applications to power production by nuclear fission or fusion. The invention of the transistor—an example of applied physics—was not prompted by its practical potentialities.

It is often difficult to distinguish between fundamental and applied science, and any considerations of this kind can lead to dangerous oversimplifications. The success of basic research derives to a large extent from the close cooperation of basic and applied science. This close relation provided tools of high quality, without which many fundamental discoveries could not have been made.

Compared to other groups the scientific community is more international or, better, more supranational because it transcends national and political differences. Personal contacts across borders are established easily between people working on similar problems; science has its own international language. The percentage of foreigners in scientific laboratories is probably greater than in any other human activity; there are some very successful international laboratories, among which CERN in Geneva stands out in the field of high-energy physics, as a model for the future United States of Europe. The international ties of science have been helpful even in nonscientific affairs; for example, in the so-called Pugwash conferences, scientists initiated a number of actions directed toward a more unified world, such as the ending of atomic bomb tests in the atmosphere and the beginning of serious talks on arms control.

Science has a peculiar relation to the traditional and the revolutionary. It is both traditional and revolutionary at the same time. Newton's mechanics and the electrodynamics of Faraday and Maxwell are still valid and alive. Current calculations of satellite orbits and of radio waves are still based on them. Revolutionary concepts, such as that of relativity and quantum theory did not invalidate the earlier ideas; they establish unexpected limitations to the old ideas, which remain valid within these limitations. On the other hand, there is a strong trend in science toward the new and the different. Technological advances and novel ways of thinking are constantly introduced to change the manner of working and the method of approach. But scientific revolutions are extensions rather than replacements. Apart from a few notable exceptions, old ideas are expanded and reinterpreted on a more universal basis. Old methods are proved not wrong, but impractical and inaccurate.

In many ways, the attitude of mind in science is opposed to some of the negative and destructive trends in today's thinking. It means being involved in activities where there is real progress; deeper and deeper insights into the natural world are continually obtained. It engenders a feeling of participation at a unique collective enterprise, the construction and improvement of a vast intellectual edifice, one of the great creations of contemporary culture. There is little dispute among scientists regarding the general value scale as to what is significant and as to the directions in which to proceed, although there are difference of opinion regarding the relative importance of different elements.

Scientific knowledge leads to an intimate relation between man and nature, to a closer contact with the phenomena derived from a deeper understanding. To know more about the laws and the fundamental processes on which the material world is based should lead to a deeper appreciation of nature in all its forms. It should show how natural events are closely interwoven and depend on each other, how almost every mineral structure and certainly every manifestation of life are unique and irreplaceable. Thus, science establishes an awareness of how the universe, the atom, and the phenomena of life coexist and are all one. It is ecology in its widest interpretation.

There are still many fascinating problems and unanswered questions at all frontiers of science. We are not yet skillful enough to deal with complexity in nature. Even the structure of liquids is not well understood. No physicist would have predicted the existence of a liquid state from our present knowledge of atomic properties. The complexity of living matter presents far greater problems. In spite of the growing insight into the fundamental processes of reproduction and heredity, we still know very little about the development of organisms, about the functioning of the nervous system, and we know practically nothing about what goes on in the brain when we think or when we use the memory. The deeper we penetrate into the complexities of living organisms, into the structure of matter, or into the expanses of the universe, the closer we get to the essential problems of natural philosophy: How does a growing organism develop its complex structure? What is the significance of the particles and subparticles of which matter is composed? What is the origin of matter? What is the structure and the history of the universe at large?

The urge to find answers to question of the nature of life and matter and to pursue the search for laws and meaning in the flow of events is the mainspring and the most important justification of science. These problems may have little to do with the immediate needs of society, but they will always be in the center of interest because they deal with the where, whence, and what of material existence.

Obligations of the Scientist

Does the actual science establishment correspond to the ideal picture of science as we have drawn it? It certainly does not appear so to many observers outside of the scientific community and even to some scientists. The human problems caused by the ever-increasing development of a science-based technology are too close and too threatening, they overshadow the significance of fundamental science as a provider of deeper insights into nature. The scientist must face the issues raised by the influence of science on society. He must be aware of the social mechanisms that lead to the specific uses and abuses, and he must attempt to prevent the abuses and to increase the benefits of scientific discoveries. Sometimes he must be able to withstand the pressures of society toward participation in activities which he believes to be detrimental. This is not an easy task since the problems are to a great extent of social nature, and the motivations are often dictated by matierial profit and political power. It puts the scientist in the midst of social and political life and strife.

On the other side, the scientist also has an obligation to be the guardian, contributor, and advocate of scientific knowledge and insight. This great edifice of ideas must not be neglected during a time of crisis. It is a permanent human asset and important public resource. The scientist who devotes his time to the solution of our social and environmental problems does an important job, But so does his colleague who goes on in the pursuit of basic science. We need basic science not only for the solution of practical problems, but also to keep alive the spirit of this great human endeavor. If our students are no longer attracted by the sheer interest and excitement of the subject, we were delinquent in our duty as teachers. We must make this world into a decent and livable world, but we also must create values and ideas for people to live and to strive for. Arts and sciences must not be neglected in times of crisis; on the contrary, more weight should be given to the creation of aims and values. And it is a great value to broaden the territory of the human mind by studying the world in which we live.

Much can and should be improved in the style and character of scientific teaching and research. The rapid increase of science acitvities during the 1950's and 1960's has left its mark on scientists and science students. Some of the positive aspects have been adulterated; in many respects, science has become an organization for producing new results as fast as possible. Changes and new perspectives are in order. One of the most dangerous aspects in today's scientific life is overspecialization. There are several trends that lead to it. One is the increasing pace of research, which does not allow the researcher enough time to be interested in other fields not directly related to his own. He has enough trouble in trying to stay ahead of his numerous competitors in his own field and cannot devote much time to anything else. Another impetus has been the general availability of research jobs in all fields; therefore the young scientist did not see the necessity of training himself in fields outside his speciality. Our educational system did not produce "physicists," it produced high-energy physicists, solid-state physicists, enzyme biochemists, and so forth. A typical symptom of this disease can be found in the manpower questionnaire that the National Science Foundation circulated among physicists, where one is asked to specify one's field which is subdivided to the extreme. For example, there are divisions of this kind: elementary particles, hadrons; elementary particles, leptons; solid-state, magnetic properties; solid-state, optical properties. . . . And people try to find a job in exactly the sub speciality of their Ph.D. thesis. What a narrow view and what a boring life with the same subfield of physics forever! A physicist should be interested in all of physics and should welcome a change of field. Most of the positive aspects of science come from an awareness of its broad range, of its universal view. The same quantum theory governs elementary subparticles and phonons or excitons in a solid.

The teaching of science must return to the emphasis on the unity and universality of science, and should become broader than the mere attempt to produce expert craftsmen in a specialized trade. Surely, we must train competent experts, but we also must bring fields together and show the connections between different fields of science. This task may be difficult because of heavy demands on the time and the intellectual capacities of those involved in modern research. But it is highly rewarding from any point of view. The teacher will get a deeper satisfaction from his work and the student will enjoy his studies more; his knowledge will be broader, it will help him to his future work, and he will have a wider choice of jobs.

I. I. Rabi says so succinctly (6):

"Science itself is badly in need of integration and unification. The tendency is more the other way. . . . Only the graduate student, poor beast of burden that he is, can be expected to know a little of each. As the number of physicists increases, each specialty becomes more self-sustaining and self-contained. Such Balkanization carries physics, and, indeed, every science further away from natural philosophy, which, intellectually, is the meaning and goal of science."

A broader understanding of science as a whole, beyond professional specialization, is a necessary condition for fostering the attitude toward nature, which should be the basic philosophy of a scientist. It is that attitude of intimacy with the universe, with its richness and its uniqueness, the feeling of special responsibility toward nature here on Earth, where we have power over it, constructive and destructive. The deeper understanding of nature as a whole leads to a duty on the part of the scientific community to be watchful and to warn against intentional and nonintentional misuse of science and its applications.

Another destructive element within the science community is the low esteem in which clear and understandable presentation is held. This low esteem applies to all levels. The structure and language of a scientific publication is considered unimportant. All that counts is the content; so called "survey" articles are understandable only to experts; the writing of scientific articles or books for nonscientists is considered a secondary occupation and, apart from a few notable exceptions, is left to science writers untrained in science, some of whom are excellent interpreters. Something is wrong here. If one is deeply imbued with the importance of one's ideas, one should try to transmit them to one's fellows in the best possible terms.

In music the interpretive artist is highly esteemed. An effective rendering of a Beethoven sonata is considered as a greater intellectual feat than the composing of a minor piece. We can learn something here: Perhaps a lucid and impressive presentation of some aspect of modern science is worth more than a piece of so-called "original" research of the type found in many Ph.D. theses, and it may require more maturity and inventiveness. Some students may derive more satisfaction from an interpretive thesis—and so may some readers.

Furthermore, the scientist helps himself by attempting seriously to explain his scientific work to a layman or even to a scientist in a different field. Usually, if one cannot explain one's work to an outsider, one has not really understood it. More concerted and systematic effort toward presentation and popularization of science would be helpful in many respects; it would provide a potent antidote to overspecialization; it would bring out clearly what is significant in current research, and it would make science a more integral part of the culture of today.

Much more could and should be done to bring the fundamental ideas nearer to the intelligent layman. Popularization of science should be one of the prime duties of a scientist. The most important instrument for spreading the spirit of basic science is education. Young people should become more familiar with the insights into the workings of nature which our age has revealed. There is more to it than the mere teaching of science. Scientific education must include active involvement in research. Students can absorb the spirit of science only if they face unsolved problems, participate in the process of analyzing facts, sift evidence, construct and test new approaches and ideas. Even at the lower levels, in elementary and high school, science activities should play an increasing role. Intelligent play involving simple natural phenomena fosters a deeper appreciation of our natural environment and transmits the job of discovery. Margaret Mead (7) expressed it most impressively:

"Any subject, no matter how abstract, how inanimate, how remote from the ordinary affaris of men, remains lively and growing if taught to young children who are themselves growing by leaps and bounds, hungering and thirsting after knowledge of the world around them. To children, an understanding of the world around them is as essential as the tender loving care that, during this century, has been so exclusively emphasized in discussions of early childhood education. The language of science will then become—for everyday use—a natural language, redundant, wide in scope, deeply rooted in many kinds of human experience and many levels of human abilities."

Epilogue

Science is involved with society in many respects. There is a broad spectrum of relations—philosophic, social, and ethical—by which science influences and is influenced by society. The significance of science becomes evident in the numerous, often contradictory, aspects in which it interacts with the affairs of men.

The philosophic significance is derived from the progressively deeper and more comprehensive insight into the workings of nature. The edifice of ideas that brought about this understanding of nature was erected during the last 300 years and is one of the most sophisticated systems of thought ever constructed by man. Its great power resides in the essential simplicity of the fundamental concepts. The infinitely complicated variety of phenomena seems to emerge from a few simple, though, subtle, laws of nature.

The social significance derives from the increasing ability to change our environment and the equality of life by applying the results of science. The changes have been both beneficial and detrimental, depending on the wisdom and intentions of those who carried them out. They have had deep and lasting effects on the social structure of society. The ethical significance derives from the recognition that the evolution of life and men on Earth is predicated on a most precarious equilibrium of physical conditions on this planet. From this recognition follows a human responsibility to protect and continue the great experiment of nature which required several billions of years to get under way. Science emphasizes the unity of human beings in the urge to gain rational understanding of the workings of nature, and in the task of caring for their natural environment. It brings people together in the search for deeper insights, on a front which to a large extent remains uninfluenced by the political and social divisions.

Science claims universality. All phenomena and all human experiences are supposed to fit into the context of natural laws and have found or will probably find a scientific description or explanation. However, the scientific interpretation of human experiences does not always shed light on those aspects that are often considered most relevant. These aspects include human emotional experiences, such as feelings and value judgments. They are decisive in the realm of human decision-making. Whenever a choice is made between actions, whenever collective or personal decisions are taken, scientific reasoning can and should provide information about predictable consequences. The actual decision, however, remains outside of science, it represents a kind of reasoning which necessarily is complementary to scientific thought.

Science contains many activities of different aims and different character—the several basic sciences with all their variety of approach from cosmology to biology and the numerous applied sciences that are spreading and involving more and more aspects of human concerns. Science is like a tree in which the basic sciences make up the trunk, the older ones at the base, the newer, more esoteric ones at the top where growth into new areas takes place. The branches represent the applied activities. The lower, larger ones correspond to the applied sciences that emerged from older basic sciences; the higher, smaller ones are the outgrowth of more recent basic research. The top of the trunk, the frontier of basic research, has not yet developed any branches. Applying this picture to the physical sciences, we would locate classical physics, electrodynamics, and thermal physics at the lowest part of the trunk with broad branches representing the vast applications of these disciplines. Higher up the trunk we would put atomic physics with well-developed branches such as chemistry, materials science, electronics, and optics. Still higher we would find nuclear physics with its younger branches symbolizing radioactivity, tracer methods, geology, and astrophysical applications. At the top, without branches, so far, we would locate modern particle physics and cosmology. There was a time, only 50 years ago, when atomic physics was the branchless top.

All parts and all aspects of science belong together. Science cannot develop unless it is pursued for the sake of pure knowledge and insight. But it will not survive unless it is used intensely and wisely for the betterment of humanity and not as an instrument of domination by one group over another. There are two powerful elements in human existence; compassion and curiosity. Curiosity without compassion is inhuman; compassion without curiosity is ineffectual.

REFERENCES AND NOTES

1. This figure is based upon an exponential increase of expenditure with a doubling time of 10 years, as it occurred during the last two decades, and a final yearly expenditure of $3 billion. The starting time is irrelevant.
2. From a contribution by H. B. G. Casimir at the Symposium on Technology and World Trade, National Bureau of Standards, U.S. Department of Commerce, 16 November 1966.
3. E. Purcell, quotation from an unpublished report to the Physics Survey Committee of the National Research Council, Washington, D.C.
4. M. Polanyi, *Personal Knowledge* (Univ. of Chicago Press, Chicago, 1958, p. 182.
5. Similar views have been expressed by T. R. Blackburn [*Science* 172, 1003 (1971)].
6. I. I. Rabi, *Science: The Center of Culture* (World, New York, 1970), p. 92.
7. M. Mead, *Daedalus* 88, 139 (1959).

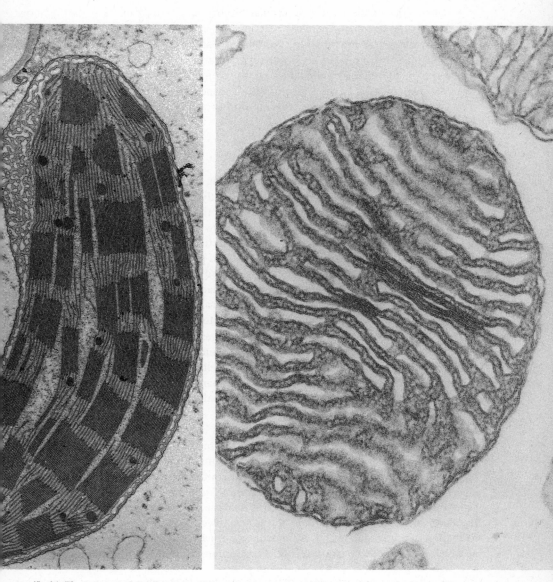

(Left) Electron micrograph of choroplast from corn (Zea mays). Courtesy Professor T. Elliot Weier, Department of Botany, University of California, Davis.
(Right) Electron micrograph of mitochondrion of beef liver muscle. Courtesy Dr. Sidney Fleischer and Mr. Akitsugu Saito, Vanderbilt University, Nashville.

2. *Molecular Mechanisms for the Regulation of Cell Function*

LEWIS J. KLEINSMITH

Discoveries in the field of molecular biology in recent years have clarified how the many complex chemical reactions in cells are regulated. Four of these mechanisms are described in the following articles: (1) transcription of genetic information from DNA to RNA, (2) translation of information from RNA into protein, (3) direct modification of a protein's activity, and (4) degradation of the protein. Dr. Lewis J. Kleinsmith is a specialist in the chemistry of the cell nucleus. At present he is associate professor of zoology at the University of Michigan.

One of the fundamental, defining features of living organisms is their adaptability, that is, their ability to change in response to constantly changing internal and external environments. Since the smallest basic unit of life is the intact cell, this adaptability must be reflected at this level in properties which allow the cell to regulate its vast array of functions. For example, during the normal functioning of a cell, regulation of metabolic pathways is continually required. If the end product of a certain pathway builds up to high concentrations, a mechanism is needed to slow down that pathway. If the product of a pathway becomes depleted, a mechanism is needed to activate that pathway. If a cell is going to divide, specific enzymes and proteins need to be activated for the process of division. Certain cells are required to perform different functions during different hormonal states of the organism. Changes in the patterns of cell function are also required during the processes of growth and differentiation, as the single-cell zygote develops into a wide variety of different cell types performing a wide variety of functions. These are just a few examples of the many diverse situations in which cellular functions need to be modified and controlled in order to insure the organism's adaptability.

Over the past decade, considerable progress has been made in helping us understand the basic molecular mechanisms involved in the regulation of cell function. This review briefly summarizes some of these mechanisms and provides a general conceptual framework within which all potential molecular control mechanisms can be categorized. Since all the functions which a cell performs are ultimately encoded in the cell's genetic information, one logical way to systematically attack the problems of how cell regulation is achieved is to look at the general levels at which the cellular flow of information can be regulated. It is immediately apparent that there are four basic levels at which control can be exerted (Fig. 1): (1) transcription of the genetic information

Source: Reprinted from *Bioscience*, *22*(6):343-348, June 1972 with permission of the publisher and author.

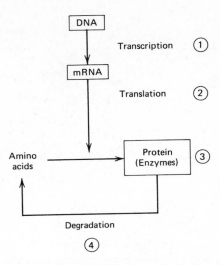

Fig. 1. Diagram of cellular information flow. The numbers indicate the four general levels at which control can be exerted: (1) transcription, (2) translation, (3) protein activity, and (4) protein degradation.

from DNA to RNA;(2) translation of the information from RNA to protein;(3) direct modification of a protein's activity and (4) degradation of the protein.

As we go through examples of how regulation is achieved at these various levels, it should become obvious that all potential ways of regulating cell function can be accommodated within this simple, general framework.

Regulation of Transcription

Probably the best understood examples of regulation at the level of gene transcription are the well-studied *repressor proteins* which occur in bacteria and viruses (Ptashne and Gilbert, 1970). The best known case is the repressor protein involved in the regulation of the so-called *lac* operon, a set of genes which contain the information coding for enzymes involved in the metabolism of the sugar lactose (Fig. 2). When bacteria containing the *lac* operon are grown in the absence of lactose, the *lac* operon DNA is not transcribed into RNA and thus the enzymes for metabolizing lactose are not produced. The reason these genes are not normally transcribed is that a "repressor" protein exists which binds specifically to the *lac* operon DNA, inhibiting transcription at this point. However, when lactose is present it binds to another site on the repressor molecule, inducing a conformational change which makes the repressor unable to bind to DNA. With the repressor no longer bound, the DNA can be transcribed into RNA. Thus, the net result is that the presence of lactose allows a specific set of genes to be transcribed which are not transcribed in the absence of this sugar.

The regulation of gene transcription of DNA-associated proteins is not limited to negative control by repressor molecules. In other cases, such as the arabinose operon,

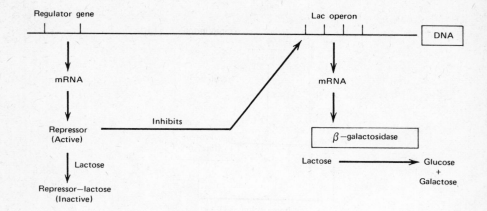

Fig. 2. Diagram of the regulation of gene transcription which occurs in the *lac* operon of *E. coli*. The *lac* operon refers to a portion of the bacterial DNA which codes for enzymes involved in the metabolism of the sugar lactose. Only one of these enzymes is shown, namely β-galactosidase, the enzyme which hydrolyzes lactose to glucose and galactose. Transcription of the *lac* operon is controlled by a repressor protein, whose activity in turn is regulated by the sugar lactose. See text for details.

gene activity is under positive control, i.e., gene transcription only occurs when a specific *activator protein* is bound to the DNA. An interesting example of an activator protein required for the transcription of specific genes is the recently discovered cyclic AMP-binding protein which regulates the genes susceptible to catabolite repression (Pastan and Perlman, 1970; Eron et al., 1971).

In cells of higher organisms, the role of DNA-associated proteins in regulating gene transcription is not so well understood. The search for specific repressors and activators, analogous to the bacterial molecules, has not yet yielded unequivocal results. The basic proteins, or histones, found associated with the DNA of higher organisms are thought to generally inhibit DNA transcription, but not in a specific fashion like the bacterial repressors (DeLange and Smith, 1971). Current investigations have led to the feeling that the specific DNA-regulatory proteins are to be found in the non-histone. or acidic protein fraction (Paul and Gilmour, 1968; Wang, 1968; Kleinsmith et al., 1970). It has also been suggested that chromosomal RNA is involved in regulating DNA transcription (Bekhor et al., 1969; Huang, 1969) although this idea is currently under dispute.

Whatever the exact mechanisms turn out to be in higher organisms, it is clear that some way exists for these organisms to regulate their pattern of DNA transcription. Different cell types, containing the same DNA, are known to be characterized by different patterns of RNA synthesis (Paul, 1970). Also, certain physiological signals, such as steroid hormones, are known to be capable of altering a cell's pattern of DNA transcription (Tomkins and Martin, 1970). Future work on the biochemistry of the

DNA-associated molecules will no doubt help clarify the mechanism involved in this regulation.

Thus far we have considered only one general way of regulating gene transcription, namely via the binding of specific repressor or activator molecules to specific regions of the cell's DNA. Another general way in which gene activity can be controlled is through effects on the enzyme responsible for gene transcription, namely RNA polymerase. Again, in microorganisms many of the details have already been worked out. RNA polymerase is known to be composed of a "core" enzyme to which regulatory subunits can be attached. Although the core enzyme shows no specificity in the genes it transcribes, addition of polypeptide chain "factors" such as the sigma or psi factor to the enzyme causes it to preferentially transcribe certain sets of genes (Travers, 1971). Thus the pattern of gene transcription can be controlled by changes in these factors associated with RNA polymerase.

In higher organisms the chemistry of RNA polymerase has not yet been worked out in such detail. However, it is known that multiple forms of RNA polymerase exist in such organisms, and that changes in the types of RNA polymerase present can change the pattern of gene transcription. For instance, one form of RNA polymerase is present in the nucleolus and preferentially transcribes the genes for ribosomal RNA, while another form is present in the chromatin and transcribes "DNA-like" RNA (Reoder and Rutter, 1970). Since these two enzymes have different requirements, regulation of their activities might be used to control the relative amounts of synthesis of ribosomal and non-ribosomal RNAs.

Thus far we have seen that gene transcription can be regulated both by altering the genes which are available for transcription and by altering the enzyme responsible for transcription, RNA polymerase. One other possibility which exists is to actually alter the amount of a particular gene present in the cell. Presumably, if more copies of a gene are present, more transcription can take place. Mechanisms for controlling the amount of a particular gene in a cell are not understood, but it is known that selective amplification of certain genes can occur. One of the best studied examples is the gene for ribosomal RNA, which is known in some cases to be amplified thousands of times to allow more ribosomal RNA to be made (Birnstiel, 1967). Current evidence (Crippa and Tocchini-Valentini, 1971; Ficq and Brachet, 1971) suggests that this specific amplification is mediated by an enzyme called "reverse transcriptase." This enzyme synthesizes DNA from an RNA template (the reverse of the usual process), and in this particular case is thought to use ribosomal RNA as a template for synthesizing multiple copies of the DNA gene coding for this RNA. It is not yet known how general a mechanism for regulating gene expression this process of amplification will turn out to be.

Regulation of Translation

Once a particular gene has been transcribed into a molecule of messenger RNA, there is still the possibility of controlling whether or not this message is translated into a protein product. Unlike the case of transcriptional control, most of the possible ways in which one could envision translational control are still essentially speculative, although it is quite clear that in a large number of instances some type of translational control is at work.

One of the best understood mechanisms for translational control is the existence of translational repressor molecules which bind to messenger RNAs and block their translation. Evidence for the existence of such translational repressors comes from systems as diverse as viruses and mammalian cells. In the case of certain RNA phages, the viral coat protein inhibits the synthesis of a specific phage protein by complexing with its messenger RNA (Stavis and August, 1970). In mammalian cells, translational repressors have been suggested to play a role in hormonal induction of enzyme synthesis, such as the specific stimulation of the enzyme tyrosine aminotransferase by glucocorticoids (Tomkins, 1969). Evidence suggests that the messenger RNA coding for this enzyme is normally not translated because it is complexed with a specific repressor, It is thought that this repressor is antagonized by glucocorticoid hormones, causing an alteration in the repressor which results in its inability to repress the translation of the messenger. Thus, the net effect in this system is that glucocorticoids induce the synthesis of the enzyme tyrosine aminotransferase.

There are several other possible mechanisms for the specific regulation of translation, although none are as well established or as well understood as translational repressors. One is the possibility of regulating the transport of messenger RNA to the ribosomes, the ultimate site of protein synthesis. It has long been known that much of the RNA made in the nucleus of cells of higher organisms is degraded before it can reach the cytoplasm. It has recently been postulated that these RNA molecules are potential cytoplasmic messenger RNAs, but that a specific regulatory mechanism exists which selects only a small portion of them for transport to the cytoplasm where they can function in protein synthesis (Harris et al., 1969; Darnell et al., 1970). Harris has accumulated evidence which suggests that the nucleolus is specifically involved in this mechanism for engaging particular messenger RNAs for transport to the cytoplasm, and that messages which are not selected by this process are quickly degraded in the nucleus. In support of this general model, recent data have shown that after estrogen stimulation of the uterus, RNA molecules which previously had been restricted to degradation in the nucleus appear in the cytoplasm (Church and McCarthy, 1970).

Another possible type of translational control which also affects the availability of messenger RNA for translation is regulation of the rate of messenger RNA degradation. It is well known that in eukaryotic cells, different messenger RNAs have different turnover rates. Some messages are very unstable and need to be continually replenished, while others are quite stable and may last for long periods of time. Thus any factor which alters the rate of degradation of a particular messenger RNA alters the steady state amount of that message present in the cell. This, in turn, determines how much of the message is available for translation.

Another possible site of control of translation are changes in the ribosome itself. The infection of bacterial cells with T4 phage causes an alteration in the bacterial ribosomes so that their ability to translate normal bacterial messenger RNAs is inhibited while they selectively translate T4 messenger RNAs (Hsu and Weiss, 1969). This finding suggests that different types of ribosomes can selectively translate different types of messenger RNAs. This conclusion is of special interest because of the recent discovery that within a given cell a heterogeneous population of ribosomes exists containing slightly different ribosomal proteins and slightly different ribosomal RNAs (Nomura, 1970). The difference between ribosome populations between different cell types may be even more striking. Since ribosomes are thus different chemically, it is conceivable that there is some type of ribosomal specificity and that some ribosomes are specific for translating particular messages (Naora and Kodaira, 1970). Although this is purely speculative at the present time, the known occurrence of ribosomal heterogeneity opens a whole new area of possible translational control mechanisms.

In addition to messenger RNA and ribosomes, another component required for protein synthesis where control might be exerted is at the level of transfer RNA. The genetic code is commonly referred to as being "degenerate," meaning that more than one codon exists for each amino acid. This means that there are several different transfer RNA (tRNA) molecules which carry the same amino acid. As a result of this degeneracy, during translation some messenger RNAs may use one particular tRNA for a particular amino acid, while other messenger RNAs may use a different tRNA for that same amino acid. The result is that the translation of different messages will require the presence of a different population of tRNAs. There are at least three ways in which this differential requirement can be employed for the regulation of protein synthesis. In one case, the synthesis of a particular species of tRNA required for the synthesis of a particular protein(s) can be altered, depending on the need for that protein. Such a control of translation via the appearance of a new species of tRNA has been shown to occur during regulation of insect growth by juvenile hormone (Ilan et al., 1970). An alternative mode of regulation is to alter the activities of the aminoacyl-tRNA synthetases, which are the enzymes required for attaching the free amino acids to their respective tRNAs. In this regard, it has been shown that during the development of the sea urchin egg, which occurs primarily via translational control (Terman, 1970), there are both quantitative and qualitative changes in the aminoacyl-tRNA synthetases present (Ceccarini et al., 1967). A final possibility for regulation at this level is an alteration in the functional properties of the aminoacyl-tRNAs caused directly by the binding of some small molecule, such as a steroid hormone (Chin and Kidson, 1971).

Thus far, we have considered the regulation of translation via controlling the properties of messenger RNA, ribosomes, and transfer RNA. In addition to the requirement for these three main components, the translation process also requires a large number of soluble protein factors. These factors include those molecules directly required for the process of protein synthesis, such as initiation, elongation, binding, and termination factors. It is probable that this list of components is still far from complete, and it is conceivable that some of these factors could turn out to be sites for the specific control of the translation of particular messenger RNAs.

Thus, we have seen the general levels at which the translation process might be regulated. Although much of this discussion is still purely speculative, it is obvious that translational control does exist and that there are a wide variety of potential mechanisms via which this control might be achieved.

Regulation of Protein Activity

Up to this point we have considered the two major levels at which the flow of cellular information for protein synthesis can be regulated, namely the transcriptional and translational levels. However, once a protein has been synthesized by a cell, there are still mechanisms available for controlling whether the protein will be active or not in performing its normal function. Since most of the proteins which cells synthesize function as the enzymes which catalyze the wide variety of metabolic reactions, it is most convenient to focus this discussion on mechanisms for regulating enzyme activity. It should be realized, however, that similar mechanisms are available for modifying the activity of other non-enzymatic types of proteins.

Although there is some overlap, one can distinguish at least 5 basic means of altering enzyme activity: (1) chemical modification; (2) cleavage reactions; (3) sub-unit interactions; (4) cofactors; and (5) allosteric control.

Chemical modification refers to the direct alteration of the enzyme by the addition or removal of a chemical group covalently attached to the enzyme. For example, the enzyme phosphorylase is only active when a specific residue in each polypeptide chain contains a covalently attached phosphate group (Krebs et al., 1966). When this phosphate is added the enzyme is activated, and when the phosphate is removed, the enzyme is inactivated. In addition to phosphorylation reactions, other chemical modifications of protein molecules, such as acetylation, methylation, and adenylation reactions are also employed in similar ways to regulate the molecule's activity.

The second type of control mechanism, namely cleavage reactions, is similar to the above in that it involves the breaking of covalent bonds. In this case, however, a portion of the polypeptide chain of the protein itself is cleaved off to convert the enzyme from an inactive to an active form. Familiar examples of proetins activated in this manner are the digestive enzymes such as trypsin and chymotrypsin, which are synthesized by the cell in inactive precursor forms (known as trypsinogen and chymotrypsinogen), and are subsequently activated by cleaving off a small number of amino acids from the protein chain (Dixon and Webb, 1964).

The third general method for controlling enzyme activity involves the association and dissociation of subunits. For an example, let us again consider the enzyme phosphorylase. In its normal form it is a dimer containing two polypeptide chains, but for the enzyme to be activated, two of these dimers must associate to form a tetramer (Krebs et al., 1966). Dissociation of the tetramer to two dimers inactivates the enzyme. Other types of situations are also known where the association of a regulatory subunit with a catalytic subunit of an enzyme renders it susceptible to some type of control influence, while dissociation of the regulatory subunit leaves the catalytic subunit to function free from this control. An example of this phenomenon is found in many of the protein kinases, which are rendered susceptible to control by cyclic AMP only when the catalytic subunit has associated with it a regulatory subunit which binds cyclic AMP (Tao et al., 1970).

The next general mechanism for modifying enzyme activity is the presence of cofactors and other required small molecules. This category includes any of the wide range of small molecules required for enzyme activity, such as coenzymes, prosthetic groups, metal ions, and even substrates. As an example of how the availability of such molecules can have relatively specific effects on the pattern of enzyme activity in a cell, again consider the two types of RNA polymerase mentioned in the discussion of transcriptional control. The nucleolar RNA polymerase exhibits optimal activity in the presence of magnesium ions, while the chromatin-associated RNA polymerase exhibits optimal activity in the presence of manganese ions. Thus, any changes in the relative concentrations of these two ions, magnesium and manganese, can affect the relative amounts of ribosomal and non-ribosomal RNA synthesized.

This brings us to the final general mechanism of enzyme regulation, namely allosteric control (Monod et al., 1963). This term refers to enzymes which have special sites which bind small regulatory molecules. The binding of such allosteric regulators alters the conformation of the enzyme, thereby modifying its catalytic activity. A wide variety of metabolic pathways are now known to be regulated by allosteric activators or inhibitors, which are usually substrates or end products which stimulate or inhibit the key enzymes in their respective pathways. One of the best known examples is aspartate transcarbamylase, a key enzyme in the pathway of pyrimidine biosynthesis. One of the end products of this pathway, cytosine triphosphate, binds to an allosteric site on the aspartate transcarbamylase, thereby altering its conformation and inhibiting its catalytic activity (Gerhart and Schachman, 1965). The net result, of course, is a feedback inhibition of the pathway.

The wide-ranging power of small regulatory molecules in influencing cell activity is probably best shown by the example of cyclic AMP, the nucleotide which has been implicated in the action of a wide variety of hormones (Greengard and Costa, 1970; Rasmussen, 1970). It has been postulated that many different hormones which act on different target cells all have the same primary effect—stimulation of the enzyme which produces cyclic AMP, namely adenyl cyclase. According to this model, cyclic AMP then acts as an activator of key enzymes in the cell, primarily protein kinases. Hormonal specificity can be explained by postulating that the adenyl cyclase in each target tissue has a specific regulatory subunit which can only be activated by the specific hormone which acts on that tissue. The fact that different cell types respond differently once stimulated can be explained on the basis that these cells already contain different protein kinases, and so the cyclic AMP simply activates the particular set of protein kinases which happen to be present in the cell being stimulated. The essential elements of this theoretical model of cyclic AMP action are summarized in Figure 3. As can be seen, cyclic AMP is a classic example of how a complex, integrated control system can be built around the simple effects of a small activator molecule.

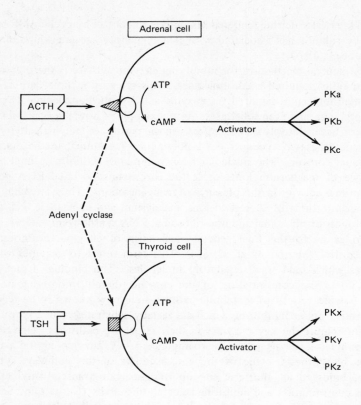

Fig. 3. Model depicting the role of cyclic AMP (cAMP) in hormone action. Two different hormones, thyrotropic hormone (TSH) and adrenocorticotropic hormone (ACTH) are shown to stimulate two different cell types because their respective target cells contain hormone-specific receptor proteins as part of their adenyl cyclase systems. However, in each case, the immediate result of hormonal stimulation is the same, namely, activation of adenyl cyclase and a resulting increase in the production of cyclic AMP. The cyclic AMP produced in each cell is then hypothesized to activate the protein kinases (PK) present in that particular cell type. Since different cells contain different types of protein kinases, the ultimate effects of the two hormones are different in the two types of target cells.

It should be obvious that none of the five mechanisms for control of enzyme activity which we have discussed are mutually exclusive. The enzyme phosphorylase, for example, is activated by chemical modification, is composed of subunits, requires cofactors, and can be allosterically controlled. Thus, many different types of control will often be at work within the same system.

Regulation of Protein Degradation

The last major level of possible regulation of cell activity is control of the rate of protein degradation. Again, the regulation of enzymes will be discussed, although it should be understood that similar mechanisms can hold for other types of proteins. This potential level of regulation has probably been the least studied of those so far considered, and as yet there is little idea as to how the degradation rate of protein is controlled.

It is easy to demonstrate, however, that such control does exist. When rats are starved, the liver content of the enzyme arginase doubles by 6 days. It can be shown that this increase in enzyme activity is not due to an increased rate of synthesis, nor does it represent activation of pre-existing inactive enzyme. Rather, the explanation is that in the starving rat, the enzyme arginase is degraded more slowly than normal, and thus the steady-state amount of enzyme increases (Schimke, 1966). This is a highly specific effect, since most liver proteins in the starving rat are actually degraded more rapidly than normal.

As to what possible mechanism can specifically regulate degradation rates of proteins, one possible analogy may be drawn from the phenomenon known as "substrate stabilization." Many enzymes are known to be less susceptible to proteolytic degradation when they are complexed with their substrate. For example, the enzyme tryptophan pyrrolase can be stabilized against denaturation and degradation in vitro by adding its substrate, tryptophan. Furthermore, when tryptophan is injected directly into animals, liver content of this enzyme is increased five-fold within 16 hours, and the increase is found to be caused by a decreased rate of enzyme degradation. Thus, it seems likely that a phenomenon similar to substrate stabilization is acting in vivo (Schimke, 1966).

Inherent differences in degradation rates of different enzymes may be very important for the overall regulation of cell activities. The steady-state level of any given enzyme in a cell is related both to its rate of synthesis and its rate of breakdown. If two enzymes differ only in their normal rate of degradation, then the one with the faster degradation rate will be present in smaller steady-state quantities and will be as a result more susceptible to control by agents which stimulate protein synthesis. For example, cortisone stimulates the synthesis of the liver enzymes tryptophan pyrrolase and arginase to the same extent. Yet, the net effect of cortisone administration is a much larger increase in the absolute amount of tryptophan pyrrolase than arginase. The explanation of this apparent paradox can be found in the inherent difference in turnover rates of the two enzymes. Tryptophan pyrrolase is degraded more rapidly than arginase, resulting in a lower steady-state level of tryptophan pyrrolase which is as a result more sensitive to increases in synthetic rate. Thus, the ultimate effect of cortisone is a differential effect on the levels of these two enzymes, even though the hormone itself is not selective in stimulating their syntheses (Schimke, 1966).

Conclusion

This brief survey of the various levels of control of cellular processes hopefully has shown that a simple, general framework can be employed to describe the wide variety of specific molecular mechanisms involved. Although more complex levels of regulation obviously exist, they can often be fruitfully analyzed in terms of the general concepts outlined above. For example, an alteration in cell activity brought about by moving a substrate from one cellular compartment to another might be analyzed in terms of changes in the activity of proteins regulating the permeability of the cellular membranes bounding the respective compartments. Thus, this general scheme has great potential for organizing our knowledge about cellular regulation.

It should also be obvious from this discussion that none of the four basic levels of regulation work in isolation from the others. For instance, the activation of a protein kinase by cyclic AMP may lead to the phosphorylation (chemical modification) of a histone which in turn may affect gene transcription. The purpose of this framework is not to claim that any particular example of the regulation of cellular activity can be explained solely via an effect at one of the four levels. Rather, the purpose of this scheme is to present a systematic overview of the various possibilities for regulation, and from this viewpoint hopefully it will serve some useful function in helping to organize our knowledge of the overall processes of cell regulation.

Acknowledgments

This review is based on a series of lectures to undergraduate biology students, and is intended as a general introduction to the field of cellular regulation. More detailed treatment of various aspects of this subject can be found in papers listed in the bibliography. The manuscript was written in part while the author was supported by a grant (GB-23921) from the U.S. National Science Foundation. The illustrations were prepared by Sandra Beadle.

REFERENCES

Bekhor, I., G. M. Kung, and J. Bonner. 1969. Sequence-specific interaction of DNA and chromosomal protein. *J. Mol. Biol., 39.* 351-364.

Birnstiel, M. 1967. The nucleolus in cell metabolism. *Annu. Rev. Plant Physiol., 18:* 25-68.

Ceccarini, C., R. Maggio, and G. Barbata. 1967. Aminoacyl-sRNA synthetases as possible regulators of protein synthesis in the embryo of the sea urchin *Paracentrotus lividus. Proc. Nat. Acad. Sci. U.S.A., 58:* 2235-2239.

Chin, R., and C. Kidson. 1971. Selective associations of hormonal steroids with aminoacyl transfer RNAs and control of protein synthesis. *Proc. Nat. Acad. Sci. U.S.A., 68:* 2448-2452.

Church, R. B., and B. J. McCarthy. 1970. Unstable nuclear RNA synthesis following estrogen stimulation. *Biochim. Biophys. Acta, 199:* 103-114.

Crippa, M., and G. P. Tocchini-Valentini. 1971. Synthesis of amplified DNA that codes for ribosomal RNA. *Proc. Nat. Acad. Sci. U.S.A., 68:* 2769-2773.

Darnell, J. E., G. N. Pagoulatos, U. Lindberg, and R. Balint. 1970. Studies on the relationship of mRNA to heterogeneous nuclear RNA in mammalian cells. *Cold Spring Harbor Symp. Quant. Biol., 35:* 555-560.

DeLange, R. J., and E. L. Smith. 1971. Histones: structure and function. *Annu. Rev. Biochem., 40:* 279-314.

Dixon, M., and E. C. Webb. 1964. *Enzymes.* Academic Press, New York.

Eron, L., R. Arditti, G. Zubay, S. Connaway, and J. R. Beckwith. 1971. An adenosine 3':5:-cyclic monophosphate-binding protein that acts on the transcription process. *Proc. Nat. Acad. Sci. U.S.A., 68:* 215-218

Ficq, A., and J. Brachet. 1971. RNA-dependent DNA polymerase; possible role in the amplification of ribosomal DNA in *Xenopus* oocytes. *Proc. Nat. Acad. Sci. U.S.A., 68:* 2774-2776.

Gerhart, J. C., and H. K. Schachman, 1965. Distinct subunits for the regulation and catalytic activity of aspartate transcarbamylase. *Biochemistry, 4:* 1054-1062.

Greengard, P., and E. Costa. 1970. *Role of Cyclic AMP in Cell Function.* Raven Press, New York.

Harris, H., E. Sidebottom, D. M. Grace, and M. E. Bramwell, 1969. The expression of genetic information: a study with hybrid animal cells. *J. Cell Sci., 4:* 499-525.

Huang, C. C., and P. C. Huang. 1969. Effect of protein-bound RNA associated with chick embryo chromatin on template specificity of the chromatin. *J. Mol. Biol., 39:* 365-378.

Ilan, J., J. Ilan, and N. Patel. 1970. Mechanism of gene expression in *Tenebrio molitor.* Juvenile hormone determination of translational control through transfer ribonucleic acid and enzyme. *J. Biol. Chem., 245:* 1275-1281.

Hsu, W. T., and S. B. Weiss. 1969. Selective translation of T4 template RNA by ribosomes from T4-infected *Escherichia coli. Proc. Nat. Acad. Sci. U.S.A., 64:* 345-351.

Kleinsmith, L. J., J. Heidema, and A. Carroll. 1970. Specific binding of rat liver nuclear proteins to DNA. *Nature, 226:* 1025-1026.

Krebs, E. G., R. J. DeLange, R. G. Kemp, and W. D. Riley. 1966. Activation of skeletal muscle phosphorylase. *Pharmacol. Rev., 18:* 163-171.

Monod, J., J. P. Changeux, and F. Jacob. 1963. Allosteric proteins and cellular control systems. *J. Mol. Biol., 6:* 306-329.

Naora, H., and K. Kodaira. 1970. Interaction of informational macromolecules with ribosomes, II. Binding of tissue-specific RNA's by ribosomes. *Biochim. Biophys. Acta, 209:* 196-206.

Nomura, M. 1970. Bacterial ribosome. *Bacteriol. Rev., 34:* 228-277.

Pastan, I., and R. Perlman. 1970. Cyclic adenosine monosphosphate in bacteria. *Science, 169:* 339-344.

Paul, J. 1970. DNA masking in mammalian chromatin: a molecular mechanism for determination of cell type. *Curr. Topics Develop. Biol., 5:* 317-352.

Paul, J. and R. S. Gilmour. 1968. Organ specific restriction of transcription in mammalian chromatin. *J. Mol. Biol., 34:* 305-316.

Ptashne, M., and W. Gilbert. 1970. Genetic repressors. *Sci. Amer., 222*(6): 36-44.

Rasmussen, H. 1970. Cell communication, calcium ion, and cyclic adenosine monophosphate. *Science, 170:* 404-412.

Roeder, R. G., and W. J. Rutter. 1970. Specific nucleolar and nucleoplasmic RNA polymerases. *Proc. Nat. Acad. Sci. U.S.A., 65:* 675-682.

Schimke, R. T. 1966. Studies on the roles of synthesis and degradation in the control of enzyme levels in animal tissues. *Bull. Soc. Chim. Biol., 48:* 1009-1030.

Stavis, R. L., and J. T. August. 1970. The biochemistry of RNA bacteriophage replication. *Annu. Rev. Biochem., 39:* 527-560.

Tao, M., M. L. Salas, and F. Lipmann. 1970. Mechanism of activation by adenosine 3':5'-cyclic monophosphate of a protein phosphokinase from rabbit reticulocytes. *Proc. Nat. Acad. Sci. U.S.A., 67:* 408-414.

Terman, S. A. 1970. Relative effect of transcription-level and translation-level control of protein synthesis during early development of the sea urchin. *Proc. Nat. Acad. Sci. U.S.A., 65:* 985-992.

Tomkins, G. M., T. D. Gelehrter, D. Granner, D. Martin, Jr., H. H. Samuels, and E. B. Thompson. 1969. Control of specific gene expression in higher organisms. *Science, 166:* 1474-1480.

Tomkins, G. M., and D. W. Martin, Jr. 1970. Hormones and gene expression. *Annu. Rev. Genet., 4:* 91-106.

Travers, A. 1971. Control of transcription in bacteria. *Nature-New Biol., 229:* 69-74.

Wang, T. Y. 1968. Restoration of histone-inhibited DNA-dependent RNA synthesis by acidic chromatin proteins. *Exp. Cell Res., 53:* 288-291.

3.   *A Multiple Origin for Plastids and Mitochondria*

PETER H. RAVEN

The recent discovery of DNA in mitochondria and chloroplasts lends support to the interesting speculation that these organelles were once free-living organisms. The following article reviews the basis for this concept and points out the many homologies between mitochondria and plastids. The author proposes that when these organelles were free-living, they established symbiotic relationships with a variety of other organisms such as red algae, dinoflagellates, green algae, and euglenoids, and became permanently incorporated in some of them. In this sense, plastids and mitochondria are postulated as having had a multiple origin. Dr. Peter Raven, director of the Missouri Botanical Gardens, is a specialist in plant taxonomy.

In view of much accumulating evidence, it now seems almost certain that the plastids and mitochondria in eucaryotic cells originated as free-living procaryotes which found shelter within primitive eucaryotic cells and eventually were stabilized as permanent symbiotic elements within them (1). This evidence, however, has been largely ignored in the construction of phylogenetic classifications of living organisms, such as the excellent one recently presented by Whittaker (2). What implications for the phylogenetic relationships between groups of organisms does a symbiotic origin of plastids and mitochondria have?

Evidence for Symbiotic Origin

The evidence for a symbiotic origin of mitochondria and plastids will be presented in two parts. First, we shall review the properties of these cellular organelles themselves, and then we shall consider the nature and occurrence of analogous symbioses that occur at the present day.

Both mitochondria and plastids have the capacity for semiautonomous growth and division which is only partially controlled by the nuclear DNA. They arise only from preexisting mitochondria (3) and plastids, respectively. Both contain unique base compositions and configurations of DNA and RNA (4, 5), both of which can be synthesized and replicated within the organelle. The DNA is known in the mitochondria of a number of multicellular animals to be present in the form of double-stranded circles, each with a molecular weight of about 9 to 10×10^6 (5). In other organisms, such as the higher plants and *Neurospora,* the mitochondrial DNA has a much higher molecular weight and has not yet been shown to exist in a circular form (5).

Source: Copyright 1970 American Association for the Advancement of Science. Reprinted by permission of the publisher and author from *Science, 170:* 1279-1283, 18 December 1970.

In yeast, circular DNA appears to be only a minor component of mitochondrial DNA (6). The amount of DNA in a mitochondrion may amount to about 0.01 of that present in a cell of *Escherichia coli,* or sometimes considerably more (5). In chloroplasts, there is nearly a hundred times as much, almost as much as in *Escherichia coli* (7). It is not known whether chloroplast DNA is divided into molecules or exists as one continuous piece (8). In bacteria, mitochondria, and chloroplasts, the DNA is histone-free and bound to membranes; in eucaryotic chromosomes, it is associated with histones and not bound to membranes (5, 9).

The extent to which plastids and mitochondria of DNA act as templates for transcription of specific messenger RNA's is currently the subject of investigation in a number of laboratories. Both chloroplasts and mitochondria are able to incorporate amino acids into proteins in vitro (5, 10). In the cell, chloroplast DNA plays a role in the synthesis of at least some of the characteristic proteins of chloroplasts, but nuclear DNA likewise participates to a large extent in these syntheses (11). One of the most significant findings has been that mutant *Euglena* strains which lack the ability to synthesize chloroplasts also lack DNA of characteristic low guanosine-cytosine content (12). In mitochondria, a few, but certainly not all (5), of the characteristic proteins are produced within the mitochondria from a template of mitochondrial DNA (13). For example, Woodward and Munkres (14) have found differences in at least one amino acid residue in mitochondrial structural protein in two "cytoplasmic" mutants of *Neurospora*. In mitochondria, the replication of DNA seems to occur independently of that in the nucleus (15). Many species of transfer RNA are found only in the mitochondria of rat livers and not elsewhere in the same cells (16). The DNA polymerase which has been found in mitochondria (17) may therefore also be produced within these organelles. Furthermore, *n*-formylmethionyl transfer RNA is known to be present only in the mitochondria of eucaryotes and in bacteria (18).

Chloroplast ribosomes from higher plants resemble bacterial ribosomes in their sedimentation behavior and the sizes of their RNA components (19), and the fact that their ability to incorporate amino acids into proteins is inhibited by chloramphenicol (20), as it is in mitochondria (21). In contrast, cytoplasmic ribosomes are larger than chloroplast ribosomes (22) and are insensitive to chloramphenicol both in vitro (23) and in vivo (24). Hybridization studies have shown that ribosomal RNA probably originates from organelle DNA, not from nuclear DNA as in eucaryotic systems (25).

Therefore, mitochondria and plastids clearly resemble entire procaryotic cells more closely than other components of the eucaryotic cells in which they occur (26). They have a higher degree of autonomy than any other cellular component. They likewise resemble procaryotic cells in size (typically 1 to 5 micrometers).

Mitochondria and plastids differ from all other cellular organelles in being bound by a double membrane. The inner membrane is convoluted into a series of folds, greatly increasing the "working area" of the cell on which enzymes are located. Similarly, in bacterial cells, the plasma membrane is often extensively folded into the interior of the cell. The outer cell wall of bacteria appears to be a specialized structure, whereas the outer layer of mitochondria, at least, may be in effect an extension of the endoplasmic reticulum (26). Photosynthetic bacteria and the blue-green algae have membranous vesicles and lamellae upon which many of the photosynthetic pigments are located, and these are presumably homologous with the inner membranes of mitochondria and chloroplasts.

A Common Origin for Mitochondria and Plastids?

DuPraw (27) has pointed out that there are remarkable similarities between mitochondria and chloroplasts, many of which have been stressed in the preceding discussion. The structural proteins of mitochondria and chloroplasts are similar—both may contain an actomyosin-like contractile protein, and both show contraction dependent on adenosine triphosphate. Both mitochondria and chloroplasts carry out the phosphorylation of adenosine diphosphate coupled to electron transport phenomena, with similar electron carriers. These considerations have led to the hypothesis that mitochondria and chloroplasts may have had a common origin.

In any case, it is virtually certain that procaryotes with the properties of chloroplasts evolved before those with the properties of mitochondria, even though mitochondria are found in all eucaryotic cells, chloroplasts only in some. Free oxygen is utilized by mitochondria as an acceptor for electrons, but chloroplasts can carry out their functions anaerobically. It now seems likely that photosynthesis had already evolved by the time of deposition of the first known fossils, at least 3.2×10^9 years ago (28). Blue-green algae appear to be at least 2.7×10^9 years old as a group. On the other hand, eucaryotic cells probably did not evolve much more than 1.1×10^9 years ago (29). Other evidence strongly suggests that all the oxygen in the atmosphere has been derived from the process of photosynthesis. The achievement of current concentrations of oxygen in the atmosphere is often coupled with the invasion of the land by plants, animals, and fungi, an event that seems to have taken place about 4.5×10^8 years ago (30). At any rate, it is likely that there was insufficient oxygen in the atmosphere to allow the evolution of cells with the properties of mitochondria until perhaps 1.5×10^9 years ago, and perhaps much more recently than that.

If plastids and mitochondria did have a common origin, then, it appears almost certain that mitochondria are derived from plastids. In that case, an autotrophic procaryote presumably became a symbiote in the cells of a larger anaerobe maintaining itself with energy from glycolysis. This symbiote was then functionally equivalent to a chloroplast in the cells of the anaerobic host. Later, modification of its properties in the course of evolution led to its acquiring the oxidative capabilities associated with mitochondria.

It appears more likely, however, that, although mitochondria and plastids did have a common origin in the sense that both are derived from procaryotic symbiotes that became stable elements in eucaryotic cells, they originated from separate symbiotic events. First, there are impressive differences between them. Chloroplasts are larger mitochondria, have a different characteristic shape, contain nearly a hundred times as much DNA, and differ greatly biochemically. Second, mitochondria occur in all eucaryotic cells, including protists, plants, animals, and fungi. Plastids, on the other hand, are found only in plants. It appears simplest to visualize the acquisition of mito-chondria by the common ancestor of all eucaryotes and the subsequent acquisition of plastids by one or more lines of eucaryotes as separate events. The simplicity with which intracellular symbioses apparently become established and are maintained argues in favor of a separate symbiotic origin of mitochondria and chloroplasts.

A Multiple Origin for Mitochondria?

Did mitochondria themselves arise from a series of separate symbiotic events? Symbioses that are observed at the present time invariably involve autotrophs and are thus analogous to the events that presumably led to the evolution of chloroplasts. All eucaryotic cells already have mitochondria, and a symbiosis involving additional mitochondrion-like particles would appear to have no selective value. In the past, this may not have been the case. If eucaryotic cells acquired characteristic mitochondria after their pattern of nuclear organization was already established, and they were in a sense already eucaryotic, one could visualize a series of symbiotic events involving mitochondrion-like organisms analogous to those involving autotrophs we see at the present day. It seems likely, however, that only one sort of procaryote was involved, because of the impressive similarities between mitochondria in all eucaryotic cells that have been investigated. In this connection, it is interesting to note that the ciliate *Paramecium aurelia,* for example, harbors a number of endosymbiotic gram-negative bacteria of uncertain adaptive value in nature, among them kappa, mu, lambda, and sigma particles (31). In any event, it is difficult to evaluate the question of a common or multiple origin for mitochondria by studying contemporary organisms.

Contemporary Symbioses and the Origin of Plastids

Photosynthetic algae are, at the present time, symbiotic in a very wide variety of organisms. Symbiotic relationships involving the procaryotic blue-green algae will be reviewed below, but first a brief survey of the symbiotic relationships of eucaryotic algae will be presented here to illustrate the diversity of these interactions.

In addition to the symbioses involving eucaryotic algae with higher plants, other algae, and vertebrates, the relationships between the algal and fungal components of lichens have been much discussed and studied experimentally. In addition to their participation in all of these kinds of systems, autotrophic algae are known to occur as symbiotes in more than 150 genera of invertebrates, representing eight phyla (32, 33). In the vast majority, if not all, of these instances, the symbiotic relationships arose independently. The symbiotes include three orders of green algae as well as the dinoflagellates and diatoms.

All stages in reduction of the cell wall in the symbiotic algae are represented among these relationships. For instance, in the dinoflagellate *Platymonas convolutae,* symbiotic in the marine acoclous turbellarian *Convoluta roscoffensis,* the symbionts lack the cell wall, capsule, flagella, and stigma of the free-living forms of the same species (31). They likewise have finger-like extensions 1 to 2 micrometers long that greatly increase the area of contact with the host cytoplasm.

A number of opisthobranch gastropods feed on siphonaceous green algae and have symbiotic chloroplasts, apparently derived from these algae, which are confined to the cells that form the hepatic tubules (32). These functional and clearly autonomous chloroplasts play an important role in the nutrition of their hosts.

The diversity of contemporary symbiotic relations attests both in the high selective value of such relations for heterotrophs and to the ease with which they are established. Thousands of such symbioses of varying age exist at the present day, and have been established even in the face of competition from established autotrophs, both unicellular and multicellular.

Symbiotic Origin of Mitochondria and Plastids

In view of the similarities between existing mitochondria and plastids, on the one hand, and procaryotic cells, on the other, it would be very difficult to conclude that they had originated separately. When the array of demonstrable symbiotic relationships between eucaryotic and, as we shall see shortly, procaryotic autotrophs and various kinds of heterotrophic organisms is taken into account, the case becomes overwhelming. Some symbiotic blue-green algae can at the present day be distinguished only with the greatest difficulty from chloroplasts—and, indeed, the distinction seems to be a false one. The case for the symbiotic origin of plastids appears overwhelming, and some of the implications of this for a phylogenetic understanding of the relationship between organisms will now be considered.

Mitochondria, like 9-plus-2 flagella and a differentiated nucleus (but see 34), appear to be characteristic of all eucaryotic cells, but they may well have had a multiple origin, as suggested above. They have apparently lost most of their DNA content subsequently. Whether the spindle apparatus of eucaryotic cells likewise had a symbiotic origin, as suggested by Sagan (35), is a separate question that will not be considered in this article.

Symbiosis and the Origin of Plastids

The numerous symbiotic associations that can be observed in living organisms, together with their diversity, strongly suggest that such relations arise with relative ease. They would have had an even higher selective value when no eucaryotic cells were autotrophic, a fact that lends credence to the view that plastids arose not once, but many times. The implications of this view for the phylogenetic relationships of the major groups of autotrophic procaryotes will now be considered.

Blue-Green Algae as Symbionts

One group of living procaryotes is biochemically and structurally (36) similar to the chloroplasts of certain eucaryotes. These are the blue-green algae, which, in addition to chlorophyll *a*, contain phycobilins, two unusual porphyrins related to the bile pigments (37) in their cells (38). Outside of the cells of the blue-green algae, phycobilins occur in the red algae (37), in several genera of flagellated Cryptophyta (39), and in the anomalous hot-spring alga *Cyanidium caldarium* (40). They function as accessory pigments, and, because of their unusual structure, it is difficult to imagine that they evolved independently in these four groups, which are markedly dissimilar when judged on other criteria—so much so that it is extremely difficult to imagine a direct phylogenetic connection between them.

Symbiotic relationships involving blue-green algae are very common. These procaryotes are frequently found as symbiotic components in the cells of amoebae; flagellated protozoa; green algae such as *Gloeochaete, Glaucocystis,* and *Cyanoptyche* that lack chloroplasts; and diatoms (41). Some are even associated with fungi such as the phycomycete *Goesiphon pyriforme* (42). In all of these organisms, the blue-green algae play the role of chloroplasts, bringing the characteristic biochemistry of their procaryotic free-living relatives with them. In the symbiont of *Glaucocysts nostochinearum, Skujapelta nuda,* the cell walls are nearly or entirely lacking, a condition that Hall and Claus (43) consider an apparent adaptation to the intracellular habitat and symbiotic association.

These relationships suggest that the simplest hypothesis to account for the biochemical similarities between the groups of algae mentioned above would be that the chloroplasts of red algae, flagellated Cryptophyceae, and *Cyanidium* are in fact blue-green algae that entered into symbiotic relationships with the ancestors of these organisms. For each group, there is evidence for and against the proposed hypothesis.

In the blue-green algae, chlorophyll *a* occurs with several carotenoids and phycobilins. No other chlorophyll is present. The presence of a cell wall in the blue-green algae and its absence in chloroplasts poses no problem for, in at least two instances (44), known symbiotic blue-green algae lack a cell wall.

In red algae, chlorophyll *d* has been reported in many, but not all forms examined. Chlorophyll *d,* when present, is usually only a trace constituent; it has never been detected in the absorption spectra of living red algal cells or thalli; and it may merely be an oxidation product derived from chlorophyll *a* in vitro (45). This hypothesis has recently been strengthened by the detection of a pigment with the spectral qualities of chlorophyll *d* in extracts derived from the green alga *Chlorella pyrenoidosa* (46). The phycobilins of red algae are slightly different from those in blue-green algae, but, all in all, there seems to be no compelling biochemical reason not to regard the chloroplasts of red algae as ancient symbiotic blue-green algae. Moreover, the structure of the chloroplasts in red algae is extremely simple. These chloroplasts appear to contain a single thylakoid (47) and thus to be virtually identical to entire cells of the blue-green algae. The blue-green algae have been in existence for some 3 billion years (48), the red algae probably for less than 650 million. If the hypothesis presented here is accepted, the relationship to be considered is not that between the procaryotic blue-green algae and the eucaryotic red algae, but between the former and the chloroplasts of the latter. Aside from their chloroplasts, the cells of red algae have nothing in common with blue-green algae in organization or biochemistry (49).

Blue-green algae, lacking cell walls, are in fact the functional chloroplasts in one member of the Cryptophyta, *Cyanophora paradoxa* (43). The flagellated Cryptophyta—cryptomonads—are essentially Protozoa with a proteinaceous pellicle and a contractile vacuole. Some of them are heterotrophic and ingest food particles. They are not plant-like in any of their characteristics except for the biochemistry of their chloroplasts.

The chloroplasts are relatively simple and similar in structure to those of the red algae (47) and to entire cells of the blue-green algae. Could these chloroplasts be ancient symbiotic blue-green algae?

The chief objection to this hypothesis appears to be the presence of chlorophyll c, together with carotenoids and phycobilins, in the chloroplasts of at least some cryptomonads (50). In structure, chlorophyll c differs widely from other chlorophylls (51). It occurs as an accessory pigment in the diatoms and some other Chrysophyta, in the dinoflagellates (Pyrrhophyta), and in the brown algae (Phaeophyta). These groups seem totally unrelated, and perhaps chlorophyll c evolved separately in the "chloroplasts" of cryptomonads. It may well be that the Cryptophyta are polyphyletic, as a number of phycologists have suggested. In a review of the algae (52), some are said to have "a cellulose membrane" and there are said to be two types of chloroplasts—"small, blue-green bodies" (symbiotic blue-green algae?) and "one or two parietal plates." As the rather numerous genera of cryptomonads are investigated further, it should be possible to determine which have chloroplasts of blue-green algae origin.

There appears to be no valid reason not to regard the large lobed chloroplast of *Cyanidium caldarium* as a symbiotic blue-green algal cell that has become stabilized in this role. These chloroplasts contain chlorophyll a, carotenoids, and phycobilins, thus closely approximating the biochemistry of living blue-green algae, to which they are likewise similar in structure. If this is the case, then biochemical evidence from the chloroplasts should not be taken into account in deciding where *Cyanidium* should be placed among the groups of organisms (53).

The Origin of Plastids in Other Algae

If similar reasoning is applied to the biochemistry of the chloroplasts in other groups of algae, certain relationships become evident (54). For example, the green algae (Chlorophyta) and euglenoids (Euglenophyta) have chloroplasts that are essentially identical biochemically. These chloroplasts resemble those of the land plants in containing chlorophyll b as an accessory pigment. In other respects, the green algae—which are "typical" plants with a cellulose cell wall—could not be more different from the euglenoids—which are "typical" flagellate protozoa with a proteinaceous pellicle and a contractile vacuole, and which at times ingest solid food particles (55). The simplest assumption to account for the similarity between the two groups is that both harbor the remnants of an ancient line of procaryotes, now presumably extinct, which had both chlorophyll a and b. These hypothetical procaryotes, then, would have been the group in which the photosynthetic apparatus characteristic of the land plants evolved (56).

The remaining algae—Phaeophyta, Chrystophyta, Xanthophyta, and Pyrrhophyta are the major groups—have chloroplasts in which chlorophylls a and c are associated with various carotenoids. If we can allow for the evolution and diversification of carotenoids after the establishment of symbiotic relationships in each case, then the biochemical similarities between the chloroplasts characteristic of these taxa might be accounted for by postulating a third group of procaryotes with chlorophyll a, in this case accompanied by chlorophyll *c*, which became symbiotic in cells ancestral to these groups (34, 57).

To summarize, the simplest way to account for the similarities and differences between the biochemistry of the procaryotic blue-green algae on the one hand and the chloroplasts of the various algal divisions on the other is as follows. First, chlorophyll *a* and a system of photosynthesis that led to the evolution of oxygen evolved in one line of procaryotes. In these, chlorophyll *a* and probably certain carotenoids that served as accessory pigments were probably arranged on photosynthetic lamellae within their cells. This line gave rise to at least three biochemically distinct derivatives: (i) the living blue-green algae, in which evolved phycobilins; (ii) the "green procaryotes," in which evolved chlorophyll *b*; and (iii) the "yellow procaryotes," in which evolved chlorophyll *c*. All of these groups entered into symbiotic associations with primitive eucaryotic cells, probably more than once, and groups (ii) and (iii) no longer exist as free-living organisms. From symbiotes of group (i) were derived the chloroplasts of the Rhodophyta, Cryptophyceae, and the anomalous genus *Cyanidium*. From symbiotes of group (ii) were derived the chloroplasts of the Chlorophyta (and through them, the bryophytes and vascular plants) and Euglenophyta. From symbiotes of group (iii) were derived the chloroplasts of Phaeophyta, Chrysophyta, Xanthophyta, and Pyrrhophyta.

Prospects for the Future

The relationships discussed here suggest several promising lines of investigation. First, the diversity of base-pair ratios in the blue-green algae (58) might provide a clue as to which might have been most likely to have given rise to chloroplasts in particular groups during the course of evolution, although the base ratios may of course have been altered subsequently by the operation of distinctive DNA polymerases. If we may take them more or less at face value, however, they may be useful; for example, the base ratios in the endosymbiote *Cyanocyte korschikoffana,* which occurs in the cryptomonad *Cyanophora paradoxa,* strongly support the placement of *Cyanocyta* (10) in the order Chroococcales of the blue-green algae (58).

Second, in searching for homologies between contemporary blue-green algae and chloroplasts, it may not be appropriate to consider the vascular plants, in which the chloroplasts are very likely not homologous with blue-green algae. When DNA-hybridization experiments are more widely applied to the problem of the origin of mitochondria and plastids and their homologies, the hypotheses presented here invite several specific avenues of attack. For example, it would seem appropriate to compare certain contemporary blue-green algae with the chloroplasts of the red aglae, preferably by DNA hybridization.

Even though it appears almost certain that mitochondria and plastids had independent, multiple, symbiotic origins, their functions are shared to varying degrees by nuclear DNA in contemporary eucaryotes. The hypothesis presented here suggests that the degree to which the nuclei have taken over what were presumably once independent functions may vary widely in different groups of eucaryotes. If DNA has actually been lost in the course of evolution from the symbiotes, perhaps becoming incorporated in the nucleus in some, this might have also altered radically the base-pair ratios in their mitochondria and plastids and might lead to unexpected results in DNA-hybridization experiments. It might eventually be possible in some systems to show that the nuclear cistrons responsible for mitochondrial or plastid functions (for example, cytochromes) might have base compositions similar to that of the organelle which they affect.

Summary

The impressive homologies between mitochondria and plastids, on the one hand, and procaryotic organisms, on the other, make it almost certain that these important cellular organelles had their origin as independent organisms. The vast number of symbiotic relationships of all degrees of evolutionary antiquity which have been found in contemporary organisms point to the ease with which such relationships can be established.

In view of this, the similarities between such totally different groups as blue-green algae and red algae, dinoflagellates and brown algae, and green algae and euglenoids can best be explained by postulating an independent, symbiotic origin of the plastids in each instance. A minimum of three groups of photosynthetic procaryotes appears to be necessary to explain the relationships among contemporary Protista and green plants: (i) the blue-green algae, which possess chlorophyll *a*, carotenoids, and phycobilins; (ii) the "green procaryotes," a hypothetical group characterized by chlorophylls *a* and *b* and a distinctive assemblage of carotenoid accessory pigments, but not phycobilins; and (iii) the "yellow procaryotes," a second hypothetical group whose members had chlorophylls *a* and *c* and various carotenoids but not phycobilins. There is, however, no reason to think that only three kinds of organisms were involved; numerous symbiotic events presumably occurred in each of these lines.

The "green procaryotes" and "yellow procaryotes" survive today only as chloroplasts from which the characteristics of the original, free-living forms can be deduced only in part. Hybridization between selected plastid DNA's may be helpful in unraveling this story, and is likewise suggested as the key to understanding the relationship between the blue-green algae and the chloroplasts of the red algae and cryptomonads.

It is postulated that the symbiotic organisms have lost various functions to the nucleus in the course of evolutionary time. If mitochondria and plastids have had a multiple origin, as suggested here, it will be necessary to examine the division of function between the two subsets of DNA for a wide variety of organisms before valid conclusions can be obtained.

REFERENCES AND NOTES

1. C. Mereschkowsky, *Biol. Zentralbl. 25*, 593 (1905); in *The Cell in Development and Heredity*, E. B. Wilson, Ed. (Macmillan, London, 1925); S. Granick, *Encycl.*

Plant. Physiol. 1, 507 (1955); A. Famintzin, Biol. Zentralbl, 27, 353 (1907); H. Ris and W. Plaut. J. Cell Biol. 13, 383 (1962); A. Gibor and S. Granick, Science 145, 890 (1964); A. L. Lehninger, The Mitochondrion (Benjamin, New York, 1964); P. Echlin and I. Morris, Biol. Rev. 40, 193 (1965); L. Sagan (now L. Margulis), J. Theor. Biol. 14, 225 (1967); M. Edelman, D. Swinton, J. A. Schiff, H. T. Epstein, B. Zeldin, Bacteriol. Rev. 31, 315 (1967); J. T. O. Kirk and R. A. E. Tilney-Bassett, The Plastids (Freeman, San Francisco, 1967); P. Borst, A. M. Kroon, C. J. C. M. Ruttenberg, in Genetic Elements, Properties and Function, D. Shugar, Ed. (Academic Press, New York, 1967), p. 81; E. J. DuPraw, Cell and Molecular Biology (Academic Press, New York, 1968); L. Margulis, Science 161, 1020 (1968); D. B. Roodyn and D. Wilkie, The Biogenesis of Mitochondria (Methuen, London, 1968). An increasingly less popular view—that these organelles have evolved independently in eucaryotes—has been defended recently by A. Allsopp [New Phytol. 68, 591 (1969)]. The arguments necessary to defend such a view in the face of contemporary evidence are far more complex than those supporting what is now clearly the majority opinion.

2. R. H. Whittaker, Science 163, 150 (1969). However, there are exceptions; see, for example, L. Margulis, J. Geol. 77, 606 (1969).
3. G. Shatz, Biochemistry 8, 322 (1969); D. J. L. Luck, J. Cell Biol. 16, 483 (1963); ibid. 24, 445 (1965); ibid., p. 461.
4. For Neurospora mitochondria see D. J. L. Luck and E. Reich [Proc. Nat. Acad. Sci. U.S. 52, 031 (1964)]; for mitochondria of the myxomycete Physarum see J. E. Cummins, H. P. Rusch, T. E. Evans [J. Mol. Biol. 23, 281 (1967)]; for Euglena chloroplasts see J. A. Schiff and H. T. Epstein [in Reproduction: Molecular, Subcellular, and Cellular, M. Locke, Ed. (McGraw-Hill, New York, 1965), p. 131].
5. References summarized by M. M. K. Nass [Science 165, 25 (1969)].
6. L. Shapiro, L. I. Grossman, J. Marmur, J. Mol. Biol. 33, 907 (1968).
7. E. J. DuPraw, Cell and Molecular Biology (Academic Press, N.Y. 1968), p. 97.
8. Chloroplast DNA has been shown by G. Brawerman [in Biochemistry of chloroplasts, T. W. Goodwin, Ed. (Academic Press, New York, 1966), vol. 1, p. 301] to differ from nuclear DNA in its much lower content of guanosine and cytosine (about 24-27 percent in green algae) or higher content of guanosine and cytosine (in most vascular plants).
9. References summarized by S. Nass [Int. Rev. Cytol. 25, 55 (1969)].
10. Evidence summarized by E. J. DuPraw [Cell and Molecular Biology (Academic Press, New York, 1968), pp. 97-98, 152-153].
11. J. T. O. Kirk [in Biochemistry of Chloroplasts, T. W. Goodwin, Ed. (Academic Press, N.Y. 1966), vol. 1, p. 301] summarizes much evidence that biosynthesis of many components of the chloroplast depends upon nuclear genes.
12. D. S. Ray and P. C. Hanawalt, J. Mol. Biol. 11, 760 (1965).
13. R. P. Wagner, Science 163, 1026 (1969); see also more indirect evidence summarized in (5).
14. D. O. Woodward and K. D. Munkres, Proc. Nat. Acad. Sci. U.S. 55, 872 (1966).
15. E. W. Guttes, P. C. Hanawalt, S. Guttes, Biochim. Biophys. Acta 142, 181 (1967).

16. C. A. Buck and M. M. K. Nass, *Proc. Nat. Acad. Sci. U.S. 60*, 1045 (1968); *J. Mol. Biol. 41*, 67 (1969).
17. E. Wintersberger, *Biochem. Biophys. Res. Commun. 25*, 1 (1966). Activity of DNA-dependent RNA polymerase has been demonstrated for both mitochondria [by D. J. L. Luck and E. Reich, *Proc. Nat. Acad. Sci. U.S. 52*, 931 (1964)] and chloroplasts (11). R. R. Meyer and M. V. Simpson [*Proc. Nat. Acad. Sci. U.S. 61*, 130 (1960)] showed that mitochondrial DNA polymerase was distinct from that located in the nucleus.
18. A. E. Smith and K. A. Marcker, *J. Mol. Biol. 38*, 241 (1968); J. E. Darnell, *Biochem. Biophys. Res. Commun. 34*, 205 (1969).
19. J. W. Lyttleton, *Exp. Cell Res. 26*, 312 (1962); N. K. Boardman, R. I. B. Franki, S. G. Wildman, *J. Mol. Biol. 17*, 470 (1966); E. Stutz and H. Noll, *Proc. Nat. Acad. Sci. U.S. 57*, 774 (1967); L. S. Dure, J. L. Epler, W. E. Barnett, *ibid.* 58, 1883 (1967).
20. E. Stutz and H. Noll, *Proc. Nat. Acad. Sci. U.S. 57*, 774 (1967); U.E. Loening and J. Ingle, *Nature 215*, 363 (1967); additional references are summarized by S. Nass (9).
21. G. F. Kalf, *Arch. Biochem. Biophys. 101*, 350 (1963); E. Wintersberger, *Biochem. Z. 341*, 409 (1965); G. D. Clark-Walker and A.W. Linnane, *Biochem. Biophys. Res. Commun. 25*, 8 (1966); *J. Cell Biol. 34*, 1 (1967).
22. Cytoplasmic and chloroplast ribosomes isolated from the same cells have been shown to differ in this respect in *Euglena* by J. M. Eisenstadt and G. Brawerman [*J. Mol. Biol. 10*, 392 (1964)] and in *Nicotiana* by R. J. Ellis [*Science 163*, 477 (1969)].
23. A. Marcus and J. Feeley, *J. Biol. Chem. 240*, 1675 (1965); B. Parisi and O. Ciferti, *Biochemistry 5*, 1638 (1966); R. J. Ellis and I. R. MacDonald, *Plant Physiol. 42*, 1297 (1967).
24. R. J. Ellis, *Phytochemistry 3*, 221 (1964).
25. Y. Suyama, *Biochemistry 6*, 2829 (1967); H. Fukuhara, *Proc. Nat. Acad. Sci. U.S. 58*, 1065 (1967); N. S. Scott and R. M. Smilie, *Biochem. Biophys. Res. Commun. 28*, 598 (1967).
26. S. Nass, *Intr. Rev. Cytol. 25*, 55 (1969).
27. E.J. DuPraw, *Cell and Molecular Biology* (Academic Press, N.Y. 1968), p. 154.
28. A. E. J. Engel, B. Nagy, L. A. Nagy, *Science 161*, 1005 (1968).
29. J. W. Schopf, in *McGraw-Hill Yearb. Sci. Technol.* 1967), p. 47.
30. The reasoning is that the development of a layer of ozone sufficient to protect land organisms from the destructive effects of ultra-violet radiation was necessary before the land could be fully occupied, and once this had occurred, the occupation may have happened rather rapidly.
31. G. H. Beale, A. Jurand, J. R. Preer, *J. Cell Sci. 5*, 65 (1969).
32. D. Smith, L. Muscatine, D. Lewis, *Biol. Rev. 44*, 17 (1969).
33. M. Droop, *Symp. Soc. Gen. Microbiol. 13*, 171 (1963).
34. This argument has recently been strengthened by the demonstration of a form of "mitosis" in the Pyrrhophyta totally distinct from that in any other organism: see D. F. Kubai and H. Ris [*J. Cell Biol. 40*, 508 (1969)]. Any direct phylogenetic connection between the dinoflagellates and, for example, the brown algae appears extremely implausible.

35. L. Sagan, *J. Theor. Biol. 14*, 225 (1967).
36. P. Echlin and I. Morris, *Biol. Rev. 40*, 193 (1965).
37. H. W. Siegelman, D. J. Chapman, W. J. Cole, in *Porphvrins and Related Compounds*, T. W. Goodwin, Ed. (Academic Press, New York, 1968).
38. Review by C. Ó hEocha, in *Chemistry and chemistry of Plant Pigments*, T. W. Goodwin, Ed. (Academic Press, New York 1968).
39. C. Ó hEocha and M. Raftery, *Nature 184*, 1049 (1959).
40. M. B. Allen, *Arch. Mikrobiol. 32*, 270 (1959).
41. M. Droop, *Symp. Soc. Gen. Microbiol. 13*, 171 (1963); L. Geitler, *Syncyanosen*, in Ruhland's *Handb. Pflanzenphysiol. 11*, 530 (1959).
42. E. Schnepf, *Arch. Mikrobiol, 49*, 112 (1964).
43. W. T. Hall and G. Claus, *J. Cell Biol. 19*, 551 (1963).
44. Summarized in N. Lang, *Annu. Rev. Microbiol. 22*, 20 (1968).
45. M. B. Allen, *The Chlorophylls*, L. P. Vernon and G. R. Seely, Eds. (Academic Press, New York, 1966), pp. 511-519.
46. M. R. Michel-wolwertz, C. Sironval, J. C. Goedheer, *Biochim. Biophys. Acta 94*, 584 (1965).
47. J. T. O. Kirk, in *The Plastids*, J. T. O. Kirk and R. A. E. Tilney-Bassett, Eds. (Freeman, San Francisco, 1967), pp. 30-47.
48. J. W. Schopf, "Antiquity and evolution of Precambrian life," in *McGraw-Hill Yearb. Sci. Technol.* (1967), pp. 47-55.
49. Arguments such as that presented by A. Allsopp [*New Phytol. 68*, 591 (1969)], who views the red algae as an intermediate group between procaryotic and eucaryotic cells, are therefore considered invalid; they are based entirely on the properties of the chloroplasts of red algae.
50. F. T. Haxo and D. C. Fork, *Nature 184*, 1051 (1959).
51. L. P. Vernon and G. R. Seely, Eds., *The Chlorophylls* (Academic Press, New York, 1966).
52. G. W. Prescott, *The Algae: A Review* (Houghton Miffin, Boston, 1968).
53. P. C. Silva, in *Physiology and Biochemistry of Algae*, R. A. Lewin, Ed. (Academic Press, New York, 1962), pp. 827-837; the carotenoid relationships are discussed by T. W. Goodwin [in *Chemistry and Biochemistry of Plant Pigments*, T. W. Goodwin, Ed. (Academic Press, New York, 1965), pp. 130-133].
54. As suggested by L. Sagan, *J. Theor. Biol. 14*, 252 (1967).
55. J. J. Wolken, *Euglena* (Appleton-Century-Crofts, New York, ed. 2, 1967); G. F. Leedale, *Euglenoid Flagellates* (Prentice-Hall, Englewood Cliffs, N.J., 1967).
56. The carotenoid pigments of most green algae and vascular plants are likewise similar; see T. W. Goodwin, in *Chemistry and Biochemistry of Plant Pigments*, T. W. Goodwin, Ed. (Academic Press, New York, 1965), p. 130.
57. The presence of fucoxanthin in Phaeophyta and Chrysophyta and its absence in Xanthophyta will likewise have to be taken into account in evaluating these relationships; T. W. Goodwin, in *Chemistry and Biochemistry of Plant Pigments*, T. W. Goodwin, Ed. (Academic Press, New York, 1965), p. 133.
58. M. Edelman, *Bacteriol. Rev. 31*, 315 (1967).
59. I thank R. B. Flavell (Cambridge Plant Breeding Institute) and A. Staehelin (Harvard University) as well as my colleagues P. C. Hanawalt, R. W. Holm, D. O. Woodward, and C. Yanofsky for their helpful comments on this manuscript. Supported in part by NSF grant GB 7949X.

4. *The Golgi Apparatus*

D. H. NORTHCOTE

The Golgi apparatus must be considered together with the endoplasmic reticulum, membrane bounded vesicles, and lysosomes as part of a complex membrane system within the cytoplasm of the cell. These organelles are always present but they are in a constant state of flux: they are constantly being formed, changed, broken down, and re-formed and they move within the cytoplasm. The space or lumen enclosed by membranes within the cell is made up of that of the nuclear envelope; the rough and smooth endoplasmic reticulum; the cisternae, tubules, and vesicles of the Golgi apparatus; the lysosomes; the pinocytotic vesicles; and the vacuole which is contained by the tonoplast in plant cells. It is also necessary to include, as part of this system, the region outside the cell beyond the plasmalemma. All these spaces may be considered to be connected either by a functional continuity or by a direct morphological connection. These membrane-bounded spaces are thus separated from the rest of the cytoplasm and nucleoplasm of the cell. They constitute a transport system throughout the cell and to its exterior and also in some cells from the exterior back into vesicles and lysosomes within the cell.

The Golgi apparatus is at an important junction of this membrane-bounded transport system between the nucleus and the endoplasmic reticulum on the one hand and the vesicles, lysosomes, tonoplast, and plasmalemma on the other [1]. Within the complete system, proteins, polysaccharides, glycoproteins, and probably membrane lipids and lipoproteins are formed and flow. The connection with the nucleus is also probably very important for the transport of material from this central control site during the growth and development of the cell.

Form of the Golgi apparatus

The existence of the Golgi apparatus has been doubted since its very first description—by Camillo Golgi in 1898—and the interpretation of images seen with the optical microscope has been constantly discussed and criticized. It can be seen in sections of fixed cells stained with metallic stains, or in the living cell using phase-contrast optics [2—5]. The organelle varies in form from a compact discrete granule or mass to a more dispersed filamentous reticulum. It is pleomorphic and its shape as well as its function varies with the metabolic and developmental state of the cell [6—8]. The advent of a thin sectioning technique for the electron microscope made possible a more detailed and definitive description of an organelle which was designated as the Golgi apparatus [9—10].

The model of the apparatus is based upon images seen in thin sections and in freeze-etched replicas of plant and animal tissues and upon negatively and positively stained cell fractions which have been prepared by centrifugation from broken cells in order to separate the Golgi apparatus from other cell organelles and membrane material [11]. It is a characteristic, easily recognizable structure and it occurs in nearly all the cells of animals and plants. It consists of a stack of disc-shaped cisternae or saccules (usually about 3–12) one above the other; they are all slightly curved, giving concave and convex faces to the whole stack. In some cases material can be seen between the cisternae of a stack and this may be part of a definite intercisternal structure. From the edges of the cisternae an anastomosing network of tubules arises and these in places swell to form various types of vesicles. The cisternae may, therefore, not be formed of continuous sheets of membranes but the membranes may be fenestrated; this fenestration can become more pronounced progressively down the stack of the cisternae (from convex to concave face) so that at the convex surface the membranes are complete and at the other they are extensively pierced by pores which fragment the cisternae into the network of tubules [9, 10, 12–14].

Formation

The development and formation of the Golgi apparatus may involve a series of different processes, such as the synthesis of a stack of cisternae, in the absence of a pre-existing apparatus within the cell; alteration in the type and number of vesicles formed; and increases in the number of stacks, in the number and size of cisternae in each stack, and in the extent of tubular and vesicular regions of the stack [6]. All these forms of growth may be part of a single process of membrane conversion and formation, but it is possible that some of the details of the separate stages might be different. For instance, individual stacks of cisternae might arise from pre-existing stacks by division or fragmentation, which would be a different process from a *de novo* formation of the apparatus. Nevertheless, a good deal of the cytological, biochemical, and chemical studies can be interpreted by postulating that there is a flow of membrane material from the endoplasmic reticulum via the Golgi apparatus to the plasmalemma. Thus a theory for the formation of the Golgi apparatus is that the cisternal membranes arise from the rough endoplasmic reticulum, which changes to smooth endoplasmic reticulum and then becomes the Golgi cisternae. These cisternae break down to vesicles which can fuse with and extend the plasmalemma.

The maintenance of the Golgi complex requires the presence of the nucleus and in its absence, or in the presence of actinomycin D, the Golgi bodies decrease in size and eventually are no longer detectable [15]. In enucleated amoebae small curved, smooth cisternae are found within 30 minutes to one hour after renucleation, and these aggregate in certain parts of the cytoplasm. A similar progressive alignment of single cisternae to form stacks has been seen in the cells of the ciliate *Tetrahymena* during mating, when Golgi bodies are formed [7]. In the amoebae the Golgi complexes grow both in size and number during the 6-24 hours after renucleation, and during this time the rough endoplasmic reticulum seems to participate in the process of membrane production. This is shown by the presence of a dense material both in the lumen of the endoplasmic reticulum and in the lumen of the cisternae of the Golgi apparatus, and also by the appearance of a direct continuity between the endoplasmic reticulum and the Golgi cisternae [15]. A.V. Grimstone [16] in an earlier study on the regeneration of the Golgi apparatus in the flagellate *Trichonympha* has also shown a close association between the endoplasmic reticulum and the Golgi cisternae; in addition, the Golgi bodies appeared sequentially after the regeneration of the rough endoplasmic reticulum during re-feeding of the starved organism.

The radio-autographic evidence of G. E. Palade and his co-workers [17, 18] demonstrated that material was transferred from the rough endoplasmic reticulum by means of small vesicles to the outer cisternae of the Golgi complex. Thus although there is no unequivocal evidence of a direct connection, the endoplasmic reticulum system is in continuity with the Golgi bodies. Numerous experimental results have confirmed these ideas and also indicate that there are transitional sections of smooth endoplasmic reticulum between the rough endoplasmic reticulum and the Golgi cisternae. Since the outermost cisternae of the Golgi apparatus could be constantly formed by the fusion of vesicles derived from the reticulum system, while the inner cisternae in any one stack could become fenestrated and pinch off vesicles, the Golgi apparatus can be conceived of as having a newly forming face at one surface and a mature or secreting face at the opposite surface. In the colonic goblet cells M. Neutra and C. P. Leblond [19, see also 20] have estimated that the entire Golgi apparatus is formed and re-formed within 20–40 minutes and that individual cisternae are replaced once every 2–4 minutes.

Chemical analysis of the membranes of the Golgi apparatus also indicate that they have a composition between that of the endoplasmic reticulum and of the plasmalemma [21]. This has also been suggested as a result of a study of the cytoplasmic membranes of the plant pathogenic fungus *Pythium ultimum* Trow, in which the overall membrane thickness, the staining intensity, and the substructural pattern of stain deposition were investigated. It was shown that at one extreme the plasmalemma and the membranes of the vesicles (free in the cytoplasm and attached to the Golgi cisternae at the distal, mature face) stained intensely, were thicker (75 Å), and were clearly unit membranes. At the other extreme, the nuclear envelope and the endoplasmic reticulum membranes stained faintly, appeared thinner (25–40 Å) and rarely revealed the dark-light-dark pattern of the unit membrane [22]. The membranes of the Golgi apparatus were differentiated across the stack so that those at the forming face appeared similar to those of the endoplasmic reticulum and nucleus whereas those at the mature face resembled the plasmalemma. Membranes in the middle of the stack were intermediate in their appearance. These results indicated that the membranes were modified within the Golgi stack and that one of the functions of the apparatus is to alter the membranes of the endoplasmic reticulum so that they resemble that of the plasmalemma. Presumably this process is irreversible. Thus the Golgi apparatus represents a one-way valve within the following system: nuclear envelope → endoplasmic reticulum → vesicles → Golgi apparatus → vesicles → plasmalemma.

The function of the Golgi apparatus

One of the functions of the apparatus is to serve as part of an internal transport system of the cell, but coupled with this function—which involves secretory and digestive processes—there are synthetic activities, not the least part of which is the formation and modification of lipoprotein membranes. The most obvious way in which this is achieved is by the incorporation of a mucopolysaccharide coat which appears inside the vesicles derived from the Golgi cisternae but which, when exported to the plasmalemma, appears at the outside of this membrane. This could imply that not only is the outer surface coat of the membrane derived and placed at the surface of a pre-existing and independently formed plasma membrane but that the whole complex macromolecular architecture of the cell membrane itself—lipoprotein and mucopolysaccharide—is assembled by the Golgi apparatus and its vesicles. This is then transferred to the growing points on the surface of the cell where the membrane is being extended [23], sometimes in a very specialized way, as in the brush borders of the intestinal absorptive cells [24].

Another general feature of the Golgi apparatus is that it is concerned with the formation and packaging of material for export from the cell across the plasmalemma by a process similar to pinocytosis in reverse. In this way cell wall material is deposited in plants [8] and secretory proteins, digestive enzymes [18, 25], and the matrix material of the connective tissue [26] are transported in animal tissues. There are also many examples of secretions derived from the Golgi apparatus which form internal cell structures such as the acrosome of the spermatocyte and spermatid [27]; certain yolk substances of oocytes [28]; and the tubular inclusions of endothelial cells [29]. Very many materials are now known to be packaged and transported through the Golgi apparatus and its associated vesicles in animals and plants [5, 30, 31]. Many of these substances are polysaccharides, or they are proteins or lipids which acquire carbohydrate moieties to become glycoproteins or glycolipoproteins. The carbohydrate is acquired in the smooth membrane systems of the Golgi apparatus and its vesicles and it may be that conjugation with carbohydrate is a prerequisite for subsequent transport of the material across the plasmalemma.

With the electron microscope, substances can sometimes be seen in the endoplasmic reticulum, in the Golgi apparatus, in the vesicles, and outside the plasmalemma; often it appears that the same substance occurs at all these sites. An apparent progression of the form of the material in the lumens of these organelles can be seen, making it very probable that a series of events and flow of products within the cell from one part of the membrane system to the other occurs [32]. I. Manton's work [33, 34] on the scale production in green and brown algae is of particular interest, since these scales, which have a characteristic shape, can be seen within the Golgi cisternae and they are progressively developed and transported in the vesicles to the outside of the cell. The scales have a dorsoventral asymmetry and are orientated in a definite way within the vesicle. Since the scale continues to develop in this asymmetric way within the vesicle before it is finally transported to the outside of the cell, it would seem that the vesicle also has a dorso-ventral asymmetry either in the lumen or on the membranes. A physical and chemical investigation of the scales produced in the Golgi apparatus of one of these algae has indicated that, in part, they consist of polysaccharide [35]. Direct experimental evidence which clearly demonstrates a flow of material and a possible modification of its structure within the membrane system of the cell can be obtained by the use of radioactive markers. The passage of the labelled material through the cell can be followed by direct autoradiography of thin sections of the tissue prepared for electron microscopy (figures 11; 12). It can also be followed by determining the kinetics of the incorporation of the radioactivity supplied as a short pulse of the precursor into the tissue and preparing from the tissue cell fractions which separate the various parts of the membrane system during the experiment.

The intracellular transport of secretory protein in the pancreatic exocrine cells has been studied by Palade and his coworkers [17, 18]. They have shown that the digestive enzymes (for example, α-amylase) or their precursors (for example, α-chymotrypsinogen) which are synthesized on the ribosomes at the outer surface of the rough endoplasmic reticulum are transferred by some mechanism, at present unknown, across the reticulum membrane into the lumen. Once in the lumen they are transported to elements of smooth endoplasmic reticulum where they are packed into small vesicles and carried into the forming surface of the Golgi apparatus. In the Golgi body the protein is concentrated into zymogen granules which are membrane-bounded and which arise from the Golgi cisternae. These granules migrate to the cell apex and discharge their contents into the pancreatic ducts, whence they pass to the intestine. The passage of the material to the outside of the cell is effected by reverse pinocytosis in which the membrane of the granule fuses with, and becomes part of, the cell membrane.

A similar series of events has been shown to occur in the goblet cells of the colon of rats by Leblond and his coworkers [19, 26, see also 32]. In this tissue a protein synthesized at the rough endoplasmic reticulum is passed to the Golgi apparatus, where it is conjugated with carbohydrate and the resultant mucopolysaccharide is transported as mucigen granules to become the mucus of the colon.

The wall of the plant cell is composed of polysaccharides which can be classified into two types: a matrix composed of pectic substances and hemicelluloses, and a fibrillar network of α-cellulose [36]. The first stage of wall formation is at cytokinesis when a cell-plate is developed at the centre of the mitotic spindle and this is progressively enlarged and thickened in an organized manner. The cell plate has around it a membrane which will become the new plasmalemma of each daughter cell. The material of the matrix of the cell plate and of the membrane is derived from vesicles, some of which can be seen to be pinched off from the Golgi apparatus and which move to the site of cell plate formation where they fuse together to give the matrix substance and the cell membrane [37, 38]. Into the clear matrix of the cell plate microfibrils of cellulose are woven at a later stage [38]. The wall is continually enlarged by the deposition of polysaccharide matrix material into it from the Golgi vesicles. The type of polysaccharide incorporated into the wall changes during its development although the method of deposition does not; this implies that the mechanism for synthesizing polysaccharides supposed to occur within the Golgi cisternae and vesicles changes during the differentiation of the cell [36].

It has been suggested that the cellulose is also synthesized within the Golgi vesicles

[39, 40] but there is other evidence which indicates that it is formed directly in the wall from the plasmalemma [36]. Since it seems likely that the plasmalemma is itself formed by membrane modification and synthesis which occurs at the Golgi apparatus and its associated vesicles, it may well be that the enzymes for the production of cellulose at the plasmalemma surface are transported there via the Golgi system.

During the thickening of the cell wall the material is often deposited in an organized manner at definite localized sites, giving rise to spiral or reticulate thickening of the wall. This phenomenon, together with others of a similar nature, indicates that it is necessary to consider not only the formation of the material and its packaging for transport but also the direction and site of transfer; this must be an organized and controlled process. In plant cells during cell plate and wall formation this directed transport can be seen, in part, to be mediated by the alignment and laying down of microtubules within the cytoplasm where the vesicles are incorporated into the cell plate [41, 42] or the wall [8, 43].

Enzymes and metabolic activity of the Golgi apparatus

The Golgi apparatus seems to effect both synthesis and modification of the material contained in the lumen of its cisternae and vesicles. In many instances this is brought about by a formation of oligosaccharide or polysaccharide; the membrane system should therefore contain enzymes—such as nucleoside diphosphate sugar transglycosylases—necessary for this synthesis. In addition, vesicles of the Golgi apparatus possibly form primary lysosomes containing hydrolytic enzymes [44]. By fusion with pinocytotic vesicles or with autophagic vesicles, these give rise to secondary lysosomes or to the digestive vacuoles of animal cells [45]. The hydrolytic enzymes are presumably first made at the rough endoplasmic reticulum; and they can pass to the smooth reticulum; this can give rise to primary lysosomes directly or the enzymes can be transferred via vesicles to the Golgi apparatus for further modification and concentration. These hypotheses about the role of the Golgi apparatus and endoplasmic reticulum in the formation of lysosomes are based on cytochemical evidence for the distribution of hydrolytic enzymes such as phosphatases, in these membrane systems [46, 47].

The Golgi apparatus and its vesicles must therefore carry enzymes of two types: firstly, those which are used for the synthesis of lipoprotein membranes and polysaccharides, and, secondly, those which are present because they are being concentrated, possibly modified, and transferred through the membrane system to be used in another part of the cell or organism. In either case the enzyme activity of the apparatus will change not only with the reversible changes which can take place in a cell's metabolism because of biochemical feedback mechanisms or supply of hormones [48, 49], but with the slower irreversible changes which are part of the development or differentiation of the cell as the organism grows and matures [36].

Cytochemical techniques adapted to the electron microscope, show that in many cells acid phosphatases are present together with nucleotide diphosphatase and thiamine pyrophosphatase. The latter may be different from the nucleotide phosphatase or (as in liver microsomes) it may be the same enzyme [50]. It has been suggested that thiamine pyrophosphatase may be a consistent marker enzyme for the Golgi apparatus [51, 52]. These studies have also demonstrated that not all the cisternae in a single stack of the Golgi body carry the same enzyme, but that the activity may be localized in certain saccules at the mature or forming surface. It has further been shown that the endoplasmic reticulum may contain enzymic activities related to those of the vesicles near the Golgi apparatus, again suggesting a functional connection between the membrane systems, in the region of the Golgi body [47].

A direct analysis of enzymic activity may be carried out on purified fractions made from broken cells [53, 54]. In Golgi bodies prepared from rat and bovine liver tissue, a high activity of an antimycin insensitive NADH-cytochrome c reductase and a NADPH-cytochrome c reductase was found. These fractions also concentrate various sugar transferase activities which may be important for glycoprotein synthesis [53, 55, 56].

Autoradiographic evidence has indicated that in plant tissues pectic polysaccharides and hemiculluloses are synthesized and transported by means of the Golgi apparatus [57, 60]. The transferases which form these polysaccharides from nucleoside diphosphate sugar compounds have been prepared from plant cells, and all have been shown to be membrane-bound [61]. H. Kauss and his colleagues [62, 63] have shown that the enzymes for the polymerization and the methylation of polygalacturonic acid are present in a membrane-bound compartment of the cell, and that this is a complex unit which allows an organized co-operation of several enzymes necessary to build up a heteropolysaccharide. This unit might well be the Golgi apparatus and its associated vesicles.

Conclusions

The Golgi apparatus is a characteristic feature of most cells having a visible nucleus. It is part of a membrane system which is concerned with intra- and inter-cellular transport, but since synthetic activities also occur within the system it is comparable more to the production line of an industrial process rather than a pipeline through which preformed material is passed. The activity of the Golgi apparatus changes during the metabolic variations of the cell and during growth and differentiation. It lies at a key position in the transport function of the system, effecting an irreversible connection between the nuclear envelope and endoplasmic reticulum system on the one hand and vesicles and plasmalemma on the other. Thus this localized part of the membrane system acts like a valve, regulating the synthesis and organized distribution of intra- and inter-cellular materials.

REFERENCES

[1] De Duve, C. In "Lysosomes in biology and pathology." J. T. Dingle and H. B. Fell (Editors). Vol. I, 3, North-Holland Publ., Amsterdam, 1969.
[2] Golgi, C. *Archo ital. Biol., 30,* 60, 278, 1898.

[3] Dalton, A. J. and Felix, M. *Nature, Lond., 170,* 541, 1952.

[4] *Idem. Am. F. Anat., 92,* 277, 1953.

[5] Beams, H. W. and Kessel, R. G. *Int. Rev. Cytol., 23,* 209, 1968.

[6] Flickinger, C. J. *J. ultrastruct, Res., 27,* 344, 1969.

[7] Elliott, A. M. and Zieg, R. G. *J. Cell Biol., 36,* 391, 1968.

[8] Northcote, D. H. *Proc. R. Soc., B., 173,* 21, 1969.

[9] Mollenhauer, H. H., Morré, D. J., and Bergmann, L. *Anat. Rec., 158,* 313, 1967.

[10] Mollenhauer, H. H. and Morré, D. J. *A. Rev. Pl. Physiol., 17,* 27, 1966.

[11] Cunningham, W. P., Morré, D. J., and Mollenhauer, H. H. *J. Cell Biol., 28,* 169, 1966.

[12] Mollenhauer, H. H. and Morré, D. J. *Ibid., 29, 373, 1966.*

[13] Flickinger, C. J. *Anat. Rec., 163,* 39, 1969.

[14] Morré, D. J. and Mollenhauer, H. H. *J. Cell Biol., 23,* 295, 1964.

[15] Flickinger, C. J. *Ibid., 43,* 250, 1969.

[16] Grimstone, A. V. *J. Biophys. Biochem. Cytol., 6,* 369, 1959.

[17] Caro, L. G. and Palade, G. E. *J. Cell Biol., 20,* 473, 1964.

[18] Jamieson, J. D. and Palade, G. E. *Ibid., 34,* 577, 597, 1967.

[19] Neutra, M. and Leblond, C. P. *Ibid., 30,* 119, 1966.

[20] Brown, R. M. *Ibid., 41,* 109, 1969.

[21] Keenan, T. W. and Morré, D. J. *Biochemistry, N.Y., 9,* 19, 1970.

[22] Grove, S. N., Bracker, C. E., and Morré, D. J. *Science, N.Y., 161,* 171, 1968.

[23] Stockem, W. *Histochemie, 18,* 217, 1969.

[24] Bonneville, M. A. and Weinstock, M. *J. Cell Biol., 44,* 151, 1970.

[25] Zeigel, R. F. and Dalton, A. J. *Ibid., 15,* 45, 1962.

[26] Neutra, M. and Leblond, C. P. *Ibid., 30,* 137, 1966.

[27] Fawcett, D. W. In "An Atlas of Fine Structure. The Cell, its Organelles and Inclusions," p. 128. Saunders, Philadelphia. 1966.

[28] Kessel, R. G. *J. ultrastruct. Res., 16,* 305, 1966.

[29] Sengel, A. and Stoebner, P. *J. Cell Biol., 44,* 223, 1970.

[30] Schnepf, E. "Protoplasmatologia, Vol. 8. 'Sekretion und Exkretion bei Pflanzen'." Springer-Verlag, Vienna. 1969.

[31] Sievers, A. In "Funktionelle und morphologische Organisation der Zelle: Sekretion and Exkretion," p. 89. Springer, Berlin. 1965.

[32] Thiéry, J. P. *J. Microscopie, 8,* 689, 1969.

[33] Manton, J. *J. Cell Sci., 1,* 375, 1966.

[34] *Idem. Ibid., 2,* 411, 1967.

[35] Green, J. C. and Jennings, D. H. *J. exp. Bot., 18,* 359, 1967.

[36] Northcote, D. H. *Essays Biochem., 5,* 90, 1969.

[37] Whaley, W. G., Dauwalder, M., and Kephart, J. E., *J. ultrastruct. Res., 15,* 169, 1966.

[38] Roberts, K. and Northcote, D. H. *J. Cell Sci., 6,* 299, 1970.

[39] Brown, R. M., Franke, W. W., Kleinig, H., Folk, H., and Sitte, P. *Science, N.Y., 166,* 894, 1969.

[40] Ray, P. M., Shininger, T. L. and Ray, M. M. *Proc. natn. Acad. Sci. U.S.A., 64,* 605, 1969.

[41] Esau, K. and Gill, R. H. *Planta, 67,* 168, 1965.

[42] Pickett-Heaps, J. D. and Northcote, D. H. *J. Cell Sci., 1,* 121, 1966.

[43] Wooding, F. B. P. and Northcote, D. H. *J. Cell Biol.*, *23*, 327, 1964.
[44] Novikoff, A. B. "Ciba Foundation Symposium Lysosomes," p. 36. Churchill, London. 1963.
[45] Cohn, Z. A. and Fedorko, M. In "Lysosomes in Biology and Pathology." J. T. Dingle and H. B. Fell (Editors). Vo. 1, 43. North-Holland Publ., Amsterdam. 1969.
[46] Novikoff, A. B., Essner, E., Goldfischer, S., and Heus, M. *Symp. Int. Soc. Cell Biol.*, *1*, 149, 1962.
[47] Novikoff, A. B., Essner, E., and Quintana, N. *Fedn Proc. Fedn Am. Socs exp. Biol.*, *23*, 1010, 1969.
[48] Hopkins, C. R. *Tissue Cell*, *1*, 653, 1969.
[49] Smith, R. E. and Farquhar, M. G. *J. Cell Biol.*, *31*, 319, 1966.
[50] Yamazaki, M. and Hayaishi, O. *J. biol. Chem.*, *243*, 2934, 1968.
[51] Dauwalder, M., Kephart, J. E., and Whaley, W. G. *J. Cell Biol.*, *31*, 25A, 1966.
[52] Shanthaveerappa, T. R. and Bourne, G. H. *Cellule Rec. Cytol. Histol.*, *65*, 201, 1965.
[53] Fleischer, B., Fleischer, S., and Ozawa, H. *J. Cell Biol.*, *43*, 59, 1969.
[54] Cheetham, R. D., Morré, D. J., and Yunghans, W. N. *Ibid.*, *44*, 492, 1970.
[55] Morré, D. J., Merlin, L. M., and Keenan, T. W. *Biochem. biophys. Res. Commun.*, *37*, 813, 1969.
[56] Ovtracht, L., Morré, D. J., and Merlin, L. M. *J. Microscopie*, *8*, 989, 1969.
[57] Northcote, D. H. and Pickett-Heaps, J. D. *Biochem. J.*, *98*, 159, 1966.
[58] Wooding, F. B. P. *J. Cell Sci.*, *3*, 71, 1968.
[59] Northcote, D. H. and Wooding, F. B. P. *Sci. Prog.*, *Oxford*, *56*, 35, 1968.
[60] *Idem. Proc. R. Soc.*, *B*, *163*, 524, 1966.
[61] Hassid, W. Z. *Science, N.Y.*, *165*, 137, 1969.
[62] Kauss, H., Swanson, A. L., Arnold, R., and Odzuck, W. *Biochim. Biophys. Acta*, *192*, 55, 1969.
[63] Kauss, H. and Swanson, A. L. *Z. Naturf.*, *24*, 28, 1969.

5. *Effects of Particles on Lysosomes**

ANTHONY C. ALLISON

Lysosomes are small membranous sacs distributed through the cytoplasm of cells. The lysosomes contain digestive enzymes with a variety of functions. White blood cells, for example, evidently use lysosomal enzymes to destroy invading bacteria. If, however, a cell's lysosomes are damaged in some way, the escaping enzymes may destroy the cell itself. This paper describes how silica particles and asbestos fibers enter cells, damage their lysosomal membranes, and bring about diseases such as silicosis and asbestosis. It appears from this type of evidence that persons working with these substances must be carefully protected. The author is head of the Cell Pathology Division of the Clinical Research Center at Harrow, England.

In the nineteenth century, when advances in microscopy made possible the study of pathology at the cellular level, it became clear that changes in diseased tissue depend on changes in the growth and function of cells. In recent years the electron microscope and the centrifuge have made for increasingly effective investigations at the subcellular level. The aim of the cell biologist is to identify the enzymes that catalyse cellular reactions and find out in what organelles of the cell each reaction takes place. A specific task for the pathologist is to identify and localize the primary error in function that may lead to a secondary disorder in other organelles and systems; in other words, to learn how disease processes may begin with a specific malfunction in an organelle. In certain disorders, for example, the primary damage is to the energy-producing reaction in the mitochondria; in others it is to the protein-synthesizing apparatus of the ribosomes.

When lysosomes were first recognized as distinct organelles, it was clear that they would be of special interest in subcellular pathology. They can digest things that enter the cell; they can also break down part or all of the cell itself or tissues outside the cell. It now appears that in doing so they are implicated in the development and the death of tissues, in diseases such as silicosis and gout, in the immune process and thus perhaps in autoimmune disease and in many other physiological and pathological reactions.

Source: Reprinted from the December 1970 issue of the *Advancement of Science* by permission of the author and the British Association for the Advancement of Science.

*Text of paper originally entitled *The role of phagocytosis in disease* delivered to Section D (Zoology) on 3 September 1970 at the Durham Meeting of the British Association.

The simplest and the most obvious role of a packet of digestive enzymes within a cell is in necrosis and autolysis, the death and self-dissolution of tissues. The membrane of the primary lysosome simply dissolves, liberating the enzymes to consume the cell. This can follow wounding or other damage to tissue or it can take place naturally in the course of development: when the corpus luteum of the ovary degenerates, for instance, or when a tadpole loses its tail. Any time a tissue or an organ is isolated from its supply of oxygen or nutrients under sterile conditions it breaks down rapidly; the large molecules such as proteins, lipids, nucleic acids and carbohydrates are digested, and there is evidence that enzymes from lysosomes play a part in the process. Rudolf Weber of the University of Berne (see Weber, 1969) found that the concentration of lysosomal enzymes in the tail of an amphibian increases before metamorphosis, and the concentration of these enzymes not consumed in the process of tissue digestion increases as the tail is resorbed.

Soon after lysosomes were first described it was suggested that the release of their enzymes into the cytoplasm or outside the cell might be the primary event that accounted for many other and different types of tissue damage. More detailed scrutiny showed that biochemical changes in other organelles sometimes preceded those that could be demonstrated in lysosomes; the conviction grew that lysosomal release was a secondary effect. It now seems that both situations can prevail. In some cases it is virtually certain that the primary event is an increase in the permeability of the lysosomal membrane, with the consequent release of hydrolytic enzymes. In others the lysosomal changes may indeed be secondary to reactions in other systems.

Effects of Silica Particles on Cells

Foreign particles, taken into the human body by inhalation, are usually innocuous, like the carbon particles that remain in alveolar macrophages more or less indefinitely. However, certain particles, such as those of several different crystalline forms of silicon dioxide or of asbestos, stimulate a severe fibrogenic reaction. It is generally accepted that the initial event in silicosis is the phagocytosis of silica particles by alveolar macrophages and consequent death of the cells. The particles so released are taken up by other macrophages which are in turn killed. In this way death of macrophages continues and stimulates collagen synthesis by fibroblasts in the neighborhood. Analysis of pathogenesis must therefore proceed in two stages: first determining how silica particles kill macrophages and, second, determining how this is related to fibrogenesis.

As Marks (1957) showed, the cytotoxic effects of silica can be conveniently reproduced in cultures of periotneal or alveolar macrophages, and the relative toxicity of different forms of silica, and of different dusts, on cell cultures agrees with the pathogenicity and fibrogenic activities of the dusts *in vivo* (Marks and Nagelschmidt, 1959; Vigliani, Pernis and Monaco, 1961). When my colleagues (T. Nash, J. Harington and M. Birbeck) and I began working on this problem in 1964, there was no satisfactory explanation of silica toxicity. King in his well-known "solubility" theory suggested that silicic acid liberated into the tissues from silica particles brings about deposition of collagen. Later observations did not support this interpretation, as King (1947) himself pointed out. Curran and Rowsell (1958) showed that silica particles implanted into the peritoneum in diffusion chambers do not induce any fibrogenic reaction, although silicic acid is liberated from the chambers. Vigliani and Pernis (1963) formulated an autoimmune theory of silicosis, but several workers were unable to obtain experimental evidence in support of this interpretation.

We therefore made a detailed study of the effects of toxic and non-toxic particles on cultures of macrophages, using time-lapse phase-contrast cine-micrography, histo-chemistry and electorn microscopy (Allison, Harrington and Birbeck, 1966). Particles of silica, diamond dust and other materials were rapidly included in phagosomes surrounded by single membranes. Lysosomes became attached to the phagosomes and discharged their enzymes into the phagosomes. So far there was no difference between the toxic and non-toxic particles. After about 18 hours' incubation, however, clear differences were apparent. The noxtoxic particles and associated enzymes were still enclosed in secondary lysosomes whereas many of the toxic particles and associated lysosomal enzymes had escaped into the cytoplasm. The macrophages that had ingested non-toxic particles were fully extended and moving about freely, whereas many of those exposed to toxic particles were rounded and immobile. Thus it was evident that silica particles, unlike non-toxic particles, can react with lysosomal membranes and make them permeable.

This appears to be a relatively non-specific reaction of silica with a variety of biological membrane systems. The simplest demonstration is provided by mixing washed erythrocytes with suspensions of silica particles or with silicic acid preparations (Stalder and Stober, 1965; Nash, Allison and Harrington, 1966). The erythrocytes are quite rapidly lysed by several fibrogenic forms of crystalline silica, whereas one form which is not fibrogenic (stishovite), and several other types of non-fibrogenic dust of comparable size and surface area, produce very little haemolysis. We have presented reasons (Nash *et al.*, 1966) for believing that the toxicity of silica is due to the fact that the particles are easily ingested and by interation with water form on their surface silicic acid which can act as a powerful hydrogen bonding agent.

There are two classes of hydrogen bonding compounds. The larger class comprises hydrogen acceptors such as ethers and ketones with active long-pair electrons on oxygen or nitrogen. The smaller class comprises hydrogen donors of which only the phenols are important among organic compounds and silicic and some other very weak acids among inorganic compounds. Compounds of one class interact with those of the other, so that it is not surprising that there is one class (the former) the members of which are compatible in living cells, sometimes in very high concentrations, whereas those of the other class are damaging.

Model experiments showed that hydrogen bonding of phenolic hydroxyl groups, of the type present in silicic acid, occurs with secondary amide groups of proteins, and this can lead to protein denaturation. However, the interaction with phospholipids is stronger, and we have presented evidence that this is more important in interactions with biological membrane systems. Evidence in support of the interpretation that hydrogen bonding is important in silica comes from experiments with poly-2-vinylpyridine-*N*-oxide (PPNO). Schlipkoter, Dolgner and Brockhaus (1963) found that this substance markedly reduces the amount of fibrous tissue formed after intravenous injection of silica. The toxic effects of silica on cultures of macrophages are also diminished in the presence of or after exposure to PPNO. We have shown (Allison *et al.*, 1966) that PPNO is taken up into lysosomes in much the same way as dextran, polyvinylpyrrollidine and other polymers (see de Duve and Wattiaux, 1966). However, PPNO is unique in having oxygen atoms which readily form hydrogen bonds with phenolic hydroxyl groups. Thus PPNO can preferentially interact with silicic acid on the surface of silica particles before the latter can attack lysosomal membranes. Mr. Nash and I synthesized and tested another compound, polyvinylpryidinio-acetic acid, which has even greater hydrogen bonding capacity than PPNO and very efficiently protects macrophages from silica toxicity.

These two facts are sufficient to explain why silica is so toxic to macrophages: the particles are taken up into lysosomes and readily damage lysosomal membranes through hydrogen bonding interactions. Various secondary reactions may occur. Thus, Munder, Modolell, Ferber and Fischer (1966) have found a considerable increase in the concentration of lysolecithin, as compared with lecithin, in macrophages damaged by quartz. This could follow activation of the enzyme phospholipase A, which catalyses the reaction lecithin → lysolecithin and which is known to be lysosomal. However, the fact that silica lyses erythrocytes (membranes of which do not contain demonstrable amounts of phospholipase A) shows that this process is unnecessary for interaction of silica with membrane systems, although the formation of surface-active lysolecithin would well accelerate damage induced by silica in macrophages. Suspensions of silica particles readily release enzymes from isolated liver lysosomes *in vitro*, as Stalder's experiments and our own have shown; the relatively low temperature coefficient for this release suggests that physico-chemical rather than enzymic reactions are involved.

Gout and Pseudo-Gout

Faires and McCarthy (1961) and Seegmiller, Howell and Malawista (1962) produced attacks of gout by injecting microcrystalline monosodium urate into gouty patients and non-gouty volunteers. The crystals are taken up by phagocytic cells, chiefly polymorphs, and the consequent inflammation seems to play an important part in the acute exacerbations of gout. Similar amounts of sodium urate in solution do not damage phagocytic cells or produce inflammation, and it seems that in the case of both silica and urate having the offending material in particulate form ensures its uptake by phagocytic cells and concentration in a vulnerable region, the lysosome.

Colchicine inhibits migration, particle ingestion and degranulation by polymorphonuclear leukocytes, and this may account for its well-known therapeutic effect in gout.

Calcium pyrophosphate dihydrate (CPPD) crystals are sometimes found in joints in association with sodium urate, sometimes alone. Their discovery accounts for the disease pseudo-gout, which can also bring about severe cartilage erosion.

Asbestosis

Inhalation of asbestos particles can produce a fibrogenic reaction (asbestosis). The basic mechanism appears to be similar to that of silicosis, although the association of groups of cells with long asbestos fibres and generation of iron-containing inclusions (ferritin) is characteristic of asbestosis. Small particles of certain types of asbestos (for example, crocidolite, or blue asbestos) also cause tumours of the lung, especially masotheliomas, tumours of the cell layer lining the pleural cavity. This is an important occupational hazard, which is now carefully watched.

Thus study of the effects of particles on lysosomes has been of academic interest and provided information about several types of human disease.

REFERENCES

Allison, A. C., Harington, J. S. and Birbeck, M. (1966): *J. exp. Med. 124*, 141.
Curran, R. C. and Rowsell, E. V. (1958): *J. Path. Bact. 76*, 561.
De Duve, C. and Wattiaux, E. (1966): *Ann. Rev. Physiol. 28*, 435.
Faires, J. S. and McCarthy, D. J. (1961): *Clin. Res. 9*, 329.
King, E. J. (1947): *Occupational Med. 4*, 26.
Marks, J. (1957): *Br. J. ind. Med. 14*, 81.
Marks, J. and Nagelschmidt, G. (1959): *A.M.A. Archs ind. Hlth, 20*, 383.
Munder, P. G., Modolell, M., Ferber, E. and Fischer, H. (1966): *Biochem. Z. 344*, 310.
Nash, T., Allison, A. C. and Harington, J. S. (1966): *Nature, Lond. 210*, 259.
Schlipköter, H. W., Dolgner, R. and Brockhaus, A. (1963): *German Med. Monthly, 8*, 509.
Seegmiller, J. G., Howell, R. R. and Malawista, S. E. (1962): *J. Am. med. Ass. 180*, 469.
Stalder, K. and Stöber, W. (1965): *Nature, Lond. 207*, 874.
Vigliani, E. C. and Pernis, B. (1963): *Adv. Tuberc. Res. 12*, 230.
Vigliani, E. C., Pernis, B. and Monaco, L. (1961): In *Inhaled Particles and Vapours*, p. 348. Ed. C. N. Davies, London: Pergamon Press.
Weber, R. (1969): in *Lysosomes in Biology and Pathology*, vol. II, p. 437, Ed. J. T. Dingle and H. B. Fell. Amsterdam: North Holland Publishing Co.

Robert Perron

6. Nutrition and Society

DAVID J. KALLEN

"Malnutrition during development leads to high infant mortality and lowered physical size. While severe malnutrition may lead to intellectual impairment, the direct relationship between moderate malnutrition and intelligence is still unknown. This is because both nutrition and intellectual development are associated with various social factors. While malnutrition is a medical problem, hunger is a social problem, complicated by the fact that the hungry are also subject to various other noxious social conditions. The negative effect of moderate malnutrition may stem from apathy in learning and other situations which relate to life success."—Dr. David J. Kallen received his doctorate in social psychology at the University of Michigan and for several years was with the National Institutes of Mental Health, and Child Health and Human Development before going to the College of Human Medicine, Michigan State University. His interests are in social organization and integration, social science and social welfare, and socialization. This article was in part given at the Conference on Nutrition and Human Development, East Lansing, Michigan, May 1969.

Recent discussion about the relationship between nutrition and human development has emphasized the presumed negative effects of malnutrition on intellectual capacity and on physiological development. Much less attention has been paid to the relationship between nutrition and society. (The large literature on nutritional experimentation in animals will not be reviewed here because the nutritional conditions under which the experimental animals were raised find their counterpart only under severe protein-calorie malnutrition in humans. A good summary of these studies is presented by Scrimshaw and Gordon.[1]) Since malnutrition is not randomly distributed in the social system, its negative impact will be felt largely by those who are subject to other noxious interferences with optimal development. The combination of poor nutrition and various social and economic dysfunction tends to lead to the perpetuation of a disadvantaged group in the society.

While some physical and psychological consequences of malnutrition hold for all individuals, others hold only for malnutrition which occurs during the physiological, psychological, and social development of the individual. The later will be considered here.

Source: Reprinted from *Journal of the American Medical Association, 215*(1): 94-100, 4 January 1971, published by the American Medical Association, by permission of the publisher and author.

Physiological and Psychological Consequences of Malnutrition

The known or postulated consequences of malnutrition during the development of the human individual include those discussed below.

Delayed Physical Maturation and Smaller Physical Size

Perhaps the best documented of all findings relating nutritional intake to human development is the reduction in growth rate.[1] Delay in bone development is part of this overall process that leads to smaller physical size. Jelliffe suggests that physical size for age, particularly height, weight, and body composition, is a good indicator of nutritional status.[2] The small-for-age child has probably been malnourished during some significant period of his development.

Viewed in more historical perspective the reverse might be a more accurate statement: high intake of nutritionally well-balanced foods (and perhaps a lowered energy expenditure) results in increased physical size.[3] It is perhaps a reflection of the American feeling that bigger is better that so much emphasis has been given to the reduction in physical size. However, other factors, such as climatic conditions and ritualized physical stress during infancy[4] (ie, circumcision, nose puncturing, etc) can also influence physical size.

Sexual maturation, particularly menses in females, appears to occur later in malnourished populations, and has been seen to occur at younger ages in succeeding generations in more highly developed nations. Here again, it may well be that the high nutritional level of the Western world is lowering the age of sexual maturation.

Interaction Between Malnutrition and Infectious Disease

The well-documented synergism between malnutrition and infectious disease is complicated by parasitic infestation.[5] A vicious circle develops in which poor nutrition and large parasitic loads combine to reduce resistance to infectious disease, which then places a heavier load on the individual who, in consequence, is unable to utilize the limited nutrients he does receive. Prolonged or frequent infant or weanling diarrhea has a particularly telling effect on early child development and nutritional status. Folk medicine and local beliefs about proper diet for the ill may further limit nutritional intake. Nutritional intervention in areas of endemic malnutrition can decrease both disease rate and infant and preschool deaths while having only a minimal impact on physical growth.[6]

The relationship between nutrition and infectious disease has not been studied in the more developed countries. Modern medical practice, better sanitary conditions, programs of immunization, and perhaps better-developed natural immunities apparently greatly reduce the effects of certain common diseases of childhood on children in the developed countries.

The relationship between malnutrition and mortality, particularly between the ages of 1 and 4, has been clearly demonstrated in the developing countries. Although the infant death rates in the underdeveloped countries are higher than those elsewhere, the difference in death rates for preschool children is even greater. It is between the ages of 1 and 4 that the child is at greatest risk for malnutrition. Prior to weaning the child is assured at least a minimal intake of a reasonably well-balanced food, mother's milk, although the amount and quality of the intake may be less than optimal towards the end of a prlonged nursing period. Weaning is usually accompanied by a sharp decrease in the quality of food, particularly protein, thus leading to malnutrition. When complicated by infectious disease or by weanling diarrhea of unspecified origin, a high preschool death rate results.

The relationship between malnutrition and infant and preschool mortality has not been well studied in the developed countries. As with infectious disease, the availability of medical care, better sanitary conditions, and lesser degrees of malnutrition attenuate the relationship. Infant malnutrition in the United States appears to be associated with maternal or family disease.[7,8] Maternal malnutrition has been implicated in fetal wastage, birth defects, and perhaps in prematurity. Small-for-age infants may be regarded as premature when in fact they may be the products of malnourished mothers. Birth weight of infants in Mexico has been linked to maternal stature, health, and to social characteristics associated with nutrition.[9]

Retarded Brain Development

The human brain develops through cell division until the child is approximately 6 months of age. Severe malnutrition after birth, resulting in Kwashiorkor or marasmus at an early age, interferes with the normal development of the brain. Monckeberg[10] has demonstrated that the brain is smaller than normal in children with these maladies; in fact, it may be smaller than head circumference would indicate, with a fluid similar to spinal fluid filling the cavity between the brain and the skull. Winick[11,12] has shown, in both animals and humans, that when malnutrition occurs during the period of cell division (ie, prior to 6 months of age) the resultant brain has fewer cells than a normal brain, presumably leading to an irreversible impairment. When malnutrition occurs after the end of the period of cell division, animal experiments suggest that the brain cells may be smaller than normal, but the change may not necessarily be irreversible. An increase in nutritional standards may restore the brain cells, and hence the brain, to normal size. Chase and Martin,[8] in a study of children hospitalized for malnutrition during the first year of life, show that those treated prior to 4 months of age, when compared to those treated later, have a greater recovery potential. While Chase and Martin attribute this to a shorter period of malnutrition, the recovery may stem from nutritional rehabilitation prior to the end of the period of cell division in the brain.

Animal studies have indicated that maternal malnutrition can also interfere with the rate of cell division in the infant. Human studies are less clear; while severe maternal malnutrition will reduce the size of the infants produced, evidence is not available per se on neurological development. Nor are the effects on the infant of prolonged chronic malnutrition of the mother clear, although in the underdeveloped countries physical size of infants of malnourished mothers does appear to be about the same as the size of infants of well-nourished populations for the first six months of life. Whether brain development is also normal during this early period is not known.

Apathy

Malnutrition is associated with apathy, listlessness, and unresponsiveness to external and internal stimuli. This apathy is characteristic of both adults and children during periods of malnutrition.

Effects of Severe Malnutrition on Intelligence

Evidence has been provided to indicate that marasmus and Kwashiorkor produce deficits in intellectual ability.[10,13-17] Studies of children who have suffered from severe nutritional deprivation early in life all indicate that the child performs less well than a normally nourished control on tests of intellectual competence. However, these results may be interpreted with caution for the following reasons:

Most of the studies have been of children who have been hospitalized for an extended period. Extended institutionalization, particularly of very young children, may produce developmental lags and interfere with intellectual development.[18] The effects of hospitalization stem both from separation from parents and from the lack of stimulation in the hospital. When the child is malnourished, lack of proper stimulation will be complicated by general lethargy and an inattention to stimulation, thus further confusing the interaction between the nutritional disease and its sequelae. In addition, all of these studies have been retrospective rather than prospective. Thus, the possibility exists that a preexisting mental incapacity leads to a relative lack of success in competition for food within the family, and therefore may make the child more vulnerable to malnutrition. Under such conditions, intellectual incompetence would be a cause, rather than a consequence, of malnutrition.

Those studies which have analyzed a number of dimensions of intellectual capacity indicate the possibility of a patterning of the deficit. Thus, for example, the children in Monckeberg's study[19] and in Chase and Martin's study[8] were more retarded in language development than in other areas.[6] Klein et al[13] indicates that rehabilitated malnourished children performed less well than normally nourished children on a series of tests which may be indicative of poor attention span or lowered motivation to learn. Alternatively, these results could come from deficits in speed of perception or from an impairment of short-term memory. The patterning of the results of tests of malnourished African children reported by Stoch and Smythe[15] does not appear inconsistent with this interpretation.

Effects on Intelligence of Mild Malnutrition

The evidence that severe malnutrition is associated with reduced intellectual capacity is reasonably clear, but the evidence for an association of mild or chronic malnutrition and intellectual incapacity is not. Mild malnutrition, which affects most surviving rural children in developing areas, may be regarded as a nutritional deficit sufficient to reduce the ultimate physical size of the child but not sufficient to endanger life. The major study, which gave rise to the hypothesis, was that of Stoch and Smythe.[20] However, they were in fact studying severely malnourished children from families with considerably greater social abnormality than the control children.

Cravioto and his associates[21] studied intersensory development of rural and urban Guatemalan and Mexican children between the ages of 5 and 11. Height was used as the indicator of nutritional deprivation in rural children; among urban children factors other than nutrition were assumed to account for differences in height. In rural children, height did correlate with differences in intersensory integration. However, these differences decreased with age, suggesting a lag rather than a permanent disability in the development of intersensory integration.

On the other hand, Ramos-Galvan and associates[22] suggest that rural residence and/ or lower-class status may be more important predictors of intellectual incompetence than nutritional status. They compared the intellectual performance at various ages of upper-class urban children, well-nourished and poorly nourished lower-class urban children, and well-nourished and poorly nourished rural children in Mexico. The upper-class children were all well nourished; they had the highest scores. The malnourished lower-class urban children in the younger age groups did not score as high as well-nourished children of the same class, but the malnourished urban slum children still had higher scores on the intelligence tests than the rural children. The scores of the rural children, which were the lowest of all groups, did not differ by nutritional status. However, by age 10, there were no differences in the scores of the well-nourished and poorly nourished lower-class urban children.

Monckeberg[10,19] studied urban children in Chile; his sample included middle-class children, lower-class children who had participated in a nutritional supplementation program, and lower-class children who had no nutritional supplementation. While the middle-class and the supplemented children had a lower rate of malnutrition, Monckeberg reports that this may be in part caused by differences in medical care and environmental motivation. In the lower class, a relationship was observed between growth retardation, low intellectual capacity, and low intake of animal protein.

In another study, the McKays and associate[23] provided protein supplement to malnourished preschool-aged children along with either cognitive stimulation, physical stimulation, or low stimulation which was similar to that the children received at home. While all groups gained physically, the cognitively stimulated group did better on intellectual tests than the physically stimulated, while the low-stimulation group did not differ from the control group. "This suggests that feeding alone is not sufficient to produce increased mental growth over a four-month period."

Children enrolled in Head Start classes were tested on a variety of mental development tests by Sulzer, who then conducted test scores as related to hemoglobin and hematocrit levels.[24] Lack of anemia was related to height and to age within age range. Although the results are not totally clear, the patterning of relationships among the various tests may reflect either a superiority of learning ability among the normals or a lower level or less-sustained level of motivation among the anemic. Since the relationship between height and anemia suggests that children currently anemic may have been malnourished earlier, Sulzer's findings are consistent with the report by Klein et al[13] that children who have recuperated from Kwashiorkor may have a lower motivation to learn. Clearly the relationship between malnutrition and motivation needs further exploration. From this brief review it is apparent that the relationship between nutrition and intellectual development is highly complicated.

Behavioral Consequences of Inadequate Nutrition

The consequences of inadequate nutrition are complicated by the fact that nutritional status is not randomly distributed in the social system. Except under most unusual situations (war is the primary example), malnutrition is most apt to be found in those segments of the society which have other characteristics which *independently* interfere with optimal development.[25] While improvement in nutrition may be a necessary condition for more optimal development, it clearly is not a sufficient condition. The task of separating the consequences of malnutrition from the consequences of the social factors found associated with malnutrition is indeed complex.

Nutrition and Social Controls

Among the consequences of malnutrition are apathy and uninvolvement in the world around one. Both apathy and uninvolvement are also consequences of anomie.[26] Anomie refers to normlessness or a state of isolation from the mainstream of society. Anomie exists in times of rapid social change,[27] that is, when the expectations and possibilities open to the individual are clearly different and changing rapidly from those he has learned to live with. Under such conditions, social control mechanisms break down because the regulatory system of the society is no longer applicable. But anomie also exists when persons or groups are placed in situations where the norms and values of the society are not completely applicable to them, that is, when there is a discontinuity between the expectations and goals taught by the society and the means available to the individual for achievement of these goals. It may also exist when individuals or groups are cut off from the mainstream of society under conditions which do not permit the full development and implementation of subgroup norms and goals. It is, of course, possible for an individual to be a member of a subgroup which is cut off from the mainstream of society, but which maintains a clear and consistent set of norms which guide the behavior of its members. In this instance anomie will not be evident in the behavior of the group members.

Anomie is much more characteristic of lower-class individuals than of middle- or upper-class individuals.[28] But malnutrition also is most apt to be found among the lower-class groups. Thus, apathy and uninvolvement may stem independently from two forces. Increasing nutritional intake will not reduce apathy if the apathy stems from both sources. But changes in social conditions will not reduce apathy if it stems from malnutrition. A dual shift, a change in both casual conditions, may be necessary to produce change in behavior or attitude. However, because malnutrition reduces the possible levels of energy expenditure, a change in nutritional status may be a necessary precondition to a change in the social position of the individual.

Nutrition and Population Control

In the developing countries, prevailing levels of nutrition may interfere with successful attempts to institute population control programs. Many nonnutritional factors, including industrialization, urbanization, and education, influence the willingness of individuals (or societies) to limit their reproduction. But the high infant and preschool-age death rate associated with malnutrition, parasitic load, and infectious disease may independently interfere with the willingness of individuals to limit family size. At least unconsciously, most women appear to have determined the ideal number of children that they wish to rear. But families in developing areas are playing a kind of Russian roulette with nature—an unknown number of children must be born in order to have at least the desired number survive. This failure to limit family size in turn contributes to malnutrition.

Malnutrition may, in some instances, contribute to the lack of industrialization by a nation. Children who must become involved in productive activities early in life cannot achieve the education required for development of an industrialized society. Thus, the cycle of poverty, subsistence production, and ignorance is perpetuated. Hence, malnutrition—or, more properly, the effects of malnutrition on health and on infant childhood mortality—contributes to the high birth rate and to other factors which retard the full development of a society. Again, while adequate nutrition will not insure the development of a nation, or even the successful implementation of a population control program, it may be a necessary precondition.

Nutrition and Infant Development

Among the consequences of malnutrition which interfere with optimal development are apathy and unresponsiveness. Beginning at an early age, the responsiveness of the infant is a determinant of the amount of stimulation that the child receives from his mother, from other persons in his environment, and from objects in the environment.[17,29] The initial mother-child interaction, which depends in part on the responsiveness of the infant as well as on the characteristics of the mother, may have an effect on later congnitive development, as well as on the exposure which the child has to various learning experiences. If the child is passive and unresponsive, a lack of interaction with the mother may interfere with adequate language development, as well as with other developmental tasks.

In fact, stimulation may effect the size of the brain. Thus, for example, Rosenzweig[30] and associates have demonstrated in rats and other experimental animals that small, but statistically significant differences in the size of the cerebral cortex result from variations in the stimulatory aspects of the environment. No information has been provided as to whether the changes are caused by differences in cell size or cell number, but the finding that mature animals can increase brain size with experience tends to support the hypothesis that the change is due mostly to a change in cell size. This set of studies also found a relationship between experience—and therefore brain size—and reversal—learning ability.

Malnutrition is not the only deterrent to a good mother-child relationship. However, malnourishment is most likely to occur in those families least able, for other reasons, to provide the infant with an optimal start in life. The mother may be unresponsive to the child because she has too many children, is physically or emotionally disorganized, or has a variety of other problems.[7] In this instance improved nutrition would not necessarily improve the mother-child relationship. But an adequate start for the child is not possible without adequate nutrition.

Pollitt[17] points out that in areas where malnutrition is endemic, the lack of responsiveness of the child is expected and normal. But in the United States, even among the poorer segments of the population, there are wide variations in the nutrition of children. If nutrition varies within a particular subgroup, the mother of the malnourished, apathetic infant may regard her child as deviant. This deviance may place the child on a deviant socialization track, in which the expectations that significant others have of his behavior are reduced because of his initial lack of responsiveness.

Nutrition and Learning

The child who lags in the *performance of his developmental tasks* will be ill prepared for the learning tasks required of him when he enters school. If he is behind when he enters, he may never have an opportunity to catch up. If the initial impression he gives is of a child who cannot fully benefit from the learning experiences provided by the school, the behaviors of his teachers towards him will reflect their expectations of his lack of performance,[31] thus reinforcing the probability of inadequate performance.

Even if the child enters school on schedule developmentally, the apathy and listlessness created by malnutrition will interfere with his learning ability. Additional interference may be created by a pervasive anxiety about food. If a child spends his time worrying about what, or even whether, he will eat that day, or if he sleeps in class because he does, not have the energy to devote to learning, he will not learn. *This may be the greatest impact that nutrition has on intellectual performance—not an effect on "innate ability" or "general intelligence" but in the utilization of the ability that the child does have.* And if the child does not learn early what the school expects him to learn, he will be forever handicapped in educational progression.

In our society, educational failure, for whatever reason, reduces the individual's life chances. Clearly, malnutrition is not the only reason why children do poorly in school or are unmotivated to achieve.[32] But for a hungry child, malnutrition may create important additional barriers to education and hence to adult achievement, and it is in those families where support and understanding for education is least that malnutrition is most apt to be found.

Nutrition and Self-Image

The self-concept[33-34]—the individual as he is known to himself—is the product of the reactions that others have to his behaviors, and of the expectations that others present for the ways in which he will behave. A series of experienced failures will lead to a negative self-image and to a self-concept in which the individual defines himself as incompetent.

But self-esteem may also be affected by a comparison of the self with culture heroes and with the standards of physical attractiveness in the society. (Thus, for example, Negro children have until recently been taught to value white exclusively and to denigrate black.[35]) Clearly, American culture values a certain body and facial conformity in females, and height and strength in males. Garn[36] reports that "the size of the individual after birth is a most conspicuous physical attribute, placing him in relation to his peers and affecting the opinions of his judges. In a culture such as ours that values sheer bigness, greater body size may be an economic as well as a social asset." To the extent that malnutrition results in reduced physical size, the male may feel that he suffers by comparison with those whose physical characteristics are nearer to the cultural ideal. In addition, physical size is frequently an important criterion for one set of culture heroes, the professional athletes. Increasingly, there is no room for the small man in athletics. At the same time, athletics remain an important pathway for upward mobility in some groups. The lower-class or minority-group athlete has more opportunities for education than his non-athletic peer of equal intellectual ability. Thus, the smaller stature of the malnourished individual may serve to cut off one potential path of upward mobility. To compound this problem, a comparison of his size with the athletic culture heroes may reinforce the negative self-image that develops from other experiences in which nutritional status is implicated.

Hunger as a Social Problem

In the affluent countries such as the United States, severe malnutrition is not a serious problem, nor is malnutrition endemic as it is in the underdeveloped Southern Hemisphere, Africa and much of Asia. But malnutrition is a problem. Although it is most apt to occur in a socially structured manner, it is a physiological condition which has social and pyschological consequences. It is associated in the United States with hunger, poverty, and their attendant evils. Our society, therefore, has defined hunger as a social problem. (A social problem is "a dislocation or dysfunction in the social system which is regarded by the society as requiring intervention by its designated agents."[37])

In an urban, industrialized society, poverty is a consequence of a division of labor which furnishes rewards on the basis of education, skill, and achievement. Because there is a division of labor, some individuals inevitably remain in the bottom positions through lack of opportunity, discrimination, an inadequate educational system, or through socially structured experiences which demonstrate that effort does not lead to reward.[38] Still other individuals are unmotivated or inept.

But America has traditionally regarded poverty not as a consequence of the organization of society but as a defect of character. Aid to the poor is provided grudgingly and sparsely. This tradition insists that if the poor did not have character defects, they would not be poor, and hence not need charity.[38] However, children, at least until they reach the attention of the juvenile courts, are held to be blameless for their condition, and the sins of the parents should not necessarily be visited upon the children. Therefore, programs of assistance are provided, although the morality holds that their levels should be scanty enough to motivate the children to become self-supporting. Welfare assistance is justified on the basis that at least some of the ineptitude of the parents is caused by early or present malnutrition and that feeding hungry children may become one way of breaking the poverty cycle. In fact, the size of the average welfare payment in the United States is such that it is inconceivable that families on welfare could provide adequate nutrition for their children. Thus, malnutrition may be one of the consequences of the social arrangements for providing for the poor in the United States.

The poor, or at least the disreputable poor, are held to a higher moral standard than either the reputable poor or the nonpoor. One of the public appeals of the present program to feed the hungry is that they provide aid in kind rather than in cash, and this insures that the aid will be used for moral purposes—feeding hungry children. The surplus food program provides aid in kind that cannot be transformed into cash, while the food-stamp program requires the expenditure of cash for controlled access to food. Aid in cash would run the risk that the cash would be spent on those "immoral" pleasures that are reserved for those who earn their way.

Even under these conditions, aid in kind is difficult to obtain. Both the surplus-commodity distribution program and the food-stamp program require the collaboration of many layers of government, which are often controlled by individuals with negative images of those to whom they provide aid. Means tests, with the consequent degradation of the individual, are still used.

In the eyes of the decision makers whose attitudes appear to have been formulated before America became an urban, industrial society,[39] the surplus-food program serves a useful purpose—it helps get rid of surplus agricultural commodities which otherwise would require tax dollars to store. Thus, the hungry may eat that for which society has no other use. There appear to be few efforts to insure good nutritional balance in the surplus-food packages, nor to integrate the contents of these packages with readily available, easily and cheaply attainable, local food-stuffs in order to insure a balanced diet.

Despite the stigmatization of the hungry, society agrees that it is better to feed them, to provide them with surplus food, than to let them starve. But we should be more concerned about the social consequences of the ways in which we do feed the hungry. Wax[40] has suggested that programs of direct feeding of the poor, particularly school lunch programs and the Head Start Nutrition program, tend to erode the role of the mother in much the same way that other welfare programs, particularly the Aid to Families with Dependent Children program, have eroded the role of the father. When children receive their primary sustenance impersonally and outside the family, the nurturant role of the mother is seriously undercut. This reduces her effectiveness as an agent of socialization, just as the destruction of the role of the father as provider reduces his effectiveness. This form of aid then interferes with a relationship of extreme importance in the socialization process, and may contribute to the mother's feelings of helplessness and inadequacy, thus further reducing her effectiveness as an agent of socialization.

Some Final Questions

It has been suggested in this communication that malnutrition, and the conditions associated with it, may interfere with the learning process, lower self-esteem, reduce social mobility, and damage the relationship between parents and children. All of these tend to produce and perpetuate within the society a disvalued and dysfunctioning group, a section of society for which the social system will have to provide as the economic opportunities for the unskilled, undereducated, and unmotivated become fewer and fewer. But there is another equally important relationship between nutrition and society.

We do not need to demonstrate intellectual impairment or any of the other possible negative consequences of hunger or malnutrition to know that permitting children to be hungry in the midst of affluence violates crucial moral norms (those norms which are necessary for the preservation of the essential forms of our society[41]). A violation of these moral norms reduces the probability of the society surviving in its present form. The persistence of hunger in the midst of affluence may well create as well as reflect attitudes and values among the affluent which are at serious variance with the values of democracy and equality. The final question then is not the impact of hunger on the hungry but on the nonhungry—the affluent. What are the consequences for an affluent society of permitting children to go hungry? The answer is unknown, but critical.

REFERENCES

1. Scrimshaw NS, Gordon E. (eds): *Malnutrition, Learning and Behavior.* Cambridge, Mass, MIT Press, 1968.
2. Jelliffe DB: *The Assessment of the Nutritional Status of the Community.* Geneva, World Health Organization, 1966.
3. Garn S: "Biological Correlates of Malnutrition in Man. Read before the Conference on Nutrition, Growth and Development of North American Indian Children. Norman, Okla, 1969.
4. Landauer T, Whiting JWM: Infantile stimulation and adult stature of human males. *Amer Anthropol 66:*1007-1028, 1964.
5. Scrimshaw NS, Taylor CE, Gordon JE: *Interactions of Nutrition and Infection.* Monograph series 57, Geneva, World Health Organization, 1968.
6. Scrimshaw, NS: Synergism of Malnutrition and infection: Evidence from field studies in Guatemala. *JAMA,* to be published.
7. Whitten C, Pettit MG, Fischhoff J: Evidence that growth failure from maternal deprivation is secondary to undereating. *JAMA 209:*1675-1682, 1969.
8. Chase HP, Martin HP: Undernutrition and child development. *New Eng J Med 282:*933-938, 1970.
9. Cravioto, J, Birch HG, De Licardie E, et al: The ecology of growth and development in a Mexican preindustrial community: I. Method and findings from birth to 1 month of age. *Monogr Soc Res Child Develop 34:*5, 1969.
10. Monckeberg F: Nutrition and mental development. Read before the Conference on Nutrition and Human Development, East Lansing, Mich, 1969.
11. Winick M, Rosso P: The effect of severe early malnutrition on cellular growth of human brain. *Pediat Res 3:*181-184, 1969.

12. Winick M: Developmental consequences of malnutrition: Neurological correlates in man. Read before the Conference on Nutrition, Growth and Development of North American Indian Children, Norman, Okla. 1969.
13. Klein RE, Gilbert O, Canosa C, et al: Performance of malnourished in comparison with adequately nourished children (Guatemala). Read before the annual meeting of the American Associaton for the Advancement of Science, Boston, 1969.
14. Cravioto J, Robles B: Evolution of adaptive and motor behavior during rehabilitation from Kwashiorkor. *Amer J. Orthopsychiat 5:*449-464, 1965.
15. Stoch MB, Smythe PM: Undernutrition during infancy and subsequent brain growth and intellectual development, in Scrimshaw and Gordon,[1] pp 278-288.
16. Brockman L: *The Effects of Severe Malnutrition on Cognitive Development in Infants,* thesis. Cornell University, Ithaca, NY, 1966.
17. Pollitt E: Behavior correlates of severe malnutrition in man: Theoretical considerations and selective review. Read before the Conference on Nutrition, Growth and Development of North American Indian Children, Norman, Okla., 1969.
18. Yarrow L: Separation from Parents During Early Childhood, in Hoffman ML, Hoffman LW (eds): *Review of Child Development Research,* New York, Russell Sage Foundation, 1964, vol. 1, pp 89-136.
19. Monckeberg F: Effect of early marasmic malnutrition on subsequent physical and psychological development, in Scrimshaw and Gordon,[1] pp 269-278.
20. Stoch MB, Smythe PM: The effect of Undernutrition during infancy on subsequent brain growth and intellectual development. *S Afr Med J,* Oct 1967, pp 1027-1030.
21. Cravioto J, De Licardie ER, Birch HG: Nutrition, growth and neurointegrative development: An experimental and ecology study. *Pediatrics 38* (suppl): 319-371 (No. 2, pt 2) 1966.
22. Ramos-Galvan R, Arturo Viniegra C, Carlos Mariscal A: Aspectos Sociales y Epidemiologicos, *Humanismo y Pediatria,* Academia Mexicana de Pediatra and Instituto Mexicano de Psicoanalisis. Fondo Editorial Nestle de la Academia Mexicana de Pediatria, Mexico DF, 1968, pp 415-457.
23. McKay HF, McKay AC, Sinisterra L: Behavioral effects of nutritional recuperation and programmed stimulation of moderately malnourished preschool age children. Read before the annual meeting of the American Association for the Advancement of Science, Boston, 1969.
24. Sulzer JL: Behavioral data from the Tulane Nutrition Study. Read before the annual meeting of the American Association for the Advancement of Science, Boston, 1969.
25. Kallen DJ: Short comment on the social environment, learning and behavior, in Scrimshaw and Gordon,[1] pp 376-379.
26. Merton RK: Social structure and anomie . . . Continuities in the theory of social structure and anomie, in *Social Theory and Social Structure,* Glencoe, Ill, Free Press, 1957, pp 131-194.
27. Durkheim E: *Suicide: A Study in Psychology.* J. Spaulding, G. Simpson (trans), Glencoe, Ill, Free Press, 1951.
28. Kallen DJ, Miller D: Some correlates of attitudes towards welfare. Read before

the Pacific Sociological Society, Long Beach, Calif, 1967.

29. Clausen JA: The organism and socialization, *J Health Soc Behav* 8: 243-252, 1967.
30. Rosenzweig MR: Environmental complexity, cerebral change and behavior. *Amer Psychol 21:*321-332, 1966.
31. Rosenthal R, Jacobsen L: *Pygmalion in the Classroom: Teacher Expectation and Pupil's Intellectual Ability.* New York, Holt Rinehart & Winston, 1968.
32. Coleman, JS, Campbell EQ, Hobson CJ, et al: *Equality of Educational Opportunity.* US Dept of Health, Education, and Welfare, Office of Education, 1966.
33. Cooley CH: *Social Organization.* New York: Charles Scribners & Sons, 1909.
34. Cooley CH: *Human Nature and the Social Order.* New York: Charles Scribners & Sons, 1922.
35. Clark KB: Desegregation: An appraisal of evidence. *J Soc Iss 9:* Oct 4, 1953.
36. Garn S: Body size and its implications, in Hoffman ML, Hoffman LW (eds): *Review of Child Development Research.* New York, Russell Sage Foundation, 1964, vol 2, pp 529-561.
37. Kallen DJ, Miller D, Daniels A: Sociology, social work and social problems. *Amer Sociol 3:*235-240, 1968.
38. Matza D: The disreputable poor, in Bendix R, Lipset SM (eds): *Class Status, and power,* ed 2, New York, Free Press, 1966, pp 289-302.
39. Hauser P: The chaotic society: Product of the social morphological revolution. *Amer Sociol Rev 34:*1-19, 1969.
40. Wax M: Social structure and child rearing practices of North American Indians. Read before the Conference on Nutrition, Growth and Development of North American Indians, Norman, Okla, 1969.
41. Angell RC: The moral integration of American cities. *Amer J Sociol 57:*1 (pt 2) 1951.

7.   *Vegetarianism: Fad, Faith, or Fact?*

SANAT K. MAJUMDER

> If today's widespread interests in vegetarianism are relevant to considerations
> of man's nutritional problems, they must rise above fallacy, fantasy, and fadism.
> Such diets must be related to human body daily requirements, food crises in
> various areas of the world, and to environmental impact both now and in later
> generations. Dr. Sanat Majumder, a native of Khulna, India, received his under-
> graduate education at Calcutta and his doctorate in plant physiology at New
> Hampshire. Formerly at Smith, he is now associate professor of biological sci-
> ences at Westfield State College, Massachusetts. His interests include radiation
> effects on pollen grains, plant regulators, stress physiology, and drought and
> salinity effects on plants.

In the context of our enhanced interest in the mechanisms of human behavior and associated research, the aphorism "You are what you eat" appears highly simplistic and
controversial. The fact remains, however, that little effort has been made to determine
the relationship between types of food on the one hand, and the mind-body complex
on the other, except with regard to administration of a specific medicine or the effects
of a specific drug. I wish to deal here with one aspect of this question—vegetarianism—
and review its relevance from historical, biological, and contemporary perspectives.

Historical observations

Vegetariansim has never been without its advocates during any period of world
history (1). Rather than a well-defined credo, it constitutes an individualistic regime
of diet, and while a strict vegetarian excludes from his diet not only meat and fish but
also eggs, there are some egg-eating "vegetarians" as well. In certain cultures this food
habit has often been determined by the prevailing attitudes toward nature, religious
beliefs, or, simply, nutrition. Needless to say, scientific observations with regard to
vegetarian diet did not begin to appear until the turn of the twentieth century.

To most of us, food is a matter of taste and habit, occasionally predicted upon such
items of controversy as cholesterol, calorie, sugar, and salt content. To many vegetarians, however, food is of the utmost importance in leading a contented, harmonious
life. Contrary to popular belief, objection to animal slaughter and consumption of
flesh per se constitutes only an insignificant part of vegetarian motivation.

One opinion holds that a well-balanced vegetarian diet encourages development of
the intellect, increases the capacity for mental labor, and promotes longevity.

Source: Reprinted by permission of the publisher and author from *American Scientist,* Journal
of The Society of Sigma Xi, *60:*175-179, March-April 1972.

"Intemperance which is the chief cause of pauperism and crime may be greatly discouraged by cultivation of vegetarianism" (2). Sir Henry Thompson affirmed that vegetarianism simplified human character and enabled the mind to enjoy more rest and, perhaps, more acuteness (3). Gandhi, the great Indian leader, made many empirical observations on dietary habits in relation to personality development. He tried various types of food and finally chose the simplest ones (fresh fruits and nuts were his favorite) for "calming of spirit and allaying animal passion" (4).

Religious arguments opposing the intake of animal flesh are quiet numerous. Monks often abstain from meat, considering it to be a luxury and, therefore, at variance with their motto of simple living. Respect for animal life is the principal reason for vegetarian diet among the Buddhists (5). Certain sects of Hindus use the same argument for dietary habit. Christian objection to meat can be lifted directly from the Bible in such quotations as "Meat commendeth us not to God," "I will eat no flesh while the world standeth," and "Be not among riotous eaters of flesh" (6). Two contemporary Protestant sects—the Bible Christians and the Seventh Day Adventists—require of their followers a vegetarian food habit.

The year 1887 was the jubilee year of vegetarian propaganda in England. In conjunction with this, A. F. Hill published a collection of essays, one of which aptly revealed his feelings: "Love underlies vegetarianism—in love, vegetarianism holds communion with God whose nature and whose name is love" (7). Anna Kingsford, a noted vegetarian of the nineteenth century and a medical practitioner, devoted much of her writing to the question of man's basic nature—herbivorous, carnivorous, or omnivorous. Her theory sought support from scientific data on comparative anatomy and pointed to the fact that the strongest animals—horses, elephants, and camels—were not carnivorous. She also referred to the amazing ability of athletes in ancient Greece who followed vegetarian diets (8).

Probably the greatest American endeavor in vegetarianism was the establishment of the Battle Creek Sanatorium in the late nineteenth century. The primary purpose of this institution was to provide medical care through diet. Significant in this connection was Dr. J. H. Kellogg's research and eventual success in the development of breakfast cereals.

Finally, we come to the emotional "back to the land" appeal among many of today's young people. This has imparted a new meaning to vegetarian habits that are being experimented with and sought by members of the "Woodstock generation" (9). The status of vegetarianism appears to be a natural outgrowth of problem-ridden industrial societies, causing many young people to look for a new economy that does not encourage abuse of technology, pure produce that does not require the use of chemical fertilizers or pesticides, and possibly altered spiritual values that sustain life.

Biology of vegetarianism

The scientifically oriented modern generation may look upon the foregoing historical observations with some degree of skepticism and limit the questions of vegetarianism to certain places and people with peculiar cultural and intellectual characteristics.

If vegetarianism is to rise completely above the alleged status of a fad or blind faith, it must satisfy three major criteria: (1) from the nutritional point of view, vegetable diets must be complete with the standard daily requirements of the human body; (2) they must alleviate, if not eliminate, the food crisis in certain parts of the world; and (3) they must make positive contributions to the enhancement of "bioethics" among the members of the next generation with regard to wild life, agriculture, and conservation of natural resources.

Biologically, it is well known that the efficiency of energy transfer from one trophic level to another is substantially lower than 100 percent. In other words, only a small fraction of the sun's energy that is transformed into chemical energy by green plants (producers) is actually passed on to the herbivores that consume plants as food. Various carnivores that consume herbivores receive, in turn, decreasing shares of the total energy budget. To use a familiar example, a tuna fish, for each pound of canned product, must have consumed 10 pounds of herring, which, in turn, required 100 pounds of prey that lived on, proportionately, 1,000 pounds of algae. The same energy distribution is applicable to ranch and forage crops that pay us dividends in the form of beefsteaks.

It is obvious that the harvest of available energy would be more efficient if more omnivores could be herbivorous. Many such animals (e.g. raccoons) change their food habit seasonally under the pressure of food scarcity, but we, the rational omnivores, can experiment with our food and respond to similar pressures with choices that are supported by cultural feasibility and nutritional studies. Scientifically, then, this argument may form the basis for vegetarianism.

The questions now arise: What are the nutritional requirements for good health? and can vegetarian diets be balanced to meet these requirements? According to Clara Mae Taylor (10), a moderately active average man of 154 pounds and an average woman of 123 pounds need 3,000 calories and 2,500 calories, respectively, in daily food intake. In addition to appropriate amounts of carbohydrate and fat, the distribution of other essential components of our diet appears as follows: protein, 60–70 g; calcium, 0.8 g; iron, 12 mg; vitamin A, 5,000 IU (International Unit); thiamine (B_1), 1.5–1.8 mg; riboflavin (B_2), 2.3–2.7 mg; ascorbic acid (vitamin C), 70–75 mg.

Needless to say, this prescription is not undisputed. The iron requirement for women is said to be greater than that for men. The controversy over the efficacy of massive doses of vitamin C in resisting the common cold is another case in point. Linus Pauling (11) claims that a larger amount of vitamin C should be recommended for the human system. Many physicians, including Dr. Frederick Stare of Harvard University, on the other hand, question Pauling's assertion because the human body readily eliminates any excess of this vitamin.

In a vegetarian diet, the choice of a protein source does indeed pose some measure of difficulty. Animal flesh is ordinarily fortified with the ten essential amino acids human beings need in their diet (12); a single vegetable source is rarely complete in this respect. Vegetable proteins are known as "second class" proteins owing to the absence of one or two essential amino acids. Besides, vegetable food is less concentrated owing to its very high percentage of water content, a fact that dictates consumption of a larger quantity for the same food value. Vegetarians, of necessity, must combine several food types to meet their daily need for protein. Such mixtures will inevitably include milk, cheese (lactoalbumin), soybean products (glycinin), Brazil nuts (excelsin), and enriched corn and wheat (glutenin). In any event, "a completely adequate protein supply can be attained solely from vegetable sources if the supply of amino acids is carefully looked after" (13). I shall return to this point below.

Obtaining an adequate amount of calcium and iron in the diet is a relatively easy task. If half the total daily calories are taken in the form of milk, cheese, fruits, and vegetables, a person will be fairly sure of a liberal calcium supply (14). Kidney and lima beans are especially high in iron content, and so are many leafy vegetables. Whole wheat bread and peanut butter are excellent supplements in this connection.

A vegetarian faces very little difficulty in finding an adequate vitamin supply. One-half cup of steamed spinach, for instance, provides over three times the daily vitamin A requirement. Turnips, dandelion greens, and broccoli are also known to be very rich in this vitamin. For thiamin (B_1), a vegetarian looks to such items as soybeans, lima beans, peas, wheat germ, and salad greens. Riboflavin (B_2), on the other hand, can be obtained from milk and vegetables such as soybeans, spinach, and asparagus.

Much research has been done on vitamin C (ascorbic acid) since the days when Captain Cook's men used citrus fruits to prevent scurvy. One medium orange or grapefruit supplies about 85 mg of vitamin C—well over the daily need. Drinks prepared and processed from rose hips are known to contain substantially more vitamin C than citrus drinks, but they have yet to be introduced commercially.

Economic and ecological implications

A vast desert of starvation and malnutrition has spread ominously in today's over-populated world. By comparison, the oases of affluence are very few and far between. The luxury of considering a balanced nutrition is indeed a cruel hoax in a society that exists at subsistence level. The fact remains, however, that with the highly improved communication and transportation systems of today, as well as our enhanced knowledge of various cultural traditions, the benefits of scientific discoveries can be made widely available to help alleviate the situation.

Of initial concern is the caloric requirement of people struggling to survive and the type of food to which they are accustomed. Until agriculture is diversified and expanded in the regions concerned, the major reliance for improving diet must be placed on cereal grains, which constitute the primary sources of both proteins and carbohydrates in almost two-thirds of the world. As indicated earlier, to raise livestock for meat, vast amounts of grains are needed as feed, and the resulting protein yield per acre is low (15). Calculations of the average pound-to-acre ratio reveal that plants have a greater efficiency for the production of eight of the essential amino acids. For instance, crops such as soybeans, peas, and beans yield 13 pounds of these amino acids per acre; others, including carrots, potatoes, cauliflower, brown rice, and cabbage, yield approximately 4.2 pounds per acre, whereas the corresponding figure is only 1.6 pounds for poultry, beef, lamb, and pork (16).

Coupled with the above observations, it should be noted that, at the suggested rate of one acre per capita, reclamation of new arable lands cannot begin to keep pace with the runaway world population, which is projected to be between six and seven billion by the year 2000 (17). Reclaimed lands in any case do not necessarily support the most desirable crops. Motivation for cultivation can be determined by "nutrition per acre" or "nutrition per dollar."

The challenge of feeding the millions is simultaneously a challenge to find new types of food and to engage in intensive research associated with human food habits. Some fundamental advances have been made in these respects by plant scientists. Corn, the staple for many people in South America and Africa, is, for example, naturally deficient in lysine and, to a lesser extent, in tryptophan. Mertz in 1963 discovered a high-lysine strain—Opaque—22—and subjected it to intensive tests in Columbia in 1969. Both Opaque—2 and Floury—2 (18) record increased lysine and tryptophan content to equal approximately 90 percent of the nutrition of skim milk. Under proper agronomic conditions, therefore, more than 200 million people who live in the tropical corn regions can derive dietary benefit, both quantitiatively and qualitatively, from these strains.

Similar work with wheat has resulted in the development of varieties that are resistant to leaf rust and have greater protein content. Nobel Laureate Norman Borlaug developed a high-yielding dwarf Mexican wheat. His meticulous selection of genes and subsequent breeding program have opened the way to a wider distribution of this variety. Many tropical countries in South America and Asia are beginning to harvest dividends from this magnificent research. In a similar effort, by crossing a short Taiwan rice with an Indonesian variety, the International Rice Research Institute in the Philippines released in 1966 the much talked-about "miracle rice," IR—8. This new variety is dwarf, early maturing, highly responsive to nitrogen fertilizer, and very high yielding. Development of this variety could not have happened at a more crucial time or in a needier region of the world (19).

The unsung hero in the area of food research is the soybean, which was originally introduced as a livestock feed. Until the 1930s, the soybean was shrouded in an "oriental mystique." The production of soybeans in the United States has increased from 4.8 million bushels in 1925 to 840 million bushels in 1965. In view of its high protein (39—46%), phosphorus, iron, and calcium content, the soybean today is being processed for human consumption (e.g. soybean "bacon," "yogurt," and "cookies"). With ingenuity and appropriate promotion, this livestock feed of yesterday—500 million bushels of which is exported annually by the United States—can legitimately prove to be both an economic and nutritional bonanza for many people today (Table 1).

The problem of malnutrition may appear to be less dramatic than death from starvation, yet it is proving to be the greatest stumbling block during the formative period of millions of children. There seems to be little disagreement among scientists that a continuous protein-deficient diet produces irreversible damage to the brain. How can we resolve the basic question of filling the stomach and reducing malnutrition among people who cannot afford the taxing economy of livestock breeding for protein or who

Table 1. *Contributions of Dried Soybeans, Other Dried Beans, and Meats (Average Portions[a]) in Percent of Daily Requirements.*

Food	Calories	Protein	Calcium	Iron	A	Vitamins B_1	B_2
Yellow soybean	10	49.3	26	48.8	1.8	72	86
Kidney bean	10	25	17	72.5	–	24	11
Navy bean	10	28	16	72.5	–	24	11
Roast lamb	7.4	32.4	1.67	15.38	–	9.64	12
Roast veal	5.8	44.25	2.61	34.8	–	20	11

[a]An average portion of dried beans is 3 ounces, or about one cup of cooked beans; an average portion of meat equals 4 ounces. (From Williams-Heller and McCarthy, ref. *23*.)

are traditionally or habitually vegetarian? The answer apparently lies in the extensive use of staples mixed with inexpensive but highly nutritive protein concentrates (20).

The ingenious approach of William D. Gray (21) is noteworthy in this connection. Mindful of the waste that occurs in the natural energy transfer, Professor Gray sought to "harvest" protein more inexpensively. He used waste substrate such as paper pulp, in which certain species of fungi grew profusely. In the presence of this "recycled" carbohydrate and added nitrogen fertilizer, the fungi readily synthesized protein. The dry, clear, tasteless, and odorless mycelium of the fungi could then be powdered and mixed with flour and other staples to augment the protein content from 2.5 to 15 percent.

"Although the carnivorous Americans are not likely to stampede for the protein-rich flour as a substitute for steaks, the prospect of its use in some developing countries cannot be overemphasized" (22). It is not as condescending for an affuent society to export protein-rich flour to people who are either vegetarians of whose dietary habit, of necessity, is cereal-based as it is to instruct those same people about developing a meat industry.

Flavored beverages enriched with protein are beginning to receive the attention of many commercial enterprises. Based on oil-seed meals, these veverages include Vitasoy of Hong Kong, Saci of Brazil, marketed by the Coca Cola Company, and the banana-flavored Puma, introduced by Monsanto. Among the mixtures of grains and oil-seed products, Incaparina (38% cotton-seed flour, 58% corn, yeast, and vitamin A) is probably the most popular example. In India, peanut flour and Bengal gram together constitute a multipurpose food. Sardiele is a soy and sesame mixture used by the Indonesians. Foods consisting of rice, wheat, soy, and peanuts have appeared in the Taiwan market. In Uganda, the common ingredients of similar food mixtures are dry skim milk, sucrose, cottonseed oil, corn flour, and peanuts.

Conclusions

What I have discussed here neither advocates vegetarianism nor does it argue against such a dietary regime. Instead, I have attempted to examine with objectivity the question of vegetarian habits in a new perspective. The present dilemma of uneven economic growth in an overpopulated world necessarily forces us to reexamine basic concepts of human nutrition and to probe for simple, inexpensive, and nutritious food.

Although substantial evidence indicates that vegetarian diets can be balanced to meet human requirements, research in this area is very limited. The scientific communities of affluent nations have an obligation to initiate such research in view of the frightening statistics which indicate that two-thirds of all the children in the world—the future citizens of our planet—suffer from malnutrition owing to protein deficiency, while meat and fish proteins are out of their economic reach.

Even less is understood about the primary effects of different types of food on human personality and behavior. I am not talking about the secondary (social) effects, such as the apparent linearity of "starchy food—obesity—self-consciousness." Some empirical observations by early vegetarians, such as "calming of spirit," "allaying animal passion," and "acuteness of mind" associated with their diets, can and should be subjected to scientific scrutiny.

There is a vast difference between people who are vegetarians by choice and those who have no other alternative, economically speaking. Both groups can benefit, however, from the assurance that their diets can be made both plentiful and healthful. Fitting the mosaic of human cultures, the multiplicity of food habits and the availability of alternate foods should be welcome in this shrinking world. The control of a precipitous population explosion and the development of inexpensive but nutritious food must be matters of primary concern for all of us. While scientists can and must provide nutritional data as common denominators for the choice of food, they can hardly solve the human paradox of knowledge and wisdom, as indicated by the following Associated Press release on June 17, 1971:

"The diet of affluent Americans able to afford any foods they choose is less nutritious than it was a decade ago. The problem: we are overfed but remain undernourished. As a result, experts say, 10 percent of the population may be anemic and 25 percent overweight."

REFERENCES

1. Alexander Bryce. 1912. *World Theories of Diet*. London: Longmans, Green and Co.

2. Charles O. Groom-Napier. 1875. *Vegetarianism, a Cure for Intemperance.* London: William Tweedie.
3. Sir Henry Thompson. 1898. "Why Vegetarian?" From *Science Paper No. 2,* The Egyptian Newspaper Press.
4. M. K. Gandhi. 1949. *Diet and Diet Reform.* Ahmedabad, India: Navajivan Publishing House.
5. E. W. Hopkins. 1906. The Buddhistic rule against eating meat. *J. Amer. Oriental Soc. 27:*455.
6. Gerald Carson. 1971. *Cornflake Crusade.* New York: Reinhart and Co.
7. A. F. Hill. 1897. *Vegetarian Essays.* London: Ideal Publishing Union.
8. Anna Kingsford. 1885. *The Perfect Way in Diet.* London: Kegan, Paul, Trench and Co.
9. *Time.* The Kosher of the Counterculture. November 16, 1970, pp. 59—63.
10. Clara Mac Taylor. 1942. *Food Values in Shares and Weights.* N.Y.: Macmillan.
11. Linus Pauling. 1970. *Vitamin C and the Common Cold.* San Francisco: W. H. Freeman.
12. W. C. Rose. 1949. Amino acid requirements of man. Federation of American Societies for Experimental Biology, *Proceedings 8:*546—52. Note that Rose lists only eight amino acids as "essential."
13. Louis Bean. 1961. *Food and People.* U. S. Congress. Joint Economic Committee. Report of the Subcommittee on Foreign Economic Policy. Washington, D. C.: U. S. Govt. Print. Off.
14. Henry C. Sherman and Caroline S. Langford. 1943. *An Introduction to Foods and Nutrition,* N. Y.: Macmillan.
15. Ralph McCabe. 1961. *Food and People.* U. S. Congress Joint Economic Committee. Report of the Subcommittee on Foreign Economic Policy. Washington, D. C.: U. S. Govt. Print. Off.
16. M. G. Lambou et al. 1966. Cottonseed's role in a hungry world. *Economic Bot. 20:*256.
17. K. W. King. 1971. The place of vegetables in meeting the food needs in emerging nations. *Economic Bot. 25:*6.
18. E. T. Mertz, L. S. Bates, and O. E. Nelson. 1964. Mutant gene that changes protein composition and increases lysine content of maize endosperm. *Science 145:*279.
19. Lester Brown. 1970. Seeds of Change: *The Green Revolution and the Development in the 1970's.* N.Y.: Praeger.
20. Proceedings of the Symposium on Integrated Research in Economic Botany VII: Protein for Food. 1968. *Economic Bot. 22:*3—50.
21. William D. Gray. 1965. *Activities Report 17.* Research and Development Associates.
22. Sanat K. Majumder. 1971. *The Drama of Man and Nature.* Columbus, Ohio: Charles E. Merrill, p. 104.
23. A. Williams-Heller and J. McCarthy. 1944. *Soybeans from Soup to Nuts.* N.Y.: The Vanguard Press, p. 12.

PART IV

VIRUSES, CANCER, AND AGING

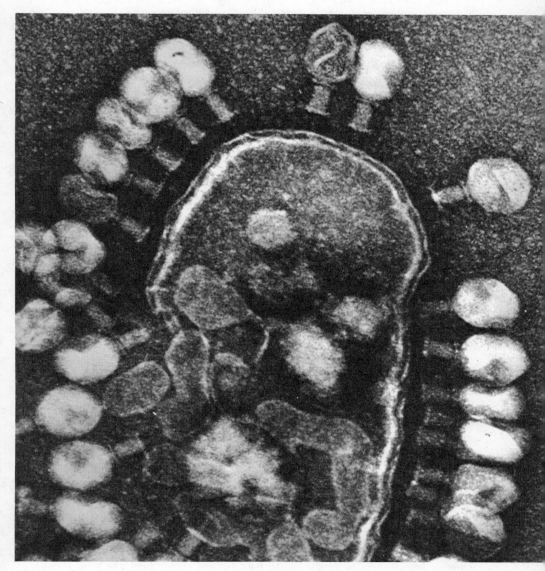

A bacterial cell E. coli *infected with and almost completely destroyed by* T_4 *bacteriophages. Courtesy of Dr. Lee D. Simon, Institute for Cancer Research, Philadelphia.*

8. *Interferon as a General Antiviral Agent*

SAMUEL BARON

For some time it has been known that animal cells when subjected to viral infection become resistant to subsequent reinfection by the same virus. The active principle was found to be a protein and was called interferon. This article reviews evidence that presents interferon not as the antiviral agent but rather as an interacting system that induces antiviral activity in the organism through the formation of antiviral components. Its relation to the array of other body defense mechanisms is considered. Dr. Samuel Baron is head of the Cellular Virology section of the Viral Diseases Laboratory, National Institute of Allergy and Infectious Diseases, N.I.H. The information given in this article was originally presented before the National Science Teachers Association-Sunoco Science Seminar in Washington, March 1971.

In the late 1950s, British researchers Alick Isaacs and Jean Lindenmann identified a nonspecific antiviral protein which is manufactured by the infected host organism within hours of viral attack. The protein was called "interferon" because it impedes virus replication. The discovery of interferon led, within a short time, to major revisions in concepts of cellular immunity and recovery of multicellular organisms from viral infections. In this paper we shall consider the implications of the interferon system as a host defense during infection and as a biological mechanism. Detailed treatment is presented in the selected references.

The interferon system is now thought to be divisible into several components. (Fig. 1) We will refer to interferon as a protein or proteins that are produced or released by cells following viral infection and certain other stimuli. Available evidence suggests that interferon is not itself directly antiviral, but rather reacts with cells to induce the formation of a new intracellular substance which mediates the antiviral activity. This antiviral component of the interferon system may be a polypeptide or a protein-containing molecule.

Production of interferon by the infected cell is thought to be preceded by derepression of the interferon cistron, leading to formation of interferon messenger ribonucleic acid (RNA) which then results in production of interferon protein. Interferon is rapidly released by cells. Production of the proposed antiviral substance by cells reacting with interferon is similarly thought to be preceded by derepression of

Source: Reprinted by permission of the publisher and author from *The Science Teacher,* *39*(1):43-48, January 1972.

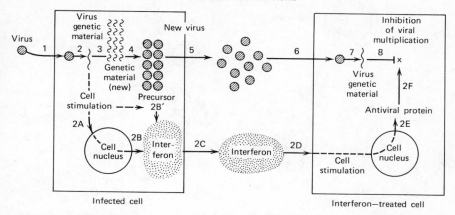

How Interferon Works

Figure 1. Virus comes in contact with the cell (1) and penetrates the cell membrane. The virus then releases its genetic material (2) which organizes the cell to make many copies of the genetic material (3). The new genetic material (4) is coated with viral protein to form completed new virus, and (5) the new virus is released into the fluid around the cells where it can now spread to other cells. During the early stages of infection, the process of viral multiplication within the cell stimulates the cell (2A) to utilize the stored information for interferon synthesis. Through a series of complex procedures, the interferon protein is made (2B) and released rapidly (2C) into the extracellular fluid. Under some conditions, the interferon may originate directly from precursors (2B') ("preformed" interferon) rather than being newly synthesized. In many instances the interferon precedes new virus to surrounding cells. The interferon stimulates the surrounding cells (2D) to utilize another set of stored information to produce an intracellular, antiviral protein (2E). The antiviral protein (2F) acts to change the cell's protein-synthesizing machinery so that it cannot be used by the viral genetic material subsequently encountered. Therefore, any infection by new virus of the interferon-protected cell (6) still leads to the intracellular release of the virus genetic material (7) but that virus genetic material is discriminated against by the cell-synthesizing machinery (8) and essential virus materials fail to be produced. Thus, virus multiplication is inhibited, and the cell is protected. The degree of protection is dependent on the level of antiviral protein.

another cistron and formation of messenger RNA which is then translated into an antiviral polypeptide.

Host Defenses During Infection

To understand the role of interferon in relation to the other host defenses during the various stages of viral infection, it is desirable to review the general functioning of the various host defenses. (Fig. 2)

The interferon system is the earliest appearing of the known host defenses, for it can be detected to be operative within hours of infection. (Fig. 3) Most viruses are able

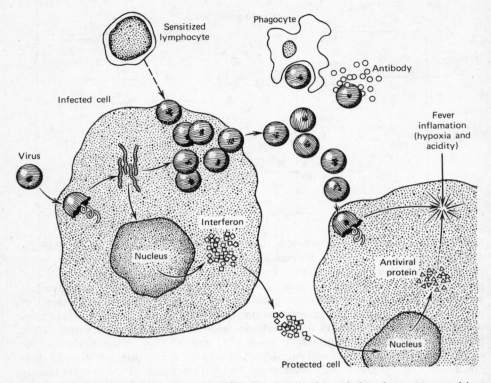

Figure 2. Host defenses against viral infection. Antibody and the phagocyte combine with and inactivate virus in the extracellular spaces. The interferon system, fever, local acidity, and local hypoxia inhibit viral replication within cells. The sensitized lymphocyte can in theory combine with viral antigens on the surface of infected cells and destroy this cell before multiplication of virus is complete.

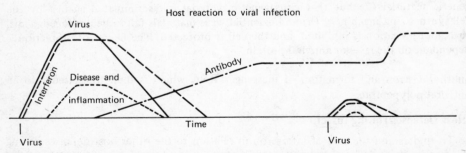

Figure 3. Sequence of appearance of host reactions to viral infection. Interferon and the other nonimmune defenses are the first to develop after the initial viral infection. Antibody and cell immunity can be detected a few days later. Reinfection by the same virus is prevented mainly by antibody which persists after recovery.

to induce at least some interferon, and most viruses are sensitive to its antiviral action, but they vary in degree of stimulation and sensitivity. The site of antiviral action of interferon is intracellular. The duration of activation of the interferon system is from one to three weeks during acute virus infections *in vivo*.

A protective antibody generally makes its appearance at three days or later after initiation of a virus infection and is first detected in the serum. It is specifically induced by the viral antigen of antigens, and it specifically acts on the inducing or related antigens. An antibody acts against a virus by combining with the virus and rendering it noninfectious. An antibody is confined to the serum and extracellular fluid and, therefore, cannot act on a virus which is multiplying within cells. A protective antibody is relatively durable in that it persists within the host for months or years and is thought to play an important role in preventing subsequent reinfection by the same virus.

Cellular immunity, which is mediated by the sensitized lymphocyte, generally makes its appearance about one day before the appearance of the antibody. As with antibodies, cellular immunity also persists for months or years. As an immunological phenomenon it is specifically induced, and it specifically acts against its antigens. Its site of action, as determined by transplantation studies, is against the living cell which contains the foreign antigen. Viral infected cells do contain viral antigens, and, therefore, sensitized lymphocytes could react with them. Thus, it is possible that sensitized lymphocytes could destroy the virus-infected cell before virus replication.

The febrile response is a nonspecific but potentially important host defense which usually makes its appearance within a few days of virus infection. It can be a generalized fever or one localized to the site of infection due to increased blood flow. It is induced nonspecifically by a wide range of stimuli, and it is nonspecific in its action against a broad range of viruses. One action of elevated temperature is to inhibit the intracellular replication of many viruses. The fever response is relatively short, lasting from a few hours to several days in most instances.

Another nonspecific defense is local inflammation at the site of virus infection, resulting in local acidity and local hypoxia, both of which have been reported to inhibit nonspecifically the intracellular replication of a broad range of viruses. Duration of the action of the acidity and hypoxia is limited to the duration of inflammation, which may last from several days to several weeks. Finally, phagocytosis may lead to the removal of viruses from the bloodstream and from body fluids before viruses spread to susceptible cells.

The possible interplay of these active host defenses will be considered here. Preexisting physical barriers and natural inhibitors will not be discussed unless pertinent to the reactive defenses.

Viral Establishment

Deposition of virus at the portal of entry of the host animal can occur either in the absence of reactive host defenses or in the presence of reactive host defenses if appropriate pre-existing infection had occurred. The portal of entry is defined as the site in the animal body where the virus is first deposited (e.g., respiratory tract for the influenza virus; alimentary tract for poliovirus). In the absence of reactive defenses, implantation of virus in a susceptible host is successful by definition. All the reactive defenses that have been mentioned here can theoretically be present at the portal of entry during attempted implantation. However, a protective antibody is the only reactive defense which can prevent infectious virus from reaching susceptible cells. The pre-existence of protective antibodies at body surfaces can completely prevent the implantation of virus or severely limit the amount of virus which is implanted. (Fig. 4)

Successful infection at the implantation site requires that the virus replicate in the initially infected cell or cells and then spread to infect other cells within the same tissue. Replication of virus in the first infected cell is probably not affected by the early appearing host defenses. For example, the interferon mechanism which is activated within the infected cell probably does not inhibit virus infection within that cell but does protect surrounding cells. This conclusion comes from the finding that inhibition of interferon production by metabolic inhibitors in virus-infected cells generally does not significantly increase the virus yield from the initially infected cells.

A virus may spread from the initially infected cells to other cells at the implantation site by means of diffusion over the external surface of the tissue, diffusion through the extracellular fluid, and direct spread from cell to cell through cytoplasmic bridges. The first reactive defense to make its appearance is the interferon system. Interferon is produced by infected cells at about the same time as the virus is produced. The interferon protein is released from cells as rapidly as it is produced and diffuses to surrounding cells, where it activates the production of the anti-viral state. In this way, the spreading virus meets an intracellular barrier to its continued replication. The degree of antiviral activity is probably dependent on the extracellular concentration of interferon so that cells in closest contact with the infected, interferon-producing cells become most resistant to virus. Protection of more distant cells does occur but to a lesser degree. The interferon defense functions in this manner not only in infected tissues at the portal of entry, but in any infected tissue.

As indicated, all antibodies, including those which neutralize virus, are confined to the extracellular space. In most tissues the newly produced antibody is first detected many days after virus infection even though it may appear first in the serum after about three or four days. Thus the early spread of the virus at the portal of entry would not be inhibited by the antibody, but later spread of the virus could be inhibited. An antibody can be expected to be more effective in inhibiting spread of a virus over the external surface of the tissue than between adjacent cells. This is because there is more time for the antibody to inactivate the virus at the surface, before the virus penetrates into cells where the antibody cannot follow. The longer time of extracellular existence of a virus at a surface, as compared with its existence between cells, is due to the larger extracellular space at body surfaces. Antibodies cannot stop the direct cell-to-cell spread through cytoplasmic bridges of certain viruses, such as the pox and the herpes viruses.

Cellular immunity, as mediated through the sensitized lymphocyte, would make its appearance shortly before an antibody but substantially later than interferon. Although it has not been directly demonstrated, cellular immunity, when produced, could prevent the spread of a virus by destroying virus-infected cells, as discussed.

Local inflammation as a result of virus-induced cell damage and immunologic reactions would give rise to the intracellularly active antiviral mechanisms of local fever, local hypoxia, and local acidity. However, inflammation would not be expected to appear as rapidly as does interferon but could occur before the immunological responses.

Successful establishment of viral infection at the implantation site sometimes leads to disease in that same tissue (e.g., influenza infection of the lung, rhinovirus infection of the nasal tract, and wart virus infection of the skin). For other viruses the established infection at the portal of entry serves principally as a dissemination site for spread to other organs and tissues (target organs) where the clinical disease first manifests itself.

Viral Dissemination

A virus may spread from the portal of entry to target organs by means of body fluid (serum or lymph) or by means of infected cells. The relative effectiveness of the various host defenses will depend upon which route the virus takes. The spread of a virus through the serum and lymph is first countered by the macrophages of reticuloendothelial system with resulting decreased viremia (presence of virus in the blood) and decreased severity of virus infection.

Interferon appears in serum within a few hours of the onset of viremia. Experiments utilizing passive transfer of interferon indicate that the circulating interferon can act both to reduce the viremia by decreasing the yield of virus from cells in contact with the blood stream and to reach target organs where it protects cells against subsequent seeding of virus from the blood stream and lymph. An antibody makes its appearance in the serum and lymph several days later and is associated with effective reduction of the level of viremia.

Cellular immunity, which makes its appearance shortly before the antibody, could theoretically inhibit the spread of the virus through fluids by destroying the cells which produce or release the virus into the circulation and also by destroying the cells within target organs which initially become infected. However, the relative contribution of cellular immunity and antibody has been difficult to distinguish.

Generalized fever is a frequent response to viremia. Those viruses which replicate poorly at elevated temperatures would, therefore, be more poorly produced at the sites seeding the circulation and would be less effective in establishing in target organs throughout the body.

The spread of a virus may also occur via infected cells, such as those in nerves, and via white blood cells (WBC). Spread of infection along a nerve has not been studied in terms of the host defenses. However, the interplay of some of these defensive factors may be surmised from other knowledge. The spread of a virus along the cellular sheath surrounding the axon would be expected to be resisted by the body defenses in the same way as is virus spread during early infection of any tissue. The spread of a virus within a nerve axon could be retarded only by intracellularly active defense mechanisms.

A virus may also spread by infected WBC, which, due to their mobility, could deposit infectious virus in target organs. Several of the host defenses have been shown to affect the ability of leukocytes (white blood cells) to support virus growth. A broad range of viral and nonviral stimuli can induce these cells to produce interferon. This has provided the first known bridge between the immunological system and the interferon system. It has been shown that sensitized lymphocytes respond to antigen (either viral or nonviral) with the production of interferon. The other intracellularly active host defenses can be expected to retard replication of a virus within leukocytes, and the antibody can be expected to help prevent dissemination of a virus which may be released from the infected leukocytes into extracellular fluids.

Host Defenses During Recovery

A virus may establish an infection in various tissues within the body, including the implantation site, tissues in contact with the vascular system, and target organs. (Fig. 4) The target organ is defined as the organ in which the major disease effect occurs. It

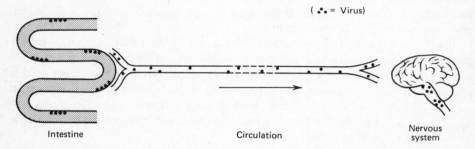

(∴ = Virus)

Intestine   Circulation   Nervous system

Figure 4. Spread of poliovirus from the portal of entry to the circulatory system, by which it is disseminated to the target organs—the brain and spinal cord.

seems probable that the mechanisms which govern the recovery of the fully infected tissue are similar at these diverse sites of established infection. The recovery process may not occur simultaneously within the various infected tissues in the body, because the time of onset of infection and production of localized host defenses is usually different for the different tissues. In this way, tissues at the implantation site or tissues in contact with the vascular system may be undergoing recovery at the time that virus infection is being established within the target organs.

Antibody During Recovery

In contrast to the effective role of the antibody in helping to prevent establishment of infection by limiting the spread of the virus through extracellular fluids, there is increasing evidence that the antibody may not be effective after virus infection has become established in a body tissue. Evidence to support this view comes from several sources. Recovery may begin before the first detectable antibody is formed during certain infections. Furthermore, the antibody may not reach certain infected tissues until many days after its appearance in serum. Immunosuppression generally does not enhance an already established virus infection within a given tissue. There are, however, a few studies in which immunosuppression by cyclophosphamide enhanced intracerebral infection with Dengue virus or with West Nile virus. Recovery from established infection was not found to be enhanced by antibody production which occurred at the time of establishment of the infection in target organs. Similarly, the transfer of physiological amounts of antibody from an immune animal to one in which virus infection was established within a target organ generally failed to enhance recovery. There are, however, two reports of instances in which transfer of an antibody may have enhanced the recovery process.

If an immunologically specific mechanism like an antibody were the major cause of recovery from a fully established infection, then the recovering organ should manifest an immunologically specific resistance to a superimposed virus infection. This has been shown not to be the case in a large number of studies. Instead, the recovering tissues and organs manifest a nonspecific resistance to a broad range of viruses. These findings indicate that if there is a specific immune component to recovery, it is masked by a more effective nonimmune reaction.

The general ineffectiveness of an antibody in enhancing recovery may be related to the intracellular site of virus multiplication where the antibody does not enter.

Cellular Immunity During Recovery

Cellular immunity could be a major factor in recovery from the established virus infection within a tissue, because infected cells could theoretically be destroyed by sensitized lymphocytes before full virus replication occurred. However, available evidence does not permit a final assessment of its actual role.

Interferon During Recovery

The finding that a major characteristic of the recovering tissue is nonspecific resistance to viral superinfection indicates that nonimmune defenses such as interferon are correlated temporarily with recovery from the established infection. Supporting evidence comes from many sources. A sufficient quantity of interferon can inhibit the multiplication of most viruses in various animal tissues *in vivo* and *in vitro*. Interferon is generally present in infected tissues of animals prior to the onset of recovery from diverse virus infections. Similar results have been obtained in studies in man. These studies have demonstrated that interferon can be produced as early as one hour after virus infection and is generally demonstrable in high titer within one to two days. In comparison, recovery, as measured by decreasing virus in the infected tissue, usually begins one or more days after interferon is first detected. These and other studies have demonstrated the presence of interferon in target organs during recovery. Apparently inconsistent findings are the low or undetectable levels of interferon in organs of mice during certain nonlethal infections. However, some of these viruses may have been very sensitive to interferon. Hence interferon produced by cells originally infected may have so reduced virus multiplication that little or no detectable amounts of interferon were stimulated.

Further evidence relating the interferon system to recovery comes from the expectation that decreased function of the interferon mechanism would increase the severity of those virus infections in which it contributes significantly toward recovery. Such were the finding in young chick embryos with an "immature" interferon system. More recent studies have extended the evidence. Decreased function of the interferon system, as caused by altered temperature, psychological stress, chemical inhibitors, and different virus strains, led to impaired recovery of animals. Carefully controlled studies of tissue cultures infected with vaccinia virus or measles viruses, herpes virus or arboviruses also indicate that impairment of the interferon mechanism hinders recovery.

Another test of the relationship of the interferon system to recovery is the transfer of interferon from extracts of infected tissues to recipient animals. Early studies demonstrated greatest protection when interferon was given prior to or at the same time as infection and not when interferon was given a significant time after infection. More recently, treatment with either large amounts of interferon or inducers of interferon production was shown to retard the development of established infections with encephalomyocarditis virus, leukemia virus, herpes simplex virus, and other viruses. In an analogous fashion, production of interferon by virus-resistant cells in a mixed tissue culture can protect the virus-susceptible fraction of the cell population.

Taken together, most of the available evidence does support a causal relationship between the interferon system and recovery from established infections. The available evidence also indicates that interferon is not the sole factor which influences recovery, and in certain virus infections it may not be the most important factor.

Other Factors During Recovery

The development of nonspecific virus resistance by recovering tissues can also be due to fever, local acidity, and local lowered oxygen tension. Further, these other nonimmune mechanisms act intracellularly and, therefore, are also possible mediators of natural recovery from virus infection. However, they do not seem to be as broadly antiviral as in the interferon system, and their duration of action is relatively short. These factors merit further study in relation to their role in viral infection.

A natural experiment which illustrates an interplay of these host defense factors concerns the human disease hypogammaglobulinemia, characterized by a poor ability to form antibodies. [6] Impairment of the production of normal amounts of antibody by patients with this disease is believed to result in increased frequency and severity of certain pus-producing bacterial infections as well as a lesser increase in severity of certain virus infections. The small amount of antibody produced by these patients is sufficient to account for resistance to spread of many viruses through the blood stream and, therefore, can account for the immunologically specific resistance to reinfection observed in these patients. Since the nonimmune and nonspecific defense factors are largely independent of the immune response, the ability of hypogammaglobulinemic patients to recover normally from established virus infection is probably due to normally functioning interferon, fever response, inflammatory responses, and perhaps in part to intact cellular immunity. The occasionally observed increased severity of virus infections in these patients (e.g., vaccinia virus and poliomyelitis infections) might be explained by delayed antibody production during the viremic spread of virus with resulting prolonged viremia and increased seeding of target organs.

Other Infectious Organisms

During *in vivo* infection of mice with chlamydiae (agents such as the trachoma and psittacosis agents), malaria, or toxoplasma, interferon production occurs. The same infectious organisms have been shown to be sensitive to the action of interferon if sufficient concentrations are used. These findings raise the possibility that the interferon system may play a role during the pathogenesis of these nonviral infections. Studies to test such a role and to help determine the mechanism of action against chlamydiae and protozoal parasites are being carried out.

Normal and Malignant Cells

Most studies of normal cells treated with interferon have not shown measurable effects on cell metabolism, cell growth, or antibody formation. More recently two studies have found that application of large amounts of interferon to cell cultures did decrease cellular protein synthesis and cell growth. Confirmation of these potentially important observations is highly desirable.

A surprising observation has been that not only does the interferon system inhibit tumor viruses and virus-induced tumors, but that interferon inducers and interferon itself inhibit the growth of transplantable tumors which were not deliberately induced by introduction of a virus. These observations raise several possibilities including (a) that the tumors which were not deliberately induced by introduction of viruses actually were transformed to malignancy by an undetected virus infection, (b) that the interferon system exerts a control over tumor cell metabolism, perhaps through a rejection of foreign nucleic acids, and (c) the interferon mechanism and interferon inducers enhance the immunological tumor rejection response.

Interferon—Evolution and Application

The limited information on the distribution of the interferon system among animals, plants, and bacteria allows only limited conclusions as to the origin and evolution of this defense mechanism. The interferon system has been shown to be operative in fish, in a poikilothermic reptile, in birds, and in mammals. There is suggestive evidence that a mechanism analogous to the interferon system may occur in plants. Unexplored as yet are the findings of a substance similar to interferon which is produced by *Pseudomonas aeruginosa* after exposure to inactivated bacteriophage. An ancient origin for the interferon system would be indicated if the plant or bacterial antiviral mechanisms are basically similar to the animal antiviral system.

If we are willing to assume that interferon reacts with cells to induce an antiviral substance, then we must accept a message-carrying function for this protein. Since the other message-carrying polypeptides and proteins within the body are generally classified as hormones, it is tempting to compare the interferon proteins with hormones. Available evidence cannot establish the hormonal nature of interferon, but it is consistent with the possibility.

Interferon holds promise of becoming an important agent for prevention and therapy of virus infections. Part of its potential utility comes from its ability to inhibit the multiplication of virtually all viruses and its lack of detectable toxicity. Theoretically, interferon could be injected into an infected tissue in advance of or supplementary to normally produced interferon. The problems to be overcome before such application include: (a) production of large quantities of interferon; (b) the requirement that interferon, because of its species specificity, must be prepared in human cells, or cells of closely related species for use in man; and (c) the tendency for protection to be localized at the site of injection and to spread poorly to distant infected sites. These problems may be largely overcome by current attempts to develop drugs such as nucleic acids and tilorone which can stimulate the production of large amounts of endogenous interferon. It has already been possible to prevent a wide variety of virus infections by prior injection of interferon into animals and, in at least one study, in man. Early studies with a nucleic acid inducer also indicate some effectiveness in prevention of experimental human respiratory infections.

Summary

The evidence relating the interferon system to the infectious process and also to normal and tumor cell function supports the view that the interferon system is an important component of the body's nonimmune defenses, which are probably the major causes of recovery from already established virus infections. In contrast, the immune defenses may not be essential for recovery but they function to prevent virus spread during primary infection and to prevent reinfection. The interferon system can also serve to limit virus spread through the bloodstream by decreasing virus shedding into the bloodstream. It can also retard viral oncogenesis by inhibiting intracellular transformation by virus, by retarding multiplication of tumor viruses, and by inhibiting tumors induced by viruses, carcinogens, and unknown causes. Interferon may be significant during infections with chlamydiae and protozoal parasites. As yet, there is only little evidence to support the possibility that the interferon system is of functional significance in the uninfected cell. This antiviral system has been shown to be operative in mammals, birds, a reptile, fish, and possibly in plants. Finally, practical usefulness toward the control of viral infections of man is being studied.

REFERENCES

1. Baron, S. In *Interferon, Advances in Virus Research.* K. M. Smith and M. A. Lauffer, Editors. Academic Press, New York. 1963. Pp. 39-64.
2. Finter, N. B., Editor. *The Interferons.* North Holland, Amsterdam. 1966 and 1972. (In Press).
3. Isaacs, A. In *Interferon, Advances in Virus Research.* K. M. Smith and M. A. Lauffer, Editors. Academic Press, New York. 1963. Pp. 1-38.
4. Levy, H. B., S. Baron, and C. E. Buckler. In *Biochemistry of Viruses.* H. B. Levy, Editor. Dekker, New York. 1969. Pp. 579-612.
5. Merigan T. C., Editor. "Symposium on Interferon and Host Response to Virus Infection." *Archives of Internal Medicine 126:* 49-157; 1970.
6. Vilcek, J. *Interferon.* Little, Brown and Company, Boston, Massachusetts. 1970.

9. The Health Menace of Tobacco

ALTON OCHSNER

A distinguished physician who first became aware in 1936 of the possible relationship between cigarette smoking and cancer of the lung discusses this now well-established link. He also describes other effects of tobacco use on man. Dr. Ochsner, President Emeritus of the Alton Ochsner Memorial Foundation, New Orleans, and former Chairman of Surgery, Tulane University has had a distinguished career in medicine. He is the author of many articles, chapters, and books that reflect clinical and research interests in thoracic surgery and disorders of the circulatory and digestive systems. This article is based on his National Lecture for Sigma Xi, given in 1968.

Tobacco use is the greatest public health hazard in the United States today. Emerson Fotte, a member of the Interagency Council on Smoking and Health, in testifying before a Congressional Committee, stated that 360,000 persons died in one year in the United States because of tobacco use. Hammond *(19)*, in 1969, writing on life expectancy studies in 447,196 men born between 1868 and 1927 and still living July 1, 1960, found that the estimated life span at age 35 was 42.4 years for men who had never smoked, 37.8 years for those who had smoked 1 to 9 cigarettes daily; 37.1 years for those who smoked 10–19 cigarettes; 36.5 for those who smoked 20–39; and 34.7 years for those who smoked 40 or more cigarettes daily. He further showed that the percentage of men aged 25 expected to die before 65 showed: nonsmoker, 22%; smoker of 1–9 cigarettes daily, 33%; 10–19, 37%; 20–39, 39%; and over 40, 46%. Ravenholt *(35)* stated, "The sum of all deaths from *accidents, infection, suicide, homicide, alcohol,* and *stomach cancer* only equals the quarter-million deaths from tobacco disease in the United States during 1962." He further states, "Cigarette disease is now the foremost preventable cause of death in the United States. For the American male, ages 35–65, who smokes a pack or more of cigarettes per day, tobacco is an environmental hazard equal to all other hazards to life combined." He added that in 1963 14% of all deaths (261,408 or 1,813,549) from all causes were excess deaths due to smoking. Because of tobacco use, particularly in the form of cigarettes, many previously rare conditions have assumed almost epidemic proportions.

One might ask if there are any advantages to the use of tobacco, because an industry which has become as large as the tobacco industry should market a worthy

Source: Reprinted by permission of the publisher and author, *American Scientist,* 59:246–252, March-April 1971, journal of The Society of Sigma Xi.

product. There is no advantage to the use of tobacco except the monetary one to the interested parties, namely the tobacco producer, the manufacturers, and the advertisers. The use of tobacco is hazardous because it produces cancer, vascular deterioration, pulmonary disease, reproductive disorders, ulcers, disabling illness and premature deaths, and tremendous economic loss.

Cancer of the Lung

Until the mid-thirties cancer of the lung was an extremely rare disease but since that time there has been a progressive increase in its incidence, the increase being more rapid than in any other cancer. In the United States in 1930, there were only 2,500 deaths from cancer of the lung; in 1950, 16,000; in 1956, 29,000; and in 1964, 43,100. The reason for the unprecedented increase in this particular cancer is that twenty years previously at the beginning of World War I, in 1914, men began to smoke cigarettes heavily, and the twenty-year lag between 1914 and the mid-thirties was just about the length of time necessary for the cancer-producing effects of cigarette smoking to become evident.

An interesting study of the incidences of cancer of the lung as determined at autopsy in Presbyterian Hospital in New York and the University Hospital in Iceland has recently been made. The study showed a parallel increase in the incidence in the two institutions, with the difference that in Presbyterian Hospital the increase began in the mid-thirties and in the University Hospital in Iceland it began in the mid-fifties. Before World War II, the Icelanders did not smoke, just as we in the United States smoked very little before World War I. Because we had bases in Iceland cigarettes were introduced into that country and the Icelander began to smoke. Twenty years later the same increase in incidence of cancer of the lung occurred among the Icelanders that had occurred in the United States approximately twenty years after World War I. In Great Britain from 1916 to 1955 lung cancer in men 45 to 64 years of age increased from approximately 10 to 120 per 100,000 population, whereas all other types of cancer decreased from 280 to 125 per 100,000 population *(40)*. Rakower *(34)*, in studies in Israel, stated that only one non-smoker in every 3,300 men aged 55 and older dies from lung cancer, whereas 1 in every 40 chain smokers dies of this disease. The chance of dying from lung cancer is thus 81 times greater in chain smokers than in non-smokers.

I first became aware of the possible relationship between cigarette smoking and cancer of the lung in 1936. In 1919, during my junior year at Washington University, a patient with cancer of the lung was admitted to the Barnes Hospital. As was usual, the patient died. Dr. George Dock, our Professor of Medicine, who was not only an eminent clinician and scientist but also an excellent pathologist, insisted upon the two senior classes witnessing the autopsy, and he stressed that the condition was so rare he thought we might never see another case as long as we lived. Being young and impressionable, I was very much impressed by the rarity of this condition. I did not see another case until 1936 when, at Charity Hospital in New Orleans, I saw nine cases in six months. Having been impressed with its rarity by Dr. Dock 17 years previously, I wondered what was responsible for this apparent epidemic.

Because all the patients were heavy-smoking men who had begun smoking during World War I and because, I learned, cigarettes were consumed infrequently until World War I, I had the temerity at that time to postulate that the enormously increased incidence was due to cigarette smoking. This observation and many subsequent ones were based upon retrospective studies which were criticized by biostatisticians as being of relatively little value. Dr. Cuyler Hammond, the chief biostatistician of the American Cancer Society, emphasized that we could not say that there was a causal relationship between smoking and cancer of the lung simply because practically all patients with cancer of the lung were heavy smokers; and until the incidence of cancer of the lung among smokers as contrasted to that among non-smokers was known, one could not say that there was or was not a causal relationship between smoking and lung cancer.

The American Cancer Society in the United States and Doll and Hill in England made prospective studies. These studies consisted of interrogating large numbers of normal individuals concerning their smoking habits and observing them over periods of time. The studies showed that the incidence of cancer of the lung varies according to the number of cigarettes smoked, the incidence being very low in the non-smoker and increasing with the amount smoked. The American Cancer Society study showed that the incidence of cancer of the lung, exclusive of adenocarcinoma, a rare type unassociated with smoking, was 3.4 per 100,000 population in non-smokers, 78.6 in the cigarette smoker, 11.4 in cigar smokers, and 28.9 for pipe smokers. Not only was cancer of the lung infrequent in the non-smoker, but its incidence increased with the amount smoked. Of those who smoked less than a half a pack a day, the incidence was 51.4; a half to one pack, 59.3; one to two packs, 143.9; and more than two packs, 217.3.

These studies further showed that if one stopped smoking, the chance of developing malignant disease of the lung became progressively less the longer the time elapsed since cessation of smoking. Whereas the incidence in the non-smoker was 3.4 per 100,000 population; in those who smoked less than a pack a day and continued smoking, it was 57.6; those who had discontinued less than 10 years, 35.5; discontinued more than 10 years, 8.3. However, in those who smoked more than a pack a day and continued smoking, it was 157.1; those discontinued less than 10 years, 77.6; and more than 10 years, 60.5. In nine such prospective studies of over 2 million individuals, the results have all been consistent. Lombard (26) showed that the development of lung cancer is influenced by the age at which smoking is begun. Of those who began smoking after the age of 20, the rate per 100,000 population was approximately 28; of those who began smoking between 15 and 19, it was 45;

of those who began smoking between 10 and 14 years of age, it was 80; and of those who began smoking before the age of 10, it was 140.

Tobacco produces cancer undoubtedly because of the contained carcinogens which have been isolated from tobacco smoke and the residue from the smoking of tobacco. Several investigators have been able to produce cancer in animals by the repeated application of the tarred residue from tobacco smoke.

Tragically, although the incidence of lung cancer is increasing more than any other cancer, it is largely a preventable disease, because, with the exception of adenocarcinoma and a few other kinds, cancers of the lung are caused by tobacco use principally in the form of cigarettes. Also, of tragic significance is the fact that whereas lung cancers are preventable, once they occur the outlook is extremely bad. The curability incidence is less than 10 percent.

Other Cancers Caused by Tobacco

Cancer of the larynx is increasing tremendously. According to Wynder et al. *(47, 48)* the incidence of cancer of the larynx in nine cities studied increased by 75 percent from 1937 to 1947. For many years otolaryngologists had recognized changes in the mucous membrane of the larynx associated with smoking. These vary from simply thickening of the cells of the mucous membrane to precancerous lesions and to real cancer. An advantage of a lesion of the larynx is that it can be observed by laryngoscopic examination and can be biopsied easily. Following total abstinence from tobacco use, the abnormal changes, even precancerous ones, revert to normal. There does come a time, however, when progression to actual cancer is irreversible.

In addition to laryngeal cancer, cancers of the lips, tongue, and mouth are increasing, and they are largely due to the use of tobacco. Vogler et al. *(46)*, in a series of 333 cases of carcinomia of the mouth, pharynx, and larynx, found 14.7% involved the lips, 16.5% the tongue, 35% the buccal cavity, 4.8% the oral pharynx, 10.8% other parts of the pharynx, and 15.9% the larynx. Of patients who had only leukoplakia (a precancerous condition) 70% used tobacco, whereas 90% of those with leukoplakia and cancer used tobacco.

In addition to smoking, other forms of tobacco use such as chewing tobacco and snuff can cause cancer of the mouth. Among 525 cases of intraoral cancer observed by Rosenfeld and Callaway *(38)*, 47% of the women in the entire group and 90% of the women patients with cancer of the gingival buccal area habitually used snuff. Chierici et al. *(9)* found that 94% of patients with cancer of the floor of the mouth, soft palate or tonsil, or oral pharynx were smokers.

According to the Surgeon-General's Report *(1)*, the chance of developing cancer of the esophagus is 3.4 times greater in cigarette smokers than non-smokers. The Report further states, "Although not as marked a gradient as in the lung cancer group, the increase in risk for esophageal cancer among smokers of more than a pack a day is greater than for laryngeal and oral cancer."

Whereas the relationship between the use of tobacco and cancer of the oral cavity, pharynx, larynx, esophagus, and lung is easily understandable, the causal effect of tobacco in cancer of the bladder is less obvious. Cancer of the skin is produced experimentally in animals by application of a carcinogen to the skin or epithelial surfaces and, additionally, there is an increased incidence of cancer of the bladder. Holsti and Ermala (21) applied tobacco tar to the oral cavity of mice, and papillomas, 10% of which were malignant, developed in the urinary bladder in 75% of the mice. Lilienfeld and associates (25), who studied the cases of carcinoma at the Roswell Park Institute in Buffalo, found that of the patients with cancer of the bladder, 84.1% used some form of tobacco. From a study of 300 men and 70 women with pulmonary cancer, Wynder, Onderdonk, and Mantel (49) concluded that the risk of bladder cancer increases with the amount smoked. Among people who smoked up to 15 cigarettes a day, bladder cancer was 1.7 times greater than in the non-smoker; among those who smoked 16 to 34 cigarettes a day, it was 3.5 times greater; in those who smoked more than 35 cigarettes a day, it was 5.7 times greater.

Kerr et al. (24) explained the incidence of cancer of the bladder in cigarette smokers by the fact that cigarette smoking interferes with the normal metabolism, so that carcinogens are excreted in the urine. They believe that cigarette smoking produces enzyme blockage, which results in lack of metabolism of the intermediate substance, which is carcinogenic. They reported that 30 determinations on six subjects when smoking showed a consistent rise in carcinogenic metabolites of tryptophan with a reciprocal fall in the end-product N'methylnicotinamide. Moreover, the level of carcinogen fell and the normal end-product rose when the subjects stopped smoking.

Vascular Deterioration

Although a great deal of emphasis has been placed upon the relationship between tobacco use and cancer (largely due to the early work instituted by the American Cancer Society), it is only relatively recently that the tremendous detrimental influence on the cardiovascular system by tobacco use has been appreciated (12). In a ten-year study of longshoremen in San Francisco, Borhani and associates (7) found a definitely higher death rate in coronary disease in cigarette smokers than in non-smokers, and this increased with advancing years. Among those from 45 to 54 years old, the death rate was 26 in smokers and 8 in non-smokers; in smokers 55 to 64 years of age, it was 72 and 16 in non-smokers; in smokers 65-74 years of age, it was 152 and in non-smokers 77.

Spain and Bradess (43) found that smoking greatly increases the risk of sudden death from coronary thrombosis. They state, "For every nonsmoker under the age of 50 who died suddenly and unexpectedly from coronary heart disease, there were 16 such deaths in those who smoked more than one pack of cigarettes per day." They also found that 95% of the smoking group died within an hour of the attack. Of 2,282 middle-aged men, observed for ten years in Framingham, Massachusetts, and 1,883 similar men in Albany, New York, observed for 8 years, Doyle and associates *(13)* found that men who smoked 20 or more cigarettes per day had 3 times as much chance of developing coronary heart disease as did non-smokers and that the risk could be eliminated if they discontinued smoking. They also reported a much higher incidence of myocardial infarction in cigarette smokers than in non-smokers (7.3 and 2.1 per 100,000 respectively). Doll and Hill *(10, 11)*, who studied 41,000 physicians in the United Kingdom, found evidence compatible with the belief that cigarette smoking is one of the most important causes of death from coronary heart disease in persons younger than 55 years of age.

Smoking is not only a major cause of heart disease but is a major cause of arteriosclerosis causing peripheral arterial disease and stroke *(5)*. The agent responsible for the vascular alterations is nicotine. Gofman et al. *(16)* have shown that the serum lipoproteins are definitely higher in smokers than in non-smokers, particularly in the 20 to 29 year level. They conclude, "The association of lipoprotein elevation with cigarette smoking will account for a sizable increase in mortality from coronary heart disease." In addition to the increase in lipoproteins and cholesterol, which are precursors of arteriosclerosis, other factors predispose to bloodclotting in individuals using tobacco. McDonald *(27)* studied the coagulability in 48 patients with ischemic heart disease and 48 healthy controls. He concluded, "A statistically significant difference was found between the patients and the controls in respect to thromboplastin-regeneration, platelet stickiness, fibrinogen-estimation, and prothrombin time. . . ; it indicated an increased coagulability of the patients' blood compared with that of the controls." Engelberg *(15)* studied the clotting time after smoking in 60 patients; in 23 there was no change after smoking, in 34 the thrombosis time was decreased, and in 3 it was increased. For the group of 60 the average decrease was 1.8 minutes, which is highly statistically significant. He concludes, "In the presence of underlying atherosclerotic disease with scarred, narrowed arteries, this accelerated thrombotic tendency may well be the major additional factor precipitating an acute coronary thrombosis . . It is further suggested that this hyperthrombotic state which appears in some individuals after smoking is a major etiological factor in the increased incidence of acute myocardial infarction in habitual smokers." Ashby and co-workers *(2)* reported that smoking 2 cigarettes definitely increased the platelet stickiness; the implication that smoking is a paramount mechanism causing coronary thrombosis is obvious.

Pulmonary Disease Other Than Cancer

The incidence of pulmonary emphysema is increasing rapidly, and the severe form is almost invariably the result of smoking, particularly cigarettes. According to Roberts *(37)* chronic respiratory diseases in 1962 caused 27,000 deaths in the United States and were contributory in an additional 43,000, making a total of 70,000. He states, "In the last 10 years the gross mortality from emphysema and chronic bronchitis has more than quadrupled and continues to grow faster than for any other disease. . . . Among all conditions for which disability allowances are awarded to males by the Social Security Administration, these respiratory diseases are outnumbered only by the cardiovascular diseases." Harris (20) states, "Virtually every investigation has demonstrated a strong epidemiologic relationship between cigarette smoking and chronic bronchitis." Auerbach, Stout, Hammond, and Garfinkel (4), in an autopsy study of patients 70 to 74 years of age, found pathologic changes characteristic of emphysema which varied according to the smoking habits of the individuals. Doll and Hill (10, 11) found that the death rate from emphysema among British doctors was 6 times greater in those smoking more than 25 cigarettes a day than in non-smokers.

Air pollution has been blamed as the cause of respiratory disease and, although there may be a slight additive effect of air pollution, it is certainly not an important factor (31). If it were, men and women should be equally affected because they both breathe the same air. Dysinger et al. (14), in studying 64,000 Seventh Day Adventists living in California over a period of time, observed 4 deaths due to emphysema as the underlying cause, whereas 22 would have been expected. An additional 14 deaths could be attributed to secondary emphysema whereas 51 would have been expected. Although smoking is contrary to Seventh Day Adventists' religious beliefs, one small group with emphysema smoked.

Air pollution is undesirable, and under certain adverse meteorological conditions may be a hazard to certain individuals with severe cardiopulmonary disease, but generally it is not a great health hazard because of the relatively low concentration of pollutant as compared with that in personal air pollution from cigarette smoking (30).

Reproductive Disorders.

It has been demonstrated that many mothers who smoke have smaller and more premature babies. Mgalobeli (29), in Germany, and Athayde (3), in Brazil, showed that women working in tobacco factories had fewer pregnancies, more abortions, and increased infant mortality. Athayde studied 1,866 women capable of procreation working in the tobacco industry. Eight hundred and forty had a pregnancy interrupted by abortion or stillbirth. In a comparative study of 112 smoking pregnant women with 1,383 non-smokers, Bernhard (6) reported an abortion rate of 22.5% in smokers as contrasted with a 7.4% in non-smokers. Moreover the rate of pre-eclampsia in smokers was 9.8% and in non-smokers, 3.4%. Russell, Taylor, and Maddison (41), in a prospective study of 2,257 pregnant women, found that cigarette smoking caused lower birth rate and an increased risk of abortion, stillbirth, and neonatal death. Hudson and Rucker (22) observed that of 580 women at term, 35% were smokers. Of the 52 who aborted, 68% were smokers. Butler *(8)* found that not only was the birth rate of babies born of smoking mothers lower but also the crude mortality rate was 40% higher. The postmortem examination of babies born of smoking mothers revealed an excess of macerated stillbirths, deaths from asphyxia in labor, from postnatal pneumonia, and from neonatal causes associated with premature delivery.

Russell *(42)*, in a series of 2,000 pregnancies, observed a fetal mortality (abortion, stillbirths, neonatal deaths taken together) of 79 per thousand in babies born of smoking mothers as contrasted with 41 per thousand of those born of non-smoking mothers. Steele and Langworth *(44)* studied 80 infants who died suddenly and unexpectedly under 4 months of age. Of the mothers of these infants, 32.6% smoked less than a pack of cigarettes a day and 23.7% one or more packs. They concluded, "There was a highly significant difference in the smoking habits of mothers of cases as compared with mothers of controls," thus clearly indicating that the expectant mother assumes a real risk for her child if she smokes. The decreased birth rate and other sequelae in infants whose mothers smoke is thought to be due to a decrease in the blood flow to the placenta, an increase in carbon monoxide content of the placental blood, as well as the effect of nicotine on the fetus. Younoszai et al. *(50)* found that the infants of cigarette-smoking mothers showed a mild metabolic acidosis as compared to the infants of non-smoking mothers.

Ross *(39)*, in a study of 17,000 infants born in one week in 1958 in Great Britain of which 89% were reexamined seven years later, found a relationship between the smoking habits of the mothers and the incidence of fits in the children. Twenty-seven per thousand had fits and seven per thousand had clonical signs suggesting permanent epilepsy. Eighteen percent of mothers smoking 10 or more cigarettes daily had children with convulsive disorders, compared with 13% who did not smoke. There was no relationship between epilepsy and abnormal delivery. Ravenholt and Levinski *(36)* stated, "Analysis of the previous reproductive experience of these 2023 mothers [in Seattle] . . . revealed that, whereas fetal deaths constituted 11.1% of deliveries of women who had never smoked, they constituted 15.9% of deliveries of women who smoked during the index pregnancy."

In addition to the deleterious effect on the fetus, smoking adversely affects the expectant mother. McDonald *(28)*, in a study of 129 unmarried primigravida white women, found that "Heavy smokers were more depressed than light smokers . . . and tended to be more depressed than nonsmokers." Goldman and Schecter *(17)*, in studying glucose metabolism in pregnant women, found that, in addition to the previously known fact that pregnancy is diabetogenic, smoking during pregnancy causes hyperglycemia of statistical significance.

In an extensive experience with individuals who for various reasons (usually health) have discontinued smoking, I have been struck by the almost universal observation (more frequently by men) that the cessation was associated with a definite increase in libido. Johnston *(23)* states, "The symptoms of tobacco smoking become apparent to the sufferer only after the disease has been arrested, and he has given up smoking. An accession of high spirits, energy, appetite and sexual potency, with recession of coughing, makes the chief symptoms of tobacco-smoking plain." Viczian *(45)* found that smoking also damages the process of sperm formation in the male.

Smoking and Ulcers

The use of tobacco is harmful to patients with peptic or duodenal ulcer. After considerable experience in the treatment of peptic ulcer, I am convinced that patients with ulcers cannot be cured as long as they smoke. Too frequently a patient with a duodenal ulcer, which usually should be treated conservatively by diet and avoidance of factors that produce increased acid, will respond well to such therapy, only to have a recurrence if he begins to smoke.

In a series of prospective studies of the expected observed deaths from all ulcers of the stomach and duodenum among cigarette smokers, it was found that in gastric ulcer 15.8 were expected but 119 were observed. In those with duodenal ulcer 69.3 were expected and 122 observed. In a larger group of cases in which both gastric and duodenal ulcers were grouped together, 117.9 were expected and 333 observed. The Advisory Committee *(1)* to the Surgeon-General concluded that "epidemiological studies indicate an association between cigarette smoking and peptic ulcer which is greater for gastric than for duodenal ulcer." Not only has the incidence increased in the two diseases but the mortality has definitely increased, particularly in gastric ulcer.

Disabling Illness and Premature Death

Many diseases are caused or aggravated by the use of tobacco, and in addition its use has caused a tremendous increase in disabling illness and premature deaths. According to Linus Pauling *(32)*, the eminent physicist and Nobel laureate, an individual aged 50 who has smoked more than a pack a day since the age of 21 has 8½ years shorter life expectancy than an individual the same age who has never smoked. He has calculated on the basis of these statistics that for every cigarette one smokes one's life is shortened 14.4 minutes.

On the basis of the Royal College of Physicians Study *(40)* completed in 1962, it is extimated that one out of every 23 heavy smokers at the age of 35 will be dead before the age of 45, as compared with one out of 90 non-smokers. They estimated that cigarette smokers have a death rate from lung cancer 30 times higher than do non-smokers. Hammond *(18)*, who studied smoking habits in the state of Maryland, found that cigarette-smoking men have the highest mortality ratios from ages 40 to 69. The death rate in cigarette smokers was 92% higher than in non-smokers. In age group 70 to 79, it was 56% higher in smokers. The risk increased with the amount smoked. A non-smoker was assigned the figure 1; for those who smoked up to 9 cigarettes a day, the risk was 1.69; for those who smoked 10 to 19 cigarettes a day, 2.13;

from 20 to 39, 2.19; and more than 40, 2.57. Hammond also determined that the earlier one begins smoking the greater the risk and that the amount of illness varied according to the amount smoked. As determined by Hammond, of his study's total sample, the percentage of patients hospitalized was: non-smokers, 12.3%; cigar and pipe smokers, 16.1%; and cigarette smokers, 17.4%. Hospitalization also varied according to the amount smoked: less than 10 cigarettes a day, 14.3%; 10 to 19, 16.8%; 20 to 39, 17.3%; and more than 40, 25.1%.

Economic Loss

It is maintained by many that although tobacco use may be harmful, it is of great economic benefit to the nation because of the tremendous amount of money involved in the industry. When one considers that the federal taxes alone paid by the cigarette industry amount to approximately 15 million dollars a day, it is evident that we are considering a multi-billion-dollar industry, probably in the neighborhood of 6 to 8 billion dollars annually. However, the other side of the coin, never mentioned, is the economic loss to the nation as a result of the detrimental effects of the use of tobacco. As stated previously, it has been estimated that 360,000 persons die annually because of tobacco use. It is difficult to calculate this loss, but it is tremendous, in addition to the anguish to the bereaved relatives. Besides the deaths *(33)* there were 77 million man-days lost from work because of tobacco use, another 88 million man-days lost because of illness associated with tobacco use, and still another 306 million man-days of partial disability. Disregarding the health and considering only the monetary side, and taking $50 a day for those totally disabled and $25 a day for those partially disabled, this economic loss to the nation in 1965 was $18,800,000,000, which is about 3 times greater than the estimated profit from tobacco.

Although there are many people who believe that the use of tobacco is justified because of the presumed pleasure it gives, one wonders if the price is not too high for a few moments of a pleasure which invariably results in tremendous increase in disabling illnesses and marked increase in premature deaths. As Michel de Montaigne (1533–1592) remarked, "Men do not usually die, they kill themselves."

REFERENCES

1. Advisory Committee to the Surgeon-General of the Public Health Service. 1964. *Smoking and Health.* Public Health Service, Publication No. 1103.
2. Ashby, P., A. M. Dalby, and J. H. D. Millar. 1965. Smoking and platelte stickiness. *Lancet 2:*158–59.
3. Athayde, E. 1948. Incidencia de abôrtos e mortinatalidade nas operarias da indústria de fumo. *Brasil Med. 62:*237–39.

4. Auerbach, O., A. P. Stout, E. G. Hammond, et al. 1963. Smoking habits and age in relation to pulmonary changes: rupture of the alveolar septums, fibrosis and thickening of walls of small arteries and arterioles. *New Eng. J. Med. 268:* 1045—54.

5. Auerbach, O., E. C. Hammond, L. Garfinkel. 1965. Smoking in relation to arteriosclerosis of the coronary arteries. *New Eng. J. Med. 273:775—79.*

6. Bernhard, P. 1962. Sichere Schäden des Zigarettenrauches bei der Frau. *Med. Woch. 104:*1826—31.

7. Borhani, N. O., H. H. Hechter, and L. Breslow. 1963. Report of a ten-year follow-up study of the San Francisco longshoremen. Mortality from coronary heart disease and from all causes. *J. Chronic Dis. 16:*1251—66.

8. Butler, N. R. 1965. The problems of low birthweight and early delivery. *J. Obstet. Gynaec. Brit. Comm. 72:*1001—03.

9. Chierici, G., S. Silverman, Jr., and B. Forsythe. 1968. A tumor registry study of oral squamous carcinoma. *J. Oral Med. 23:*91—98.

10. Doll, R., and A. B. Hill. 1956. Lung cancer and other causes of death in relation to smoking. A second report on the mortality of British doctors. *Brit. Med. J. 2:*1071—81.

11. Doll, R., and A. B. Hill. 1964. Mortality in relation to smoking. Ten years' observation of British doctors. *Brit. Med. J- 1:*1399—1410.

12. Doyle, J. T., T. R. Dawber, W. B. Kannel, et al. 1962. Cigarette smoking and coronary heart disease: Combined experience of the Albany and Framingham studies. *New Eng. J. Med. 26:*796—801.

13. Doyle, J. T., T. R. Dawber, W. B. Kannel, et al. 1964. The relationship of cigarette smoking to coronary heart disease. *JAMA 190:*886—90.

14. Dysinger, P. W., F. R. Lemon, G. L. Crenshaw, et al. 1963. Pulmonary emphysema in non-smoking population. *Dis. Chest 43:*17—25.

15. Engelberg, H. 1965. Cigarette smoking and the in vitro thrombosis of human blood. *JAMA 193:*1033—35.

16. Gofman, J. W., F. T. Lindgren, B. Stresomer, et al. 1955. Cigarette smoking, serum lipoproteins, and coronary heart disease. *Geriatrics 10:*349—54.

17. Goldman, J. A., and A. Schecter. 1967. Effect of cigarette smoking on glucose tolerance in pregnant women. *Israel J. Med. Sci. 3:*561—64.

18. Hammond, E. C. 1964. Smoking habits and health in Maryland and neighboring states. *Maryland Med. J. 13*(11):45—49.

19. Hammond, E. C. 1969. Life expectancy of American men in relation to their smoking habits. *J. Nat. Cancer Inst. 43:*951—62.

20. Harris, H. W. 1963. Chronic bronchitis, emphysema and asthma. *Amer. J. Public Health 53:*Sup. 7—15.

21. Holsti, L. R., and P. Ermala. 1955. Papillary carcinoma of the bladder in mice, obtained after peroral administration of tobacco tar. *Cancer 8:*679—82.

22. Hudson, G.S., and M.P. Rucker. 1945. Spontaneous abortion. *JAMA 129:*542—44.

23. Johnston, L. 1952. Cure of tobacco-smoking. *Lancet 263*(2):480—82.

24. Kerr, W. K., M. Barkin, P. E. Levers, et al. 1965. The effect of cigarette smoking on bladder carcinogens in man. *Canad. Med. Ass. J. 93:*1—7.

25. Lilienfeld, A. M., M. L. Levin, and G. E. Moore. 1956. The association of smoking with cancer of the urinary bladder in humans. *Arch. Intern. Med. 98:*129—35.

26. Lombard, H. L. 1959. *Cancer Fact Book.* Massachusetts Department of Public

Health.
27. McDonald, L. 1957. Coagulability of the blood in ischemic heart disease. *Lancet* *273*(2):457—60.
28. McDonald, R. L. 1965. Personality characteristics, cigarette smoking and obstetric complications. *J. Psychol. 60:*129—34.
29. Mgalobeli, M. 1931. Einfluss der Arbeit in der Tabakindustrie auf die Geschlechtssphäre der Arbeiterin. *Monatschr. f. Geburtsh. u. Gynak, 88:* 237—47.
30. National Advisory Council on Cancer. 1970. *Progress Against Cancer.*
31. Ochsner, A. 1967. The hazards of air pollution—fact or fiction. *Medical Tribune 8:*11, 14.
32. Pauling, L. 1961. Aging and death. Address presented at the University of Toronto, March 30, 1961.
33. The Health Consequences of Smoking. 1967. *Public Health Review,* 1967.
34. Rakower, J. 1965. Smoking habits and lung cancer in Israel. *Harafuah 68:*115.
35. Ravenholt, R. T. 1964. Cigarette smoking: magnitude of the hazard (Letters to the Editor). *Amer. J. Publ. Health 54:*1923.
36. Ravenholt, R. T. and M. J. Levinski. 1965. Smoking during pregnancy. *Lancet 1:*961.
37. Roberts, A. 1965. Public Health Service activities in chronic respiratory diseases. *Public Health Rep. 80:*336—38.
38. Rosenfeld, L., and J. Callaway. 1963. Snuff dipper's cancer. *Amer. J. Surg. 106:*840—44.
39. Ross, E. M. 1970. In *Third European Symposium on Epilepsy.* Helsingor: Denmark.
40. Royal College of Physicians. 1962. *Smoking and Health.* London: Pitman.
41. Russell, C. S., R. Taylor, and R. N. Maddison. 1966. Some effects of smoking in pregnancy. *J. Obstet. Gynaec. Brit. Comm. 73:*742—46.
42. Russell, C. S. 1968. Smoking and pregnancy. *Lancet 2:*1190—92.
43. Spain, D. M., and V. A. Bradess. 1970. Sudden death from coronary heart disease. Survival time, frequency of thrombi, and cigarette smoking. *Chest 58:*107—10.
44. Steele, R., and J. T. Langworth. 1966. The relationship of antenatal and postnatal factors to sudden unexpected death in infants. *Canad. Med. Ass. J. 94:* 1165—71.
45. Viczian, M. 1968. The effect of cigarette smoke inhalation on spermatogenesis in rats. *Experientia 24:*511—13.
46. Vogler, W. R., J. W. Lloyd, and B. K. Milmore. 1962. A retrospective study of etiological factors in cancer of mouth, pharynx, and larynx. *Cancer 15:*246— 58.
47. Wynder, E. L., I. J. Bross, and E. Day. 1956. A study of environmental factors in cancer of the larynx. *Cancer 9:*86—110.
48. Wynder, E. L., I. J. Bross, and R. M. Feldman. 1957. A study of etiological factors in cancer of the mouth. *Cancer 10:*1300—23.
49. Wynder, E. L., J. Onderdonk, and N. Mantel. 1963. An epidemiological investigation of cancer of the bladder. *Cancer 16:*1388—1407.
50. Younoszai, M. K., A. Kacic, and J. C. Haworth. 1968. Cigarette smoking during pregnancy: The effect upon the hematocrit and the acid-base balance of the newborn infant. *Canad. Med. Ass. J. 99:*197—200.

10. *Biological Theories of Aging*

ALEX COMFORT

Among the theories discussed here is one that presents aging as an information loss, chiefly at subcellular levels. Information loss in man is thought to have been at a stable rate throughout recent times, to involve feedback mechanisms, and to be subject to his manipulation if concerted efforts of study are pursued. Dr. Comfort received his degree in biochemistry from the University of London where now he is Honorary Research Associate in Zoology and Director of Research in Gerontology. He has authored an impressive array of novels, poems, plays, essays, as well as scientific texts. His most recent book, *The Joy of Sex,* is a best-seller.

Many of us believe that biological gerontology will be to the medicine of the 70's what chemotherapy was to the medicine of the 40's and 50's, and what immunology may be to the medicine of the 60's—with the difference that its implication is even wider. For the first time we are in the position to say that if medicine applies itself with proper backing to this subject, then within a relatively short time, and for a relatively small research expenditure, judged by the standards of other technological programs, it could present man with the option of realizing one of his oldest fantasies: that of controlling his lifespan, in part at least. At the very least it could, in ten years of serious work, answer the question whether or not a practical measure of control is likely to be fairly easily obtainable. The odds on success, by a useful amount, are about two to one against. The option is there, but it depends on whether we can persuade a major scientific power, one with the necessary tradition of critical biology and the necessary manpower and hardware, to take the project seriously.

"Hobbes said that the life of uncivilised Man was nasty, brutish, and short. The sanguine among us will say that technology has made it less nasty and less brutish; we have to try to put out of our minds the fact that it is still short. That fact has been obscured by our successes in not dying young."

The credentials of gerontology, as a study which merits the sort of effort I personally think it will soon command, include the diagram shown in Fig. 1. It charts the phenomenon of lifespan in Man. The corollary of this diagram is clear: we are already within sight of the practical limits of public health and of social advance, so far as lifespan is concerned, and we can see what those limits are. It has been calculated that the total prevention or cure of the three

Source: Reprinted by permission of the publisher and author from *Human Development, 13:* 127-139, 1970.

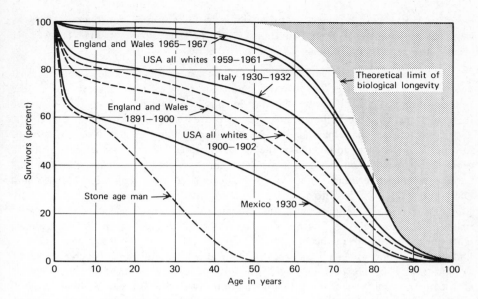

Fig. 1. Life tables indicating the effect of social and general medicine on life span.

main causes of death in the United States—heart disease, vessel disease, and the malignancies—would increase the overall expectation of life by less than five years, and the expectation of life at 65 by much less. Given the pattern of information loss from the human body with age, no amount of extra effort in conventional medicine, in cancer research, in sparepart surgery, or in welfare— good as all these are— is going to do more than make the distribution of age at death rather more regular about a mode of 75 or 80 years. Underprivileged countries are moving toward this norm and will reach it if war, overpopulation, and economic injustice do not stop them. Privileged societies are already pushing that limit. Today, nearly all our intractable medical problems that are not psychosomatic are geriatric.

"I think that perhaps—and I do not say this critically—prosperity makes us resent and fear death and the humiliation of aging more than ever before, and we rightly fear aging rather than death. How many of us can really face the emotional and personal implications of a visit to an old folk's home? It shows us that throughout human history we have been putting up with the intolerable. These women were once beautiful, or at least young; these men were once people. This is the limit to our achievement as people. It may be Faustian for prosperous nations to think in this way, but we do. We think, as Yeats did, about
  'The death
  Of every beautiful eye
  Which made a catch in the breath.'

We have put up with the idea of aging as a fact of life, but only so long as we saw no chance of doing anything about it. The time for that is over. If we see the possibility, we can change it, or at least try."

The information loss which defeats us in age is highly stable, and it must be timed by something. The age of senility is now exactly what it was in the time of Moses. The difference is that in privileged cultures more of us survive to reach it. On the other hand, if it proves feasible to tamper with the 'clock', whatever that may be, it is highly probable on the basis of current information that we could move the onset of most if not all the major deteriorations and malignancies to a higher age—that we could roll back the actuarial barrier which determines that we will decline at the age when Moses and Pharaoh declined. That this can be done in mammals, sometimes by relatively simple interventions, shows that the lifespan of mammals is in principle modifiable. Actuaries who estimate annuities work, quite rightly, in the confidence that no conventional medical advance can seriously shift the biological "gate"— that however good our medicine, most of us will get old and die between the ages of 75 and 85. They are perfectly right, but only so long as we make no fundamental attack on the timing mechanism of information loss itself. This can be done. Even if one makes full allowance for experiments in which life prolongation is really due to getting rid of a single predominant pathology in a pure strain or in a species, there are still enough examples where relatively simple intervention seems to affect the age of onset of age-associated changes in mammals en bloc. The conviction of gerontology is that in spite of great difficulties it is time to look intensively for ways of doing this in Man.

Aging is a loss of information; the old organism lacks the homoeostatic resources to remain as viable and stable as it once was. Most investigators start from this proposition, and many make the second assumption that the loss is occurring, partly at least, at the primary cellular level. If this second assumption is also correct, it leads to two subsidiary questions. First, is the loss predominantly in fixed cells? (By this I mean cells such as those of the nervous system, which do not divide, and which live as long as we do.) Or is the loss predominantly in dividing cells—in clones that produce successive generations during our life? If the second, how do these successive generations change? And how do the new cells of an old man differ from those of a baby? That is one group of questions. The second group is this: Is the leading process of information loss one of noise accumulation (using noise in the sense of random error, the communication engineer's sense) or is it secondary to differentiation? Does it, in other words, depend on the irreversible switching-off of genes in the process of development toward adulthood? Clearly, if we are dealing with a phonograph record that becomes increasingly unintelligible through the accumulation of random scratches, we might be able to lubricate the needle and reduce the random noise injection. If on the other hand we are dealing with a record that comes to an end and cannot be replayed, then in order to make it last longer we must run it more slowly, though not so much more slowly as to spoil the music. Of course, neither of these models is exclusive or exhaustive. We may have to do first one, then the other. If switching-off is involved, we are still probably dealing with information loss by random damage to molecules or structures which can now no longer be replaced. Moreover, both noise accumulation and programmed loss of capacity are prima facie susceptible to interference with their rate.

These are the basic theoretical questions. One can incorporate them in a logic diagram (Fig. 2). In fact, they may prove to be the last questions about aging that we can answer exhaustively. I think it is marginally more likely that we shall be able to answer them by discovering techniques which empirically alter the rate of aging than that we shall alter the rate of aging by first reading back its exact nature and site. All of the theories that exercise us today concern the factors which might contribute to error accumulation and information loss, and the site at which these might be expected to operate. In whole-animal research we can very rarely discriminate between these aims. At the same time, in order to experiment, we need hypotheses that fit the largest number of observed facts.

We know that the most critical cellular information system is that which leads from DNA through RNA to the production of protein, and that certain proteins, the synthetases, being themselves machine tools, are informationally critical. If they are misproduced, all subsequent production will be affected. The body is a factory whose end product is a capacity for self-maintenance. At a fixed rate, this self-maintenance fails. Are the errors in the blueprints themselves? Do these fade or become illegible? Are the errors in the copies made from them—the RNA? Or are they in the machine tools which the factory itself produces to service all of these—the synthetases and the organelles? We know that the chemical errors which affect materials—crosslinking in molecules, the kind of free-radical attack which leads to perishing of rubber, radiation injury, enzyme spoilage—are relatively nonspecific. They could act at any level of the process. But to induce runaway deterioration, which is what we see in mammalian aging, it seems necessary both mathematically and logically to require some kind of feedback process—one which, once begun, is self-aggravating. Apart from verifying the model as a whole, we need to determine the size and possible sites of such feedback loops. Yet the error may not be in the hardware but in the management of the factory—in the relation of cells to one another. MacFarlane Burnet's idea that mutation produces altered lymphocytes, this in turn leading to autoimmune civil war within the body, is a model of this kind. Here the error is in the hardware and the feedback is in the management. But let us start with the simplest models, realizing that we may need to combine them.

The simplest hypothesis is still that information is being lost from cellular DNA—by mutation, by macromolecular damage such as cross-linking, and other

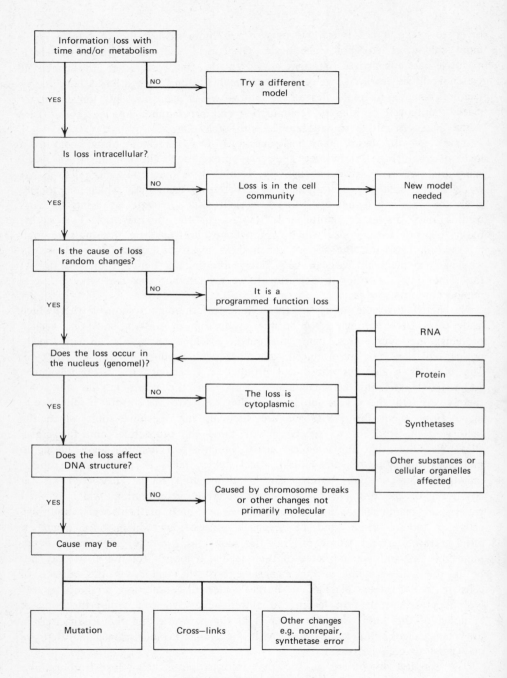

Fig. 2. Aging of mammals.

deteriorative changes we see in non-living long-term molecules. This is a historically august theory that stems largely from the attempt to explain the life-shortening effects of radiation. Simple point mutation throughout the cell community seems now to be ruled out, both by the mathematics of the process, and by the fact that not all mutagenic chemicals shorten life. Long-term damage to macromolecules by cross-linking, irreparable strand breakage, etc., are more acceptable. This would square with radiation studies, with the effects of chemical mutagens, and with what we know of large molecule deterioration generally. The damage is to the nuclear furniture and the function of the DNA strand, rather than to the actual genes. We have rather lost sight of the idea, popular with early biochemists like Marinescu and Ruzicka, that long-term cell constituents undergo the kind of damage we see in artificials such as rubber, leather, or plastics. If they don't they must be specially protected in some way. The return of this concept may be important, because it brings in the wide experience of food and materials chemists in preventing this kind of attack. DNA in general *is* specially protected, or we should not be here. All of these attack processes could operate equally at other levels. There would seem to be a major difficulty in locating aging damage in the DNA molecule, unless it is assumed that the information loss is confined to fixed cells only. We know that there are such things as stable clones—cells which can divide indefinitely without loss of stability. If there were not, we should have no bananas or William pears. We also know that in long-term cells capable of division, such as the ova, the damage rate with storage is not very high. There is some evidence of damage; Down's syndrome is a case in point. The large wastage of ova in Man may represent a screening-out of damaged cells of a kind which cannot take place, e.g., in brain cells. But if DNA is the site of agewise information loss due to random damage, we have to explain why this occurs about fifty times faster, judged by mortality, in mouse DNA than in that of Man. So far we have not been able to compare old with young DNA, at least to my satisfaction, and what work has been done suggests chiefly an increase in protein binding capacity in old DNA. It is now much more likely that error-injection, if it occurs, takes place further down in the chain of command—perhaps in the RNA or the synthetases.

Alexis Carrel taught that if fibroblasts were cultured from a chick, they would divide indefinitely, producing an immortal cell culture from a mortal organism. Until recently most biologists accepted this view. It now appears that there are two kinds of cell-lines or clones without sexual reproduction: stable and unstable: that cells from a vertebrate body are not truly stable clones; that some at least are intrinsically limited to roughly the number of cell divisions they would have undergone during the life of the animal from which they were taken (with a bit to spare); and that the number of further divisions they can undergo depends on the age of the donor. Since the last congress of Gerontology, which devoted a lot of time to possible aging mechanisms involving DNA, an important new field has been opened in the examination of aging in clones, starting with Hayflick's demonstration that cultured body-cells lines may have fixed lifespans. This work really started in 1956 with the work of Rizet and Marcou on fungi, and it has been brilliantly extended to Muggleton and Danielli in Buffalo, and by Holliday at the National Institute of Medical Research.

Muggleton and Danielli took stable clones of *Ameba* and put them in conditions which strongly inhibited protein synthesis over long periods of time. Under these conditions, irreversible changes took place in the division capacity of the cells. In some cell lines, they found a behavior like that of the cells which generate the horny layer of the skin; at each division one product died, the other dividing again. In other lines, division continued tree-wise, but all of the products died simultaneously after a fixed number of generations. This is exactly the behavior Hayflick had found in his vertebrate cell cultures. By some extremely elegant transplatation experiments, Muggleton and Danielli showed that while stem-cell behavior could be produced by implanting a 'spanned' nucleus in normal cytoplasm, Hayflick-type behavior could be induced in normal Ameba by very small injections of spanned-clone cytoplasm. This is exactly what Rizet had found in fungi—that senescence in a cell line can be cytoplasmically transmissible. Thus, in the cytoplasm of such cells whose life is limited, there is a constituent that can generate escalating disorder in normal cells.

If this is shown to be equally applicable to mammalian cells, it has great importance for our search for the site of error accumulation. In 1963 Orgel, now at La Jolla, pointed out that the one place in the process of cellular information flow where error introduction would be bound to escalate if it once began was in the production of synthetases. These enzymes are the machine tools governing the correctness of all subsequent cellular function. If errors occurred here, either in the specifying RNA or in its transcription into enzyme, it would be self-aggravating in precisely the way required by the mathematics of somatic aging. And this error should be cytoplasmically transmissible, like a virus, producing more of itself. With this in mind, Holliday has carried out experiments on fungi (Neurospora and Podospora) which show that raising the rate of protein error with nonsense amino acids accelerates clonal aging, and that the cytoplasmic factor present in aging clones decreases the accuracy of protein specification when injected into stable clones. He used a very elegant method based on nutritional requirements in particular strains.

None of this work has yet been applied directly to mammalian material. Mammals, including ourselves, do not die of old age through simple failure of cell division; old men heal normally. But the work suggests that if error accumulation is a major factor in senescence, it may well be occurring, not, as was thought, at the blueprint level, in the nuclear coping system—or not there alone—but lower down the chain of replication, at the point where the enzymes that service further copying are specified. This, if correct, is an important change of emphasis. The work of Wulff and his co-workers at Utica, who have been looking at DNA, RNA, and other cell constitutents for evidence of nonsense-material production, or of run-away synthesis to compensate for it, seems to me to point in the same direction.

A different approach to the error question leaves aside its site. We know that radiation shortens life, and we think it may do so by inducing random damage to a variety of structures, chiefly by the agency of free radicals—the things which sour lard or perish rubber automobile tires. If any structure is attacked in this way, it ought to be possible to slow the rate of attack by means of substances similar to those which the rubber and food chemists add to their products. This idea has now been tried, by Harman at Omaha. He has fed mice a number of substances which are either radioprotectants or antioxidants used in the food industry. These act as peroxide scavengers. With three of these, 2-mercaptoethylamine; BHT (ter-butylhydroxytoluene), which is added commercially to rubber and to lard; and ethoxyquin, which is added to broiler chick feed to prevent toxic peroxidation ('crazy chick disease'), he has produced significant increases in mouse lifespan. It is clearly critical to know whether he has shifted the biological "gate" of lifespan upward, as happend with food restriction, or only removed one important cause of presenile death, such as lipid intoxication. In my own experiments with ethoxyquin I have found a big effect on weight gain, so we may only be spoiling the animals' appetites and in this way lengthening their lives. But if antioxidants increase resistance to error accumulation, this would be both a theoretical and a practical advance of wide importance. Whether free radical attack is the main source of error-accumulation in cells, and, if so, where it acts, is not yet known. But a 40% increase in longevity is an important finding. Experiments are currently underway to determine whether antioxidants affect the appearance of error and limitation in clones.

Another important theory which we are on the verge of being able to incorporate into our picture of aging is that of autoimmune damage. For many years evidence has been accumulating that with increasing age the defence mechanisms of the body, which cause it to reject grafts and invaders, more and more frequently attack their proprietor's body cells, with consequences very like those of graft rejection and of the degenerative pathologies, such as amyloid disease, which we see in old animals. If errors occur anywhere in cell copying, the diversity of our cells might well increase with age to the point at which the body's self-recognition is challenged. Walford of UCLA has for many years collected evidence that such a process contributes to aging of the mammalian kind. In the past, workers have looked for mutation either in cells generally or in the immunity-controlling lymphocytes and plasmacytes; cells mutate to become criminals, or policemen mutate to arrest the law-abiding, and society founders. We might now look equally to the other types of error

Table I. *Substances Which Might Modify Rate of Ageing*

Nature	Theoretical Basis	Findings
Antioxidants	Scavenge free radicals, prevent attack on DNA or some other system	'Prolong life' in mice but do not appear to shift specific age (Harman)
Radio-protectants	Assume ageing similar in nature to radiation damage	Protect against radiation— life shortening
Protein synthesis inhibitors	Break 'vicious circle' if faulty synthetases involved	
Lysosome stabilizers	Prevent escape of enzymes (including lysosome DNAse)	
Immuno-suppressants	Abolish any ageing effects due to immune divergence	Imuran prolongs mouse life slightly (Walford). Prediction that ALS will limit some age changes but increase clonal divergence by stopping clone deletion
Anti-crosslinking agents	Ageing due to crosslinks in long-term molecules	Under test (LaBella, Kohn and Leash), β-amino propionitrile, penicill-amine
Hormonal agents	Modify chemical allometry, to produce discrepancies between process rates and retard senescent program	
Anabolics	Prevent decline of protein storage and muscular strength	Effective clinically (in prevention of decline of protein storage and muscular strength)
Somatotrophin	Maintain "young" pattern of protein synthesis	Carcinogenic in rat (Moon *et al.*) fails to prolong rat life (Everitt)
Prednisolone	Program slowing, anti-autoimmune?	Doubles lifespan in short-lived strain of mouse (Bellamy)
17-ketosteroids	Decline closely parallels human senescence	
Antimetabolic drugs	Delay program, simulate calorie restriction, induce active diapause, i.e. lowered metabolism without reduced physical and mental activity	Proposed (Strehler) but not tested

described above. With the break-through in immunology and the development of antilymphocytic sera, we may be on the verge of testing this theory. At one time one might have hoped that effective methods of immunosuppression would delay aging by cutting out such processes. But it seems equally likely that if the basic senile error is that wrong cells are being produced, immunosuppression will increase the discord. Early evidence suggests that it already increases tumor incidence. We may suppress some reactive manifestations of aging by suppressing autoimmune response, but it might equally be that we need to strengthen rather than weaken the body's filter in order to reject malspecified cells.

This paper cannot cover every theory that has been offered to explain aging and which is stimulating experiment today. A reviewer of theories can only select those signposts which seem to point the same way, and draw his map from that. However hopefully we draw such maps, the possibility always remains that some single and simple process has been at work all the time: the restriction of blood supply to a vital organ, an accumulation of 'clinkers' in cells, a leakage of information from the nervous system, a change in cell membranes due to crosslinking, leakage of enzymes from lysosomes, etc. and that an elegant experiment will subvert all of our theories. That may yet happen. In the meantime we must do our best, leaving colleagues who see these possibilities to investigate them. In spite of this, a picture does seem to emerge today which at least does not contradict the work which has been done and the phenomena we see. Aging appears to be a loss of information from the organism. That loss may well be cellular, at the DNA loci or elsewhere. It could represent random attack which injects error or failure at some vital point, so that the mischief is self-aggravating and homoeostasis fails. That process could be lifelong, or it could supervene only when reparative capacities have been switched off in the course of development at the DNA loci or elsewhere; or when other processes, postponed by natural selection to the end of the reproductive period, catch up with us. It may affect one group of cells sooner than another, leading to a train of predominant mischief which we could remove if we protected or replaced those cells. What we do know is that the mischief in aging is stable, so that the human lifespan has changed little in history; that it appears to involve feedback, and that it is almost certainly accessible to interference in Man if we care to apply ourselves to the subject. Simple chemical steps such as those which we see in the time-linked deterioration of foods and materials may well be involved, whence our interest in molecular crosslinking, free-radical attack, radiation damage, and the spoilage of enzymes. At the moment the point at which error capable of inducing cumulative breakdown seems most likely is perhaps the step of RNA transcription into synthetases. But it may occur elsewhere—in DNA; in molecular repair, in cell membranes and organelles, in recognition processes between cells. If random chemical error is important, we can probably slow its accumulation.

With DNA and other longterm program molecules, our task would be like that of conservative dentistry: to slow down tooth loss. We could probably not replace lost teeth or DNA molecules. If, however, the main error is at the protein level, this has the important consequence that while it might be very difficult to prevent escalation once error has occurred, it might be possible to scrub out the entire erroneous protein crop and return to the original specification, if that is still undamaged. Nobody has yet restored a spanned clone of Ameba or a fungus to unlimited vigor, but they may yet do so. Orgel's hypothesis, if it is right, opens the possibility not only of slowing aging, but of reversing it.

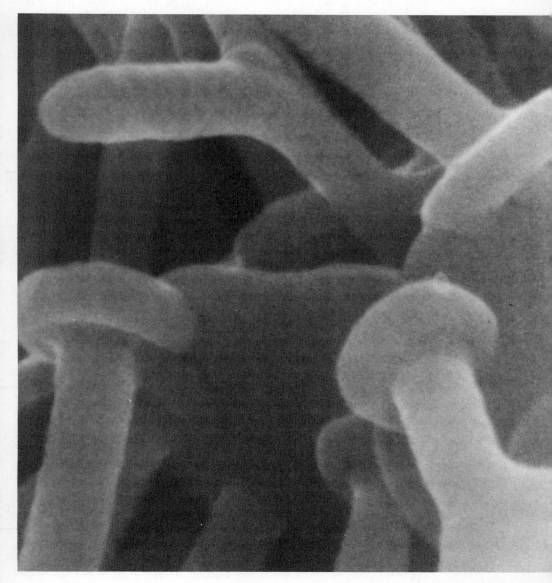

Electron photomicrograph of synaptic knobs in the abdominal ganglion of the sea hare Aplysia *(Dr. E. R. Lewis).*

11. *The Brain of Pooh: An Essay on the Limits of Mind*

ROBERT L. SINSHEIMER

Everyone should recall from the children's stories of A.A. Milne the character of Edward Bear better known as Winnie-the-Pooh. The loveable but bumbling bear once was heard to declare, "For I am a Bear of Very Little Brain and very long words bother me." According to Dr. Sinsheimer, developments in genetics will someday enable man to alter his central nervous system and overcome shortcomings not unlike those of Pooh. Chairman of the Division of Biology, California Institute of Technology, Dr. Sinsheimer is a scholar and academician of renown, the author of many publications, and a member of the National Academy of Sciences and the American Academy of Arts and Sciences.

The inviolate principle of causality—that a precisely determined set of conditions will always produce precisely the same effects at a later time—underlies our entire scientific perception of the universe. And yet in the smoothly flowing channels of natural causality there has always been in our conception one seemingly irrational, unordered, swirling eddy—the human mind. Increasingly now we cannot avoid this vortex, nor can we continue to skirt around it, for herein is the ultimate perceiver and herein form the shapes of surmise. And so, as in our dreams where we are surprised by that which we ourselves have conjured, the perceiver must in wonder inquire, "How do I perceive?" and the mind ask, "What is thought?"

The great discoveries in genetics and our enlarged understanding of the biochemistry of heredity have led to increasing discussion of the possibility of the designed change of human beings—not only of the repair of overt genetic defects but also of the longer range enhancement of the capabilities of man. Naturally, much of this discussion has concerned the improvement of man's finest and most precarious quality, his mind.

To consider this issue in any serious way one must first inquire as to what qualities of mind are considered to be genetic in origin (and are thus susceptible to genetic modification) and to what extent these qualities limit the performance of man—and what might be the consequence of their modification. In a philosophical sense such an endeavor—man trying to improve his own capacity—is clearly a bootstrap project, an adventure in positive feedback. And yet this is what we have done all the way from the jungle. What we consider

Source: Reprinted from *Engineering and Science* magazine, January 1970, by courtesy of the author and editor. Published at the California Institute of Technology.

now is but an extension, albeit in a new dimension. What can we honestly say about the mind from our present knowledge? I do believe that such a presentation can be useful in the same sense that the sixteenth-century maps of the world were useful, essentially as a rough chart of what it is we need now set out to learn, bearing in mind that the enterprise may well require as many years as were needed to fill in those ancient maps.

Further, in this *special* case there is special merit in such a projection of knowledge that we may hope to have concerning the human brain and thus concerning its, and our, future potential. For this effort to see how our brain came to be and how it might be advanced can serve to provide us a valuable perspective in which to view our present reality, in which to see more clearly our present limitations and, therein, the origins of some of our most basic dilemmas.

The very opening lines of *Winnie-the-Pooh* provide my theme.

"Here is Edward Bear, coming downstairs now, bump, bump, bump, on the back of his head, behind Christopher Robin. It is, as far as he knows, the only way of coming downstairs, but sometimes he feels that there really is another way, if only he could stop bumping for a moment and think of it."

Now Edward Bear, or Winnie-the-Pooh as he was known to his friends, was of course a bear of very little brain. But nonetheless I often think that these lines constitute a splendid parable to man and his whole scientific enterprise— that we perforce go bump, bump, bump along the paths of scientific discovery when had we but the acumen, the brain power, we could immediately deduce from the known facts the one right and inherently logical solution. This seems particularly true in biology wherein all extant phenomena have for so long been subject to and ordered by the harsh disciplines of natural selection, and wherein the right answer, when we find it, always does seem so inevitably right.

And yet of course we don't have the acumen and we can't immediately deduce the right solution because, like Pooh, our brains too are really very limited compared to the complexity about us and the frequent immediacy of our tasks. And in simple fact what else can we sensibly expect when we are apparently the first creature with any significant capacity for abstract thought? Indeed, even that capacity developed primarily to cope with stronger predators or climatic shifts, not to probe the nature of matter or the molecular basis of heredity or the space-time parameters of the universe.

A physicist friend of mine frequently remarks on how much more difficult it seems to be to teach a 17-year-old a few laws of physics than it is to teach him to drive a car. He is always struck by the fact that he could program a computer to apply these laws of physics with great ease but to program a computer to drive a car in traffic would be an awesome task. It is quite the reverse for the 17-year-old, which is precisely the point. To drive a car, a 17-year-old makes use, with adaptation, of a set of routines long since programmed into the primate brain. To gauge the speed of an approaching car and maneuver accordingly is not that different from the need to gauge the speed of an approaching branch and react accordingly as one swings through the trees. And so on. Whereas to solve a problem in diffraction imposes an intricate and entirely unfamiliar task upon a set of neurons.

I think the computers first made us aware of one of the more evident limitations of the biological brain, its millisecond or longer time scale. Computers flashing from circuit to circuit in microseconds can readily cope with the input and response time of dozens of human brains simultaneously or can perform computations in a brief period of time for which a human brain would need a whole lifetime.

Similarly I believe that we will come to see that our brains are limited in other dimensions as well—in the precision with which we can reconstruct the outside universe, in the nature and resolution of our concepts, in the content of information that may be brought to bear upon one problem at one time, in the intricacy of our thought and logic—and it will be a major contribution of the developing science of psychobiology to comprehend these limitations and to make us aware of them, to the extent that we have the capacity to be aware of them.

For I think it is only logical to suppose that the construction of our brains places very real limitations upon the concepts that we can formulate. Our brain, designed by evolution to cope with certain very real problems in the immediate external world of human scale, simply lacks the conceptual framework with which to encompass totally unfamiliar phenomena and processes. I suspect we may have reached this point in our analysis of the ultimate structure of matter, that in various circumstances we have to conceive of a photon as a wave *or* as a particle because these are the only approximations we can formulate. We, and I mean we in the evolutionary sense, have never encountered and had to cope with a phenomenon with the actual characteristics of a photon.

And likewise with the subnuclear particles. I was intrigued to learn that the latest attempt to formulate a theory of subnuclear particles is a bootstrap or self-consistent field theory, which as I understand it is a bit like saying it is there because it is there and it has to be there. To my mind this is in effect a bold attempt to adapt the concepts available to the human mind to an intractable and perhaps unimaginable reality. As Einstein so well said, "The most incomprehensible thing about the universe is that it is comprehensible."

Similar problems of concept may well arise on the vast scale of the universe or, more to the point, in the intricate recesses of the mind. Our problem will be somehow to shape a mirror to the mind such that we can comprehend its reflection.

I have tried to think how we might approach this problem of the limits to thought inherent in the structure of our brain and therefore potentially extensible by genetic modification. One approach would be comparative or phylogenetic. If we could trace the detailed chemical and structural changes in the central nervous system as evolution has progressed through the vertebrate species, and if we could correlate these changes with the changes in the reactive and conceptual capacities of these species, we would have one basis for future extrapolation.

Now the comparative approach to phylogenetic evolution has been somewhat in disfavor in this recent era of biochemical ascendance, and for good reason. The biochemistry of all living creatures is really so similar. Hardly anyone would venture to suggest the differences between man and monkey are a consequence of a novel and major innovation in biochemistry. Indeed the biochemistry of man and a yeast cell are astonishingly similar. It is evident that almost all of the most basic processes of biochemistry must have been elaborated in some very remote time of evolution.

Rather, then, the differences between man and monkey must derive largely from some elaborations of structure, and thereby function on a cellular and multicellular level and primarily in the central nervous system. And these innovations must have arisen through the usual genetic mechanisms. How many genetic changes were there, literally? It's clear they did not require any major addition to the genome. The haploid DNA content of man and monkey is identical within the precision of measurement—a few percent.

Our abilities to compare and homologize or differentiate the DNA's of different species—say, man and monkey— are as yet very crude. The DNA-DNA hybridization experiments of Roy Britten and D. Kohne indicate that, on the average, the DNA sequences of man and monkey are highly homologous. Comparative measurements of the thermal stability of human DNA, chimpanzee DNA, and test-tube hybrids of these DNA's suggest that in the fifteen or so millions of years since these species have diverged there have developed about 1.6 nucleotide changes per 100 nucleotide pairs, or about 5 changes per 300 base pairs—which is equivalent to 100 amino acids of protein sequence. Since, because of redundancy, about 20 percent of random nucleotide changes will not result in an amino acid change, we might expect a mean evolutionary distinction between these two species of about 4 amino acids per sequence of 100 in the absence of selective bias.

However, the interpretation of such homologies has since been complicated by the recognition that these experiments as they have been performed to date can only concern or involve certain fractions of the DNA, specifically those fractions that are made up of large families of molecules, or closely related sequences represented literally tens or hundreds of thousands of times in the genome. These represent about 40 percent of primate DNA.

Under the conditions of these experiments, sequences represented less often

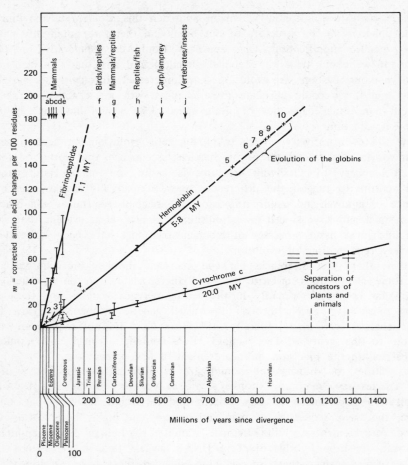

Fig. 1. The number of changes in amino acids between the same protein from two different species is plotted here against the time in the past at which the two species' ancestors diverged. The unit evolutionary period is the average time required for one difference to show up per 100 residues. Molecules such as cytochrome c, which interact closely with other macromolecules, have longer unit evolutionary periods than such non-specific proteins as the fibrinopeptides. (Adapted from R. E. Dickerson and I. Geis. *The Structure and Action of Proteins.* New York: Harper & Row, 1969).

simply never find a partner with which to hybridize in any reasonable time. The existence of these large families of closely related sequences, which may in total comprise some half of the genome, is both a surprise and a conundrum in itself, but in addition it does at present clearly limit the quantitative significance of statements about DNA homology between species, for we can say as yet very little about the possible homology of the less frequent DNA species.

Studies of the available rates of genetic mutation, as evidenced by changes in the amino acid sequences of particular proteins, suggest that the time of divergence of man and present-day monkeys from a presumed common ancestor *has been sufficient* to allow significant changes. The observable changes in amino acid sequence in any special protein are of course strongly biased by possible, and generally unknown, selective pressures that limit permissible change. Thus the alpha hemoglobin in the gorilla differs in only one amino acid from that of man. And that of the chimpanzee is identical to that of man. In an over-all sense, the rate of acceptable mutation in the globins is only about 1 amino acid per 100 residues per 6,000,000 years. For other proteins, such as cytochrome c, the allowable rate proves to be even less: 1 in 20,000,000 years. But a more accurate measure of the possible rate of amino acid replacement may be obtained from the fibrinopeptides which appear to serve no other function than to be excised from fibrinogen, when it is converted to fibrin in the formation of a blood clot, and then to be degraded. In these the apparent rate is 1 amino acid change per 100 residues per 1 million years. These numbers are in reasonable agreement with the averaged estimate from nucleotide change—approximately 4 replacements per 100 amino acids per 15,000,000 years.

It is thus possible to suggest that in the last several million years a considerable number of the proteins of man could have undergone mutational changes in one or two amino acids. But a *major* change in a particular protein would be highly unlikely—at least by the mutational processes leading to the changes so far studied.

Now of course the body undoubtedly has mechanisms whereby the consequences of even a single amino acid change in a strategic protein can be greatly amplified. But the conclusion I tend to draw from this admittedly loose argument is that the genetic distinction between a man and a monkey is, in a quantitative sense, not a great one. Hence, there is a greater chance that in time we will be able to define and understand this change and conceivably recapitulate it in the laboratory. In this connection it would certainly be of great value to have phylogenetic comparisons of specific brain proteins as well as of hemoglobin.

In addition to his enhanced capacity for conceptual thought, man exceeds other primates in his enlarged consciousness, his power of speech, and undoubtedly in such underlying functions as memory and capacity for numeration. What changes provide the bases for these qualities?

If we compare the brain of a man and, let us say, that of a rat, we find the rat brain weighs a little over 1/1,000th that of the man: 1.6 grams vs. 1,450 grams. Yet the rat is a rather complex organism. It can learn intricate mazes; it can fight or defend itself; it reproduces; it has, particularly in the wild, quite intricate behavior. After observing the rat for a while, one begins to wonder what the other 99.9 percent of the brain is doing in man. If one compares the volume of the cerebral cortex, the ratio becomes even greater: 5,000 to 1 (500 cubic centimeters to 0.1 cubic centimeter).

Of course size of brain is a rather crude indicator. The brain of a chimpanzee weighs 450 grams, that of a man, 1,450 grams. A dog has 80 grams of brain, a rabbit 10. But the brain of man is not all that extraordinarily large. The brain of the dolphin weighs 1,700 grams. It rivals that of man in structural complexity and proportions. What is it doing? The brain of an elephant weighs 5 kilograms, a whale 6 to 7 kilograms.

If we examine animals at various levels of phylogenetic development, one trend is, clearly, that more and more information is brought into the central nervous system. Thus, in man somewhat over 2,000,000 sensory fibers bring information to the brain, about half through the cranial nerves (optic, auditory, etc.) and half through the spinal cord.

If we compare man with the rat, we find that 12 times as many sensory fibers enter the spinal cord and 10 to 12 times as many fibers carry auditory and visual information. But most of this increase in informational capacity has already developed by the evolution of the primate. The principal difference between the primate and man appears to be in the elaboration of structures for the analysis and integration of the sensory input. If we compare the number of cells in area 17 of the visual cortex in man and in the macaque, or of the areas 17, 18, and 19 of the visual cortex in man and the orangutan, or the number of cells in the auditory cortex in man and the chimpanzee, we find large increments in man over the primates. And, of course, even larger differences are found in the volumes of the frontal cortex, the functions of which are still disturbingly poorly understood. We are only now, in experiments such as those of David Hubel and T. Wiesel, beginning to learn some of the ways in which networks in these areas of the cortex analyze the sensory input in monkeys; we have no information yet as to how these means of handling sensory data may differ between the lower primates and man. [See Tables 1 and 2.]

Table 1. *A Better Indicator of Brain Capacity is the Amount of Information Brought into the Central Nervous System. A Comparison of Man and the Rat Shows that 12 Times as Many Sensory Fibers Enter the Spinal Cord and 10 to 12 Times as Many Fibers Carry Auditory and Visual Information.*

	Sensory Fibers into Spinal Cord	Auditory Fibers	Optic Nerve Fibers
Man	1,000,000	30,000	1,000,000
Macaque	550,000	30,000	1,000,000
Rat	80,000	3,000	80,000

Table 2. *The Principal Difference Between the Primate and Man Appears To Be in the Elaboration of Structures for the Analysis and Integration of the Sensory Input. Large Increments in Man over the Primates Are Shown in this Comparison of the Number of Cells in Area 17 of the Visual Cortex in Man and in the Macaque, or of the Areas 17, 18, and 19 of the Visual Cortex in Man and the Orangutan, or of the Number of Cells in the Auditory Cortex in Man and the Chimpanzee.*

	Man	Primate
Cells, area 17, visual cortex	540,000,000	150,000,000 (Macaque)
Cortical surface, area 17	26 cm^2	18.7 cm^2 (Orangutan)
Cortical surface, area 18	39 cm^2	14.5 cm^2 (Orangutan)
Cortical surface, area 19	39 cm^2	14.2 cm^2 (Orangutan)
Cells, area 41, auditory cortex	100,000,000	10,000,000 (Chimpanzee)

Consciousness

One of the most obvious distinctions between man and the lower animals is in the quality and quantity of his consciousness. Man can escape from the here and now; he can compare alternate responses and originate new actions by internal imagery. In the nature and origin of consciousness is one of the most profound of mysteries.

What determines the modality of consciousness? How do certain stimuli cause pain, others color, others tone or taste? What defines the spectrum of color sensation? Why are there no more colors or no other tones? Clearly there are structural and very likely chemical, and therefore, genetic, bases for these phenomena.

There are individuals, for instance, who are genetically insensitive to pain. In some instances this defect is peripheral. The nerve receptors in the skin which are usually considered to be the sensors of pain are lacking. In others the sensory cells and sensory fibers, at least as far as can be seen, appear to be intact, and the defect may be central—an indifference to pain. An interesting point is that these people, lacking a sensory modality, do not appear to know what pain is. It is absent from their consciousness, which thus seems, in part at least, discrete. It is of interest that such people often can distinguish temperature quite normally but there is no pain associated with hot or cold. This condition is most often disastrous to the individual. It is also of great interest that in two cases siblings from first-cousin marriages have shown this trait, suggesting that it may be a consequence of a fairly simple genetic alteration.

Now I personally rather doubt that we have the conceptual capacity to really comprehend the origin of consciousness; but I do expect that we will learn that consciousness of various modalities may be associated with circuits of the brain connected in diverse ways, possibly with diverse chemical transmitters and effecters, all programmed genetically, and that by modifying these programs we may indeed in a true sense expand consciousness into unknown sensations and into undreamt intensities. If this sounds absurd, consider that many vertebrates have no color vision at all. By changing their genetic program an entire new sense has been added. We might be able to build chemical switches into various sectors of consciousness so that pain specifically could be turned off for surgery or a widened sense of taste or color turned on for enjoyment. Conceivably new receptors—for electric fields or radio waves, for ionizing radiation or what have you—could be developed to go with new modalities of consciousness.

Whatever may be the basis of conscious thought, it is clear that much of the operation of the brain cannot be brought to consciousness; it is, somehow, inaccessible or screened. There is very likely much merit in the automation of many activities. Yet, as we know, conficts and distortions on the subconscious level can produce grave disturbances of the psyche and are most difficult to detect and analyze. If more of the unconscious could be made at least selectively accessible, it could be a very considerable boon.

Language

Of course one of the major distinctions of man is his ability to communicate, particularly through speech. Remarkable as this capacity is, it must be recognized that it is a limited device. There are very real limitations of language and communication. Can we truly express everything we experience or conceive in speech? There are problems of precision, of connotation, and of association. We frequently have to coin new words for new concepts, and still it is difficult to convey their meanings to others. The expression of feelings and emotions is particularly difficult and seems to interweave several dimensions of emotionality. One can sum up a whole complex of emotions by an analogy (such as an Oedipus complex or a messianic complex) which is extremely hard to decompose analytically in words.

The average person is said to know some 20,000 to 60,000 basic words (dependent somewhat upon the definition of "know") and perhaps 100,000 derivatives of these. In ordinary speech he uses 2 to 3,000 basic words; in ordinary writing, maybe 10,000. (This difference between stored information and effective information is curious. It is of interest that there is a similar difference between the overall sensory input—2 to 3,000,000 fibers—and the overall motor output—about 350,000 fibers in man).

The rate of direct communication is typically about 150 words per minute. These it may be estimated contain at most 2,000 bits of information. Of course that depends a little upon the speaker.

Speech is probably genetically one of the newest of nature's inventions and obviously one of major importance for the development of inter-individual communication, the consequent development both of group behavior and properties, and the transmission of knowledge and culture from one generation to the next. Yet there is no reason to believe this relatively recent innovation is perfected. Indeed, as we have indicated, there is good reason to believe speech is a very imperfect device for communication.

If we could manage a significant improvement in the potential precision and speed of our vocal communication, this could be of major consequence. We could, for instance, use many more of the potential phonemes and thereby markedly increase the potential information density.

I think it is interesting that our friend Pooh, although of little brain, used language with considerable precision and economy—as in the time he was hanging onto a balloon suspended in the air and, wanting down, he asked Christopher to shoot the balloon. So Christopher aimed very carefully and fired.

'*Ow!*' said Pooh.

'Did I miss?' Christopher asked.

"'You didn't exactly *miss,*' said Pooh, 'but you missed the *balloon.*'"

One well-known indicator of the limitations of our capacity for speech is our frequent inability to bring to mind the right word for an object or a person or a concept. Pooh also suffered from this all-too-human failing—as when Christopher says:

"'What do you like doing best in the world, Pooh?'

"'Well,' said Pooh, 'what I like best—' and then he had to stop and think. Because although Eating Honey *was* a very good thing to do, there was a moment just before you began to eat it which was better than when you were, but he didn't know what it was called."

I am of course assuming here that our command of language and indeed the structure of language, whatever language it is, are at least in large part a consequence of genetically determined neuronal structure. I think this is very reasonable. And along these lines I would like to return to the concept I developed earlier—that we can learn to do certain things rather easily because, in effect, approximate programs for these operations are built in.

Learning

Could we not extend this? Could we not build in through the proper circuitry certain packets of knowledge so that every generation need not learn these anew, such as a language, or the periodic table, the Krebs cycle, etc. Migratory birds evidently have genetic programs that enable them to recognize stellar constellations. Other birds innately recognize rather complex songs. It does not seem inconceivable.

This is only an extension, although certainly in another dimension, of the wise ideal so well expressed by Whitehead, who wrote: "It is a profoundly erroneous truism—that we should cultivate the habit of thinking what we are doing. The precise opposite is the case. Civilization advances by extending the number of important operations which we can perform without thinking about them."

Statistically at least it is clear that there are changes in the human brain with aging. There are times of optimum ease of learning such matters as language or mathematics. There are optimum periods for creative work, and these seem to differ in the different sciences. In early childhood there are critical periods for mastering certain skills, and if these are past, the effect may well be nearly irreversible. Also we know there are at various times in life irreversible hormonal influences on parts of the brain. We know the number of cells in the brain does not increase after six months or a year of age and, indeed, decreases after 30 to 40 years of age.

If we understood these matters, we could perhaps control these factors. We might keep open and extend critical periods of learning. We might learn to reverse untoward hormonal effects or even to increase the number of brain cells and thus permit continued increase of information and counteract senility.

Matters of learning clearly involve the intake of information and the storage of memory. We do not understand these matters well. Numerous studies of varied design indicate that the rate at which we can abstract information from our sensory presentation—visual, for example—is highly limited by a narrow channel capacity. Various studies have been made and, despite some variation of interpretation, there seems to be a general agreement that while some 40 to 50 bits of information may be taken in visually in a flash and held for somewhat less than a second, at the most 10 bits of information per second can be abstracted from a presentation and used to control an output or be relayed to a memory bank. This channel capacity is certainly a major parameter in the determination of the speed and quality of the working of the brain.

The limited capacity of the brain to abstract information from a visual display underlies the McLuhan fallacy and explains why people still read books. Could this rate of information - handling be markedly increased? If so, we could enter the McLuhan era.

Further, it seems likely that the limitation upon the bits of information we can process at one time is related to a deeper question. How many data can we hold in our mind at one time, how many can we bring to bear upon a particular conceptual problem? Surely this is limited, and this in turn restricts our ability to cope with problems of great complexity except by overabstraction and oversimplification. Conceivably we might be able to increase this quantity.

Memory

Now, of course, it is easy to list various qualities and suggest independently improving this or that one. But properly one needs to consider and needs to be able to consider the effect of changing any one facet of intellectual performance upon an individual's whole personality. Personality is like a network with more-or-less-balanced tensions and strains; modification anywhere can affect the whole. Consider what one might first think to be a purely mechanical element such as memory. Upon reflection, memory is easily seen to be a central element in the whole cerebral process. With a little reflection I think it becomes obvious that the quality of memory, its extent, its rapidity, its precision and acuity must influence the whole life pattern through our perception of and response to any situation.

We know all too little about memory. It has become known that there is a short-term memory for the relatively brief storage (on the order of seconds) of information, and a longer term, more enduring memory, of a qualitatively different nature. I suspect that we do not yet begin fully to grasp the significance and function of these distinct memories. As we learn more about the roles of these separate memory systems, we may find that the existence of erasable short-term memory provides an essential gap that permits a distinction between our internal and external worlds. It provides a transient recording that permits us to respond to the immediate yet not to be constantly overwhelmed by the immediate, so we may select from it the important and the general. Without such a buffer we could not plan, we could not withdraw sufficiently from immediate reality.

It is even possible that our sense of time and of time passing is related to the rate of decay of our short-term memory. In our subconscious and in our internal world there is little sense of time. A past event can seem as real as the present. Drugs which affect our sense of time may do so through their effects upon these processes.

If these speculations have any validity, then the ability to alter physiologically or genetically the rates and extent of these processes of memory could have profound effects upon our perception of the world.

I might insert at this point that to a biochemist one of the major impediments to research upon many of these questions is the existence of the so-called blood-brain barrier. This is a poorly understood physiological mechanism that stringently restricts the transport of foreign substances into the central nervous system. Presumably this was designed to provide a specific neuronal environment and to protect the brain against physiological vicissitudes and not just to frustrate biochemists. But certainly one major contribution that genetics could make would be to alter this barrier—optimally, perhaps, by incorporation of some biochemical switch whereby it could be opened or closed so as to permit biochemical investigation.

Human Genetic Variants

Another and different approach to the potentials inherent in further development of the brain is by a consideration of the attributes of individuals with special gifts of one character of another. It is clear that, presumably by genetic circumstance, individuals arise with marked asymmetries of talents. It is also clear that in accord with the concept of interdependence of various cerebral functions the hypertrophy of one talent is often accompanied by major, even disastrous, consequences to others, although we are at present unable to trace the causal connections.

The so-called idiot savants who have a general mental age of two or three years but can, with great rapidity, perform extraordinary numerical feats are an extreme example. One of these, given the series 2, 4, 16, immediately continued to square each successive number into the billions. Similarly, given the numbers 9-3, 16-4, he proceeded to do square roots of numbers into 3 and 4 digits. Another class of feeble-minded individuals is known to have extraordinary talents of mimicry.

Of a less drastic and more desirable nature are the special talents we associate with musical genius, such as a Mozart who composed significant works at the age of four, or artistic genius, or literary genius, or extraordinary skill at chess. There are individuals who are extraordinarily articulate; there are others with extraordinary ability in three-dimensional visualization and spatial orientation far beyond the corresponding talents of normal people.

The capacities of these individuals indicate levels of achievement that could become commonplace, beside which we may feel like Pooh who was somewhat weak in this matter of spatial orientation and symmetry.

"'I *think* it's more to the right,' said Piglet nervously. 'What do *you* think, Pooh?'

"Pooh looked at his two paws. He knew that one of them was the right, and knew that when you had decided which one of them was the right, then the other one was the left, but he could never remember how to begin.

"'Well,' he said slowly—'"

A particular case of an extraordinary development of the faculty of memory has recently been described in considerable detail by A.R. Luria in the book *The Mind of a Mnemonist*. This analysis is of particular interest because Luria is especially concerned not only with this unusual mnemonic talent but with its consequence for the whole personality of the man who had it. This man's memory in truth could not be saturated and was apparently imperishable. He could quickly, in two or three minutes, learn a table of 50 numbers or a list of 70 words which he could then repeat or just as easily present in reverse order, or, if given an intermediate word, go forward or back from this. He memorized a nonsensical formula

$$N \cdot \sqrt{d^2 \times \frac{85}{vx}} \cdot \sqrt[3]{\frac{276^2 \cdot 86x}{n^2 v \cdot \pi 264}} \cdot$$

$$n^2 b = sv \; \frac{1624}{32^2} \cdot r^2 s$$

in a few minutes, and when asked 15 years later, without warning and with no intervening exposure, he was able to reproduce the earlier test situation and the formula without error. He literally never forgot or lost anything once committed to memory.

As Jerome Bruner suggests in his foreword to this book, it is as though the metabolism responsible for short-term memory was defective in this man and everything experienced was transferred into the long-term memory. His world was one of intense visual imagery. He was never able to develop and grasp or project ideas and generalities. He was, in effect, overwhelmed by an endlessly increasing store of perceptions.

As another corollary, the man had significant difficulty in distinguishing between the internal and the external world. He had great difficulty in planning. He could not withdraw enough from the immediate reality. Furthermore, his sense of time was often faulty. For this man the past was as real as the present. He had no childhood amnesia and seemingly could remember impressions to very early childhood.

This man was also remarkable in another way. He had a strong synesthesia. As I have pointed out, to most of us our senses are quite distinct. Sight, sound, taste, smell, touch, pain are all uniquely stimulated, except when under the influence of certain drugs which appear to facilitate sensory interaction. In this man almost all the senses seemed fused. Every sound also had an image in color, often a taste and a touch and a smell as well. (It is conceivable that this effect is also related to a short-term memory defect. The persistence of a sensory input may permit it to spread and involve other perceptual centers.) He said:

"I recognize a word not only by the images it evokes but by a complex of feelings the image arouses. It is not a matter of vision or hearing but some overall sense I get. Usually I experience a word's taste and weight, and I don't have to make an effort to remember it. But it is difficult to describe. What I sense is something oily slipping through my hand. Or I am aware of a slight tickling in my left hand caused by a mass of tiny lightweight points. When that happens I simply remember without having to make the attempt. . . . Even when I listen to works of music I feel the taste of them on my tongue. If I can't, I don't understand the music. This means I have to experience not only abstract ideas but even music through a physical sense of taste."

I think it is obvious that for such a person the world would be a very different place than it is for us.

His strongest reaction was imagery. He lived very much in a world of images. Obviously this could create very serious problems. For some words, for example, the images the sound of a word created would fit its meaning, but for others there would be a conflict and confusion. Many words we know have multiple meanings (fast, for example). This created great difficulty for him. He could not comprehend metaphors at all.

"Take the word nothing. I read it and thought it must be very profound. I thought it would be best to call nothing something. I *see* this nothing and it is something. If I am to understand any meaning that is fairly deep I have to get an image of it right away. So I turned to my wife and asked her what nothing meant. But it was so clear to her that she simply said nothing means there is nothing. I understand it differently. I saw this nothing and thought she must be wrong. If nothing can appear to a person then it means it is something. That's where the touble comes in."

It's interesting that Pooh had the same difficulty with abstractions—as when Christopher says:

"'. . . what I like *doing* best is Nothing.'

"'How do you do Nothing?' asked Pooh, after he had wondered for a long time.

"'Well, it's when people call out at you just as you're going off to do it, What are you going to do, Christopher Robin, and you say, Oh, nothing, and then you go and do it.'

"'Oh, I see,' said Pooh.

"'This is a nothing sort of thing that we're doing now.'

"'Oh, I see,' said Pooh again."

The Two Realities

There is one other aspect of this man's unusual mental and psychical structure that should be mentioned. His poor distinction between external and internal reality was perhaps reinforced by an extraordinary control over his automatic functions. He could increase his pulse rate from 70 to 100 by imagining he was running and then reduce it to 64 by imagining he was lying quietly in bed. He could raise the temperature of his right hand by two degrees and then later lower that of his left hand by one degree. How did he do this? He said,

"There is nothing to be amazed at. I saw myself put my right hand on a stove. Oh, it was so hot. So naturally the temperature of my hand increased. But I was holding a piece of ice in my left hand. I could see it there and began to squeeze it and of course my hand got colder."

He claimed also to be able to alter his sensitivity to pain at will.

"Let's say I'm going to the dentist. You know how pleasant it is to sit there and let him drill your teeth. I used to be afraid to go but now it's all so simple. I sit there and when the pain starts I feel it. It's a tiny orange-red thread. I'm upset because I know that if this keeps up the thread will widen until it turns into a dense mass. So I cut the thread, make it smaller and smaller until it's just a tiny point and the pain disappears."

It was demonstrated that he could vary his eye adaptation by imagining himself to be in rooms of varying levels of illumination.

His strange memory and his synesthetic experience created in this man a critical difficulty in distinguishing between the world of his imagination and the external world. Lacking a clear distinction, such as we know is observed in certain drug states, his fantasies could be as real or more real to him than the external world.

"This was a habit I had for quite some time. Perhaps even now I still do it. I look at a clock and for a long while continue to see the hands fixed just as they were and not realize time had passed. That's why I'm often late."

All of which may bear importantly on the major question of how we make this critical distinction between internal and external.

I have gone into detail because this individual provides such a powerful illustration of the interlocking and interdependent character of our various mental and psychological attributes, and thus of the extensive consequences of what are undoubtedly a few strategically placed genetic alterations. Conceivably, they might amount to little more than an altered metabolism leading to the localized endogenous synthesis of an unusual substance with certain LSD-like properties.

I have thus far been principally concerned with the more cerebral and operational aspects of central nervous system function. Another most important field for genetic intervention is our motivational and emotional states. It seems all too clear to me that we are the victims of a variety of emotional anachronisms, of internal drives no doubt essential to our survival in a primitive past, but quite unnecessary and undesirable in a civilized state. We have surely more than we need in aggression. Could we not lower aggressiveness, bearing in mind that we must be on guard for possible corollary consequences?

Pessimism and depression are perhaps necessary in a world that merits suspicion, but their exaggeration has little merit. This is illustrated splendidly in the Pooh stories, where Eeyore, the donkey, one of Pooh's friends, is the embodiment of depression. One day Eeyore finds his tail is missing.

"'You must have left it somewhere,' said Winnie-the-Pooh.

"'Somebody must have taken it,' said Eeyore, 'How like Them,' he added."

But in a more humane world such qualities might be of little use. We will undoubtedly continue to have need of compassion. There is in the Pooh stories another episode in which after a period of intense rain and general flooding of the premises Edward Bear and Christopher Robin are impelled to rescue their close friend Piglet, who is stranded on a tree branch not much above the rising water. But how to accomplish this? After both are stumped for some time, Pooh has an idea which certainly far exceeds his normal cortical limitations. He suggests that they invert Christopher Robin's umbrella and use it as a boat. Christopher is so awed by this unexpectedly brilliant and, I might add, successful invention that he later names this worthy craft *The Brain of Pooh*.

I like to think that driven by necessity or even better by compassion we too will learn to exceed our normal cortical limitations and we too may tap talents yet unseen.

So much of what we see, so much of what we perceive, so much of what we experience is in truth what we conceive. It is contributed by the mind of the beholder and thus must depend in detail upon the innate structures and functions of the mind, upon its accumulated experiences, upon its physiological state, and even in a regenerative manner upon how the mind conceives of itself. And our view of the mind, even the very concept that we may at some future time be able to augment and improve our capacities, may react upon our behavior long before we achieve these visions.

For a number of the most strategic and salient structural elements of the mind there is already evidence of significant genetic determination. These genetic factors, and they may not be so many in number, define our intellectual and conceptual limits. I propose that through phylogenetic studies and through studies of the rare human genetic variants we can learn much concerning their basic cerebral components, in preparation for the day when we wish to begin to move back their limits.

And so perhaps, when we've mutated the genes and integrated the neurons and refined the biochemistry, our descendants will come to see us rather as we see Pooh: frail and slow in logic, weak in memory and pale in abstraction, but usually warm-hearted, generally compassionate, and on occasion possessed of innate common sense and uncommon perception—as when Pooh and Piglet walked home thoughtfully together in the golden evening, and for a long time they were silent.

"'When you wake in the morning, Pooh,' said Piglet at last, 'what's the first thing you say to yourself.'

"'What's for breakfast?' said Pooh. 'What do *you* say, Piglet? '

"'I say, I wonder what's going to happen exciting *today*?' said Piglet.

"'Pooh nodded thoughtfully. 'It's the same thing,' he said.''

Interaction between two adult baboons showing grooming behavior while an offspring nurses. In primates the tactile stimulation of grooming appears to lessen tensions among group members.

12. *Gypsy Moth Control with the Sex Attractant Pheromone*

MORTON BEROZA AND E. F. KNIPLING

The gypsy moth, an alien to this country since 1869, has been a serious pest of the northeastern forest for many decades. If not stopped, it could very well extend its range southward and westward and infest millions of acres of forest and shade trees. Containment of the moth to its present range may be possible by use of the synthetic and highly specific sex attractant, disparlure, and by use of sterile males. These measures are very expensive but would more than balance out the losses of timber resources and shade plantings. The authors are both with the Agricultural Research Service (ARS) of the United States Department of Agriculture. Dr. Morton Beroza, a physiological and analytical chemist, is an investigations leader of the Pesticide Chemicals Research Branch of ARS. Dr. Edward Knipling is an entomologist and a science advisor with ARS. He pioneered much of the work on gypsy moths and sex attractants.

The gypsy moth [*Porthetria dispar (L.)*], a serious defoliator of forest, shade, and orchard trees in northeastern United States, is spreading rapidly to the South and gradually to the West and threatens to become a national problem.

There is deep division among scientists, administrators, environmentalists, and public officials about whether its spread can be stopped or should be stopped. On some occasions we read of citizens and township officials begging for relief from the moth's depredations; at other times it is claimed that after the initial flareup damage can be small, and that we should learn to live with the insect and find ways to minimize the damage rather than attempt to halt expansion of the present infestation. The way in which the gypsy moth problem has been handled has been criticized *(1)*, but the critics have not come up with practical and ecologically acceptable solutions.

In this article we describe the problem and discuss the possibilities of using the recently identified sex pheromone of the gypsy moth *(2)* to combat this insect.

History

The gypsy moth, a native of Europe, Asia, and North Africa, was brought to Medford, Massachusetts, in 1869 for the purpose of producing silk for local industry; unfortunately, some insects accidentally escaped. The moth became established, but was largely unnoticed until 20 years later when there was a devastating population explosion. The following comment of a local resident is typical *(3)*: "In 1889 the walks, trees and fences in my yard and the sides of the house were covered with caterpillars. I used to sweep them off with a broom and burn them with kerosene, and in half an hour they were just as bad as ever. There were literally pecks of them. There was not a leaf on my trees. . . . The stench in this place was very bad."

Though local infestations were gradually brought under control by natural forces and such efforts as burning egg masses, treating them with creosote, banding trees to catch the caterpillars, and spraying or dusting with chemicals such as lead arsenate, spread of the moth into new areas continued. For a long time the moth was confined to New England, but the cost was high— averaging $1.7 million a year for the 33 years preceding 1940 *(4)*. When DDT (dichlorodiphenyltrichloroethane) became available in the late 1940's, it served as a powerful weapon against the insect *(5)*; but its use was phased out after 1958, and the pest has spread rapidly as far south as Virginia. Widely scattered finds of the insect have been made recently in North and South Carolina, Alabama, Florida, Ohio, and Wisconsin.

The Moth and Its Threat to the Environment

The gypsy moth has one generation a year. In the Northeast the larvae or caterpillars emerge from overwintering eggs in late April or early May (usually over a period of 2 to 4 weeks) and begin to feed on suitable hosts. Although oak, willow, poplar, speckled alder, basswood, gray birch, river birch, and apple are favored, many other trees and ornamentals are attacked, including evergreens. Many young caterpillars spin down on silken threads that break off and act as sails *(6)*, allowing the wind to carry the tiny insects off, usually for a few hundred meters but sometimes for more than 40 kilometers away *(7)*, where they may start new infestations. The caterpillars feed voraciously on leaves, normally passing through five and six instars (molts) for the males and females, respectively, and attaining a length of 4 to 7 centimeters. In late June or early July they change to the pupal stage, usually for 10 to 14 days. The brown adult male moths, which start emerging from pupation before the females, are strong fliers and are capable of mating several times. The off-white female, with its abdomen filled with eggs, does not fly. To lure the male for mating, she releases a sex attractant, which the male detects with great sensitivity; he then flies upwind to find the female *(8)*. The female normally mates once, and then lays from 100 to 800 eggs in a buff-colored, hair-covered mass, from which the larvae merge the following spring. Adult moths do not feed and they live only a short time after mating. Damage by the insect is thus limited to the larval stage.

The efficient means by which the sexes find one another for mating and propagation, the large number of eggs laid, and the voracious appetite of the larvae account for the explosive population buildups of this insect and the great harm it does. A single 5-cm caterpillar eats about 0.1 square meter (1 square foot) of leaf surface a day *(9)*. A single defoliation has been known to kill white pine, spruce, and hemlock *(10)*. Two successive defoliations can kill most hardwoods *(10)*. As an example, 3 years after the use of sprays was banned in the Morristown (New Jersey) National Historical Park in 1967, a survey showed that one-third of the park's oak trees, average age 100 years, had been killed by gypsy moth defoliation *(9, 11)*. With oak forming the natural ground cover of much of the Northeast, the ecological implications are inescapable.

In 1970, this pest defoliated 800,000 acres of forest. In 1971 this figure rose to 1.9 million acres (1 acre is equivalent to 0.4 hectare). Although non-commercial forests and parks have suffered most, commercial forests of the Appalachian and Ozark mountain ranges and of the South are now threatened.

Cities and towns are not spared. The complaint of one resident of Shirley, New York, was quoted in the *New York Times (12)*: "Our children cannot go out. Our pools are finished for the summer. It's a question of survival—the caterpillars or us."

Current Control Measures

Intensive efforts have been made in the past to solve the gypsy moth problem by the use of natural biological agents, and many species of parasites and predators have been imported and released for this purpose *(13)*. Native parasites and predators also attack the pest at various stages. Vertebrates feeding on it include white-footed mice, cuckoos, blackbirds, and grackles. Among the invertebrates *Calosoma* beetles kill the caterpillars and a chalcid wasp, *Ooencyrtus kuwanai,* parasitizes the eggs.

Attempts are being made to develop means of using a virus causing "wilt disease" in the larvae to suppress populations *(14)*. Also, a strain of *Streptococcus faecalis* and a commercially available bacterial insecticide, *Bacillus thuringiensis,* are being investigated for control *(15)*.

Carbaryl (Sevin), the insecticide now most frequently used against the gypsy moth, is applied at the rate of 1 pound (0.45 kilogram) of active material per acre to prevent large infestations from killing trees. Egg counts are made before the insecticide is applied to determine if the population density is sufficient to cause defoliation *(16)*. The cost of effective treatment with insecticide averages $3 to $5 per acre for a season, and the time at which the insecticide is applied is critical.

The foregoing measures are meant to minimize the damage caused by the moth rather than to eliminate it, and none is considered capable of doing more than slowing the moths' rate of spread. With the insecticides now available there is little hope of our preventing the spread of the pest throughout its potential range, which may include western forests.

Left to its own devices, the gypsy moth would probably continue to spread at its rather slow natural rate. In recent years vastly increased trade and traffic have enhanced greatly the chances of artificial spread of the moth, particularly because egg masses or pupae can pass undetected on mobile homes and camping trailers. Federal-state quarantine measures have been able to minimize the artificial spread, but they have been ineffective in preventing local natural movement.

Authorities face two immediate problems in combating the moth. One is to prevent excessive damage in areas of severe infestation (defoliation of commercial forests and highly valued shade trees and harassment of residents by the caterpillars). Present plans call for an integrated approach involving various combinations of the control methods cited. These plans include more extensive use of traps baited with the sex attractant to detect new infestations and to assess moth abundance in known infestations. The number of moths trapped can be used to signal the need for application of pesticides and other control measures; the use of pesticides can then be made more efficient and restricted to those areas where treatment appears necessary, thus minimizing pollution from this source.

The other problem, more pressing at the moment, is to prevent further spread of the pest by finding and eliminating light or incipient populations beyond the areas now generally infested.

Now that the highly potent sex attractant pheromone of the gypsy moth, cis-7, 8-epoxy-2-methyloctadecane or disparlure (2), is available in ample quantities, means of using it to prevent spread of the pest are being intensively investigated. Because action of disparlure is very highly specific to the gypsy moth and it is effective in very low concentrations, its use is expected to pose no hazards to people and to be ecologically acceptable from all standpoints.

Disparlure in Detection and Survey

The isolation, identification, and synthesis of disparlure culminated a search for this attractant pheromone that started at Harvard University in the 1920's (17) and was taken up by the U.S. Department of Agriculture (USDA) in 1940 (18). As a crude extract of the last two abdominal segments of the female (called a tip), the sex lure was used in traps to detect moth infestations since the 1940's (19, 20). Female pupae were laboriously collected in the field and the tips clipped into benzene 24 hours after the females emerged. The tips were extracted with benzene, the solution concentrated, and the extract hydrogenated to stabilize the lure (18, 21). Survey traps were baited with extract equivalent to ten tips.

In 1960, the sex attractant of the gypsy moth was reported to be 10-acetoxy-7-hexadecen-1-ol (called gyptol) (22); a homolog called gyplure (12-acetoxy-9-octadecen-1-ol) was also reported to be active (23). Both compounds were later found to be inactive (24), and use of the natural extract of the moth was resumed.

Collections of tips frequently had to be made in distant countries (Spain, Yugoslavia, French Morocco) because moth populations available in the United States, especially when DDT was used extensively, were often low or uncertain. Costs of collections to bait the 60,000 survey traps used by USDA and the states were about $25,000 a year. Bioassay and chemical analysis indicated that the last extract of tips collected for the USDA in Spain in 1969 contained the equivalent of about 0.2 nanogram of sex pheromone per tip.

In 1969, the sex pheromone of the gypsy moth was identified and synthesized (2). One of our first preparations of disparlure was a 30-gram lot. At the rate the USDA had been using attractant to bait its traps, this amount was enough for the next 50,000 years. The eventual cost of disparlure is estimated at 30¢ per gram.

With a synthetic lure readily available, survey traps are now being baited with 100 micrograms of disparlure to increase detection efficiency. This tiny quantity is about 50,000 times the amount of lure present in traps baited with the extract of insects collected in Spain in 1969. But the greater amount of lure used is only part of the story. Disparlure is formulated with "keepers," which are volatile or nonvolatile diluents, to regulate its volatilization (25) and thereby prolong the action of the lure. A variety of these compounds were tested, and the best (trioctanoin) was used in the traps. Typical data, given in Table 1, show the great superiority of the disparlure-trioctanoin combination over the natural moth extract in both intensity of attraction and persistence (25). In 1970, baited traps were set out in mid-April, early June, and mid-July, that is, 12, 6, and approximately 1 week, respectively, before the flight

Table 1. *Numbers of insects captured with attractant aged 1, 6, and 12 weeks (25).*

Amount of Attractant per Trap (μg)	Approximate Age of Attractant (weeks)		
	1	6	12
Natural extract[a]			
	11	6	0
Disparlure[b]			
1.0	127	146	138
0.1	155	109	126
0.01	88	90	40
0.001	69	20	7

[a]Equivalent to ten tips per trap, the amount used in standard survey traps.
[b]Trioctanoin, 5 mg, added as keeper in each trap.

of the moth. In this way aged and fresh materials were compared simultaneously under identical conditions. Traps containing 1.0 and 0.1 μg of disparlure at the three different ages were so effective they were actually saturated with moths in the moderately infested test area because of the low trap capacity (about 20 moths maximum). Traps with as little as 1 ng of disparlure plus 5 milligrams of keeper still caught moths after being exposed for 12 weeks in the field.

The Animal and Plant Health Service of the USDA and the cooperating state agencies currently use a weather-resistant cylindrical cardboard trap 5 cm in diameter and 10 cm long with clear plastic ends having 2.5-cm openings. Males responding to the disparlure (on a cotton wick) enter the traps and get stuck on a gummy material within. In intensive surveys the traps are placed at 7/8-mile (1.4-kilometer) intervals in lines about 1 mile apart (1.6 kilometers). The spacing of survey traps that will provide assurance of detecting new infestations at the lowest practicable level has not yet been determined with the new lure.

In the 1970 survey, hundreds of traps baited with 1 to 10 μg of disparlure were interspersed among thousands of traps baited with the natural extract; reports from the states in which the traps were used showed that captures per trap with disparlure were 9- to 37-fold greater than those with the natural lure.

In 1971, survey traps baited with 20 μg of disparlure plus keeper were used for an entire season for the first time. The fact that gypsy moths were found in many outlying areas where they had never been found before is no doubt due in part to the high degree of efficiency of the new lure. Furthermore, about $50,000 a year was saved because the new traps, unlike the old ones, did not require rebaiting in mid-season.

The high potency of the lure and its exceptional persistence greatly exceeded our expectations, and the use of disparlure as a control measure began to appear feasible.

Disparlure for Control

The major objective in the proposed use of the sex pheromone is to prevent male moths from finding females and mating, thereby preventing their propagation. The proposed techniques are regarded as suitable only for light incipient populations, such as those now being found in areas where the moth has recently spread in the mid-Atlantic, southern, and midwestern areas of the United States. If high populations can be suppressed by other means, the attractant can then be employed to maintain suppression or to eliminate infestations.

To exercise control, the pheromone must disrupt communication between the sexes. Toward this end two methods are under investigation. One method consists of luring males to their destruction in some simple but suitable trapping device, and the other consists of "confusing" males with pheromone vapors dispensed into the atmosphere.

The information needed for the traps to be used effectively includes (i) an assessment of the attractant power of the pheromone in traps as opposed to the attractant power of the competing virgin females in the natural populations; (ii) the number and distribution of traps required relative to the number and distribution of gypsy moths in the population to be controlled; and (iii) the growth rate of gypsy moth populations, which determines the degree of control required to reduce their numbers.

To apply the confusion method, it will be necessary to provide a sufficiently high vapor concentration to prevent normal responses or proper orientation of males to the pheromone produced by the females. As with traps, the degree of mating inhibition required to achieve suppression of populations will depend on the insects' normal potential for increase in the absence of control.

An understanding of the dynamics of an insect population is basic to the development of suitable strategies for its control, regardless of the methods employed. Such understanding is particularly important in the use of the sex pheromone as a means of control.

The gypsy moth may spread by two means. Young larvae can become airborne and drift for some miles before settling on host plants in a new site (7), and the insect can be carried to new sites by the movement of egg masses or other immature stages on vehicles or timber, or by other means of transport (9). Adults of the larvae spread by air can be expected to be scattered downwind in a somewhat random fashion. The males must then locate the scattered individual females, and the nonflying mated females will each deposit a cluster of eggs. Therefore, beginning in year 2, the larvae and subsequent adults will tend to exist in colonies. Insects emerging from individual egg masses transported to uninfested areas will similarly tend to exist in such colonies.

The rate of growth of gypsy moth populations has been a subject of study by entomologists for some years. While each female is capable of depositing several hundred eggs, insect mortality is naturally high, and few of the potential progeny survive to mate. In one study, there was a 6.4-fold average increase in the population per generation (26); in another study there was a 7.5-fold increase (27). Although the rates of population increase can be expected to vary from place to place and from year to year, they are likely to be higher in low-density populations than in moderate- to high-density populations because of the suppression forces that are dependent on population density. Some segments of a population may increase at a higher-than-average rate, while the rate of increase of other segments may be less than average. From the standpoint of developing effective control methods, it seems prudent to assume rates somewhat higher than average to ensure that the suppressive measures used will be sufficiently effective to achieve near-maximum results. Therefore, we propose a tenfold rate of increase as being a reasonable growth rate for incipient gypsy moth populations in favorable environments.

With this rate of increase per generation (or year), we would expect a population to grow as shown in Table 2, until it reaches a density that will be adversely affected by normal density-dependent suppression forces. The projected rate of increase would lead to substantial defoliation in localized areas by year 6 or year 7. Thus, to avoid damage, a population has to be suppressed at an earlier period in its growth.

Table 2. *Rate of Increase per Generation (or per Year) in a Population of Gypsy Moths in a Favorable Environment. A Tenfold Rate of Increase per Year Is Assumed, as Is the Presence of a Single Mated Pair in Year 1.*

Successive Years	Number of Adults in Population
1	2
2	20
3	200
4	2,000
5	20,000
6	200,000

If the tenfold growth rate were realistic, the reproductive capability of 90 percent of an incipient population would have to be nullified each year merely to keep the population from increasing. The reproductive potential of each generation would have to be reduced more than 90 percent if elimination were to be achieved.

Early detection of new gypsy moth infestations is vital to the successful use of the attractant to suppress or eradicate new infestations. Studies are required to correlate the data obtained from trapping with the size of incipient populations. The assessment of population sizes in absolute numbers will be vitally important in estimating the number of pheromone traps required to achieve suppression or elimination.

Mass Trapping to Eliminate Gypsy Moth Populations

The principles of insect suppression through the use of sex pheromone traps were developed by Knipling and McGuire *(28)* who used models of postulated insect populations. The validity of these principles was confirmed by recent field-trapping studies on the boll weevil *(29)* and the red-banded leaf roller *(30)*.

The parameters that influence the absolute efficiency of such traps in practical control are many. Because of the continuous release of the synthetic pheromone from traps as opposed to the intermittent release of the pheromone by the females, the traps should be considerably more efficient than the females in attracting males. The efficiency of trapping could, however, be limited in large-scale trapping programs where the traps are distributed at random over large areas while concentrations of males and females are emerging from colonies. The dispersal of adult males, following emergence and before they seek females, could minimize this effect. Although it is not possible to give accurate estimates of the effects of all the parameters on the efficiency of a trapping system, it is possible to estimate with reasonable confidence the general magnitude of the results that would be expected from the use of various ratios of traps to competing females.

In the calculations that follow we will assume that each trap is as attractive as a single unmated female. This assumption is probably conservative for the gypsy moth. Thus, Fig. 1 shows that disparlure-baited traps containing from 1 to 6 μg of the pheromone in 5 mg of trioctanoin caught approximately the same numbers of males as did traps baited with single virgin females (31). Because of the use of 100 to 500 μg of disparlure in each trap is contemplated, the attractant power of a trap should be substantially greater than that of an unmated female.

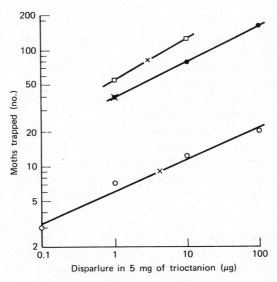

Fig. 1. Numbers of released moths captured in traps baited with live virgin females (crosses) compared with the numbers captured in traps baited with several concentrations of disparlure in three simulated field tests (data of different tests designated by squares and by closed and open circles). One female moth is equivalent to 1 to 6 μg of disparlure in 5 mg of trioctanoin (31).

The height of the traps relative to the height at which virgin females emerge could influence the competitiveness of the traps. Traps dispersed by air are apt to drop to ground level. Since the numbers of males captured by traps placed at 0, 3, 6, and 12 feet (1 foot is equivalent to 0.3 meter) above ground were 27, 22, 34, and 16, respectively [20], traps on the ground can be expected to be effective in capturing male moths. However, some means of suspending traps at different heights above ground could assure greater competition with females in the environment and a greater reduction of the moth population.

The results of the studies in Table 1 indicate that the traps will probably retain their attractant power throughout the emergence and mating period of the moths.

The longevity of females is an important consideration in male annihilation techniques. If females are unable to mate on a given day because males are absent, they may survive to the next day and be successful in luring a male. The longer the time that mating can be prevented, the more likely it becomes that the females will die before they can mate and deposit fertile eggs. A natural mortality rate of 25 percent of the unmated female population each day has been assumed in population suppression models for other insects [28]. This figure seems reasonable for the gypsy moth since the average life expectancy of unmated females is less than 1 week [3]. With a mortality rate of 25 percent per day, an uncontrolled insect population will stabilize at a level approximately four times the daily emergence rate [28]. If a high percentage of females cannot mate each day because the males have been captured in traps, there will be a gradual accumulation of unmated females, and the ratio of traps to unmated females will decline steadily until the unmated female population stabilizes. The shifting ratio of traps to competing females must be taken into account and allowances made in estimating the number of traps needed for various natural population densities.

Traps suitable for a mass-trapping program of the gypsy moth can be of simple design and can be produced at low cost. Because their usefulness will probably be limited to low populations, trapping devices large enough to hold two or three males should be adequate. The cost of such traps, including pheromone and adhesive to snare the moths, is estimated at 2¢ each. The cost of distributing traps in wooded areas will probably be considerable, but should be no higher than that of applying pesticide sprays. (Studies to improve the efficiency of available trap designs and distribution systems are under way.)

Utilizing the various elements discussed, we will develop population models to indicate the effects of mass trapping on incipient populations of the gypsy moth. Starting with a hypothetical population 3 years after its origin from ten single isolated egg masses distributed over a 10-square-mile area (1 square mile is equivalent to 259 hectares), we can expect 100 males and 100 females at each location. Although the total population of 1000 males and 1000 females would normally be expected to emerge over a period of several weeks, we will further simplify the model by assuming that emergence occurs during a 10-day season. This stipulation will make the expected results more conservative since a given number of traps competing with the same number of insects emerging over a period of 20 days would be essentially twice as effective in preventing mating because half as many insects per day would emerge. This model could be representative of small established colonies; for a population originating from larvae airborne from a heavily infested area, the model would be similar except that the insects would tend to be randomly distributed throughout the 10-square-mile area. Although the distribution of traps in relation to the distribution of unmated females can be expected to influence the efficiency of the traps, the theoretical effect of traps will be calculated without regard to this spatial factor.

A total of 5000 traps in the 10 square miles (500 traps per square mile) will be provided to compete with the 1000 unmated females in the area, and we will make the following assumptions: (i) The insects emerge at the rate of 100 males and 100 females per day. (ii) Each trap is equal to a virgin female in attractant power. (If the traps are more competitive, for example, three to five times as competitive as females, the number of traps required could, of course, be reduced.) (iii) The mortality factor is 25 percent per day for both the males and females. (iv) Females mate once and then are no longer attractive to the males. (v) The males can mate once per day on each day during their lifetime. (vi) Unmated females, as well as the traps, are attractive continuously during the mating period. (Since the "calling period" of the female is not continuous, this would be another conservative assumption that could compensate for unknown factors that might make expectations too liberal.)

The calculated effects of the hypothetical trapping program are shown in Table 3. The ratio of traps to unmated females starts at 50 : 1. Because the unmated females accumulate while trap density remains constant, the ratio of traps to unmated females drops to about 14 : 1 by day 10. With this ratio, the degree of mating inhibition is still theoretically 93 percent, which is adequate for the suppression of a population having a net potential increase rate of 10.

In an uncontrolled population the degree of mating success is assumed to be 100 percent, so that 1000 females mate. In the controlled population the total number of females expected to mate is 54. Therefore the overall suppression due to the capture of males is 94.6 percent, or approximately 95 percent.

If the level of suppression is 95 percent (the figure is rounded to simplify calculations) and there is a tenfold net increase in the population, during the following year one would expect the population to decline by one half. With the same number of traps and an emerging population of only half the original number per day, a theoretical ratio (traps to unmated females) of 99 : 1 would be achieved initially, and the overall reduction in mating would be 97.5 percent in year 2. In year 3 the population should be reduced an additional 75

Table 3. *A model showing the theoretical effects of traps baited with disparlure being used to intercept males before they can mate with females in an incipient population. The number of females mating in an uncontrolled population would be 1000: the total number of females mating in the controlled population is 54. Thus mating is controlled by 94.6 percent (32).*

Day	Number of Traps	Unmated Females	Males	Ratio of Traps to Unmated Females	Males Captured (No.)	Matings (No.)	Unmated Females Remaining	Daily Suppression (%)
1	5000	100	100	50:1	98	2	98	98
2	5000	174	102	28.7:1	99	3	171	97
3	5000	228	102	21.9:1	98	4	224	96
4	5000	268	103	18.7:1	98	5	263	95
5	5000	297	104	16.8:1	98	6	291	94
6	5000	318	105	15.7:1	99	6	312	94
7	5000	334	105	15:1	98	7	327	93
8	5000	345	105	14.5:1	99	7	338	93
9	5000	354	105	14.1:1	99	7	347	93
10	5000	360	105	13.9:1	99	7	353	93

percent to give an emergence rate of about 12 moths of each sex per day. Again, with the same number of traps, the initial ratio of traps to virgin females would rise to about 400 : 1, and the population should be practically eliminated during year 3.

The increasing efficiency that is characteristic of the pheromone trapping system as the population declines is analogous to the effects obtained when sterile insects are released for insect control *(33)*.

For the model shown in Table 3 we used an isolated population and did not take into account the larvae that would continue to invade trapped areas adjacent to existing high populations. In actual suppressive programs conducted in areas adjacent to high populations, at least 5000 traps per square mile would probably be employed. The extra traps (ten times the number in the hypothetical model) would be expected to provide adequate compensation for larval incursion, localized concentrations of emerging females, and unknown factors not taken into account in the model.

The use of biodegradable traps and innocuous adhesives should pose no hazards to people or to animals, although the presence of large numbers of such traps in populated areas may be objectionable from an aesthetic stand-point. Preliminary toxicological tests indicate that disparlure has very low mammalian toxicity.

New finds of the moth or of egg masses well outside the generally infested area would also have to be dealt with. The placement of traps by hand or aircraft drops in and around the areas of such finds should be a simple and inexpensive means of preventing the moth from becoming established.

Monitoring Results of Trapping with Pheromone

A major problem in the control or elimination of pests is the assessment of the results obtained by the suppression efforts. This is particularly true when pest populations are low, and the methods used for detecting the pest are not very efficient. Fortunately, disparlure provides an excellent means of detecting the gypsy moth. Nevertheless, when widely scattered incipient populations occur and efforts are made to eliminate or keep the populations suppressed, measurements of the effects of such containment programs are generally uncertain or unreliable.

Pheromone traps used for suppression have a built-in system for monitoring the results of their use. This aspect is worthy of discussion because the data obtained by trapping relative to the degree of control of the moth can readily be misinterpreted. In contrast to the conclusion that most people would come to, the greater the number of moths caught per trap, the more unsuccessful will be the control system. The validity of the foregoing statement can be illustrated with some simple hypothetical examples.

If 100 traps, each equal in attraction to a female, were in operation in an area having 1000 males and 1000 unmated females on a given day, one would expect the traps to capture 1/11 or 91 of the males, which amounts to 0.91 male per trap. The females would attract 10/11 of the males and 909 matings would result. When the average rate of capture per trap per day approached (or exceeded) one, as in the example given, at least 90 percent of the females would be expected to mate, and no significant suppression could be expected.

If the moths that were captured were well distributed and on a given day 100 traps captured 0.1 male per trap, we could, by applying the same reasoning, calculate that approximately 90 percent of the males were captured and that 10 males (and consequently 10 females) were in the area. This may be considered satisfactory control, but it is hardly enough to accomplish suppression.

With only occasional moths being taken in large number of traps and the captured moths being well distributed, the degree of control should be high. For example, an average capture rate of 0.01 male per trap per day would suggest a reduction in mating of approximately 99 percent, enough to effectively suppress populations of low density.

In the foregoing discussion we emphasize the need for monitoring a high proportion of traps to ascertain the progress of suppression programs, especially when capture rates are low. The capture of large numbers of moths per trap indicates that the number of competing females must also be high, and we cannot expect control to be effective. Should the trap be made equivalent in attraction to many females, then somewhat higher average rates of capture could still be consistent with a high degree of control. The attraction of traps baited with pheromone has not yet exceeded by more than several times the attraction of competing unmated insects.

"Confusion" Method

If the atmosphere is permeated with disparlure, or if many individual sources of the chemical are distributed in an area, the ability of male moths to locate females in that area is greatly impeded, presumably because the synthetic sex odor is everywhere and is indistinguishable from the natural female scent. Furthermore, the odor reception system of the males may become saturated or habituated and thereby insensitive or less sensitive to the attractant as a result of continuous exposure to it. The so-called "confusion" method, originally suggested in 1960 (34), has shown promise in small trials with some insect species (35), but has not yet been fully investigated for any insect. The mechanism of action in confusing the males and the influence of population density on effectiveness of the method have not been established.

In 1971 two tests of the confusion method were conducted with laboratory-reared gypsy moths before the mating flight. The results were encouraging (36). In the more pertinent test, pieces of disparlure-treated paper measuring 0.06 square inch (0.4 square centimeter) were uniformly distributed by aircraft over 40-acre plots; about 5000 pieces of paper were used per acre, containing in all less than one drop (20 mg) of disparlure per acre. This would represent about 3.2 million pieces of treated paper and about 12.8 grams of the attractant per square mile. Males released periodically in these plots were unable to find special traps containing unmated females or disparlure (about one trap per acre) for 6 days. In contrast, males were captured by such traps in untreated plots. During the next release of males, 21 days after dropping the disparlure-treated papers, some males were captured by the traps containing females or disparlure within the treated area; however, the number caught was still two-thirds less than the number captured in the untreated plots.

If we assume that the receptors of the males were not adversely affected by the pheromone and that the males continued to search for females, the presence of the artificial sources of pheromone can be regarded as a "numerical confusion" method. We believe that during the first 6 days of the test, each treated piece of paper had a great amount of attractant power, and in view of the many sources of the attractant (5000 papers per acre) from which the lure emanated, the males were repeatedly diverted from the few females (one per acre) or the disparlure-baited traps (also one per acre). By the 21st day the attraction of the treated papers apparently had declined sufficiently to allow some males to locate traps containing females or disparlure.

If the many sources of the synthetic attractant confused the males because of numerical superiority, we would expect this technique to be more effective against low than against high populations. If the average population per square mile is as low as 100 of each sex and if there are 1 million artificial sources of the attractant, each fully competitive with a female, the initial ratio of attraction would be 10,000 : 1. If the average population is 1000 males and 1000 females, the ratio would drop to 1000 : 1. With a population of 10,000 of each sex, the ratio would drop to 100 : 1. If we assume that the males respond to pheromone sources, the efficiency of the technique will depend largely on the average number of responses that the males will make in their lifetimes. When first released in the presence of the lure, males become highly excited, a condition that is likely to hasten their demise. Also, continuous exposure to the pheromone does appear to dull their response.

Although we have much to learn about the mechanism and efficiency of the confusion method, we are encouraged in this new approach by the finding that mating was inhibited for 6 days with a released insect population corresponding to 1000 per square mile, and that some inhibiting action persisted for 21 days. Two key questions are: How many sources of the lure and what amount of the artificial attractant will be required? How can the confusion effect be prolonged to encompass the entire duration of the mating flight? Although the main mating flight usually lasts 10 to 14 days, some insects appear before and after this period. We estimate that the confusion period should continue for 4 to 5 weeks, provided the sources of the lure are distributed to coincide with the start of moth flight. This time is indefinite and varies from year to year depending upon local climatic conditions. It is therefore important to achieve maximum persistence of the attractant so that applications can be made well in advance of the beginning of adult emergence.

Persistence of the lure can be increased by starting with much larger amounts (for example, 1 gram per acre), by multiple applications, and by improved formulations. It might be possible to include microencapsulated lure, various chemical diluents (keepers), additives to employ chemical stability, or to use lure carriers other than paper. Chemical analyses and laboratory bioassays of various formulations, aged or exposed under simulated conditions (rain, weathering), are being made to select the most persistent formulations for field testing.

Lure-treated papers falling on trees eventually drop to the forest floor, undoubtedly aided by wind and rain. Disparlure, known to be vulnerable to acid, may be degraded by contact with the usually acid forest floor. The effectiveness of papers falling in low spots or under forest debris is also diminished because the heavy attractant vapors have little tendency to rise and spread. Means of suspending the lure in the trees (for example, disparlure on strings or mixed with a sticker) are therefore needed to improve both the stability of the lure and effectiveness of the confusion method.

If, in the final stage of finding a female, males become oriented with the aid of sight, as suggested by Doane *(37)*, visual factors could also influence the effectiveness of the confusion method. The likelihood of males coming within visual range of females would be increased at high population densities. However, it seems reasonable to assume that at low densities, vision alone would not be a major detection mechanism for the gypsy moth. The existence of a very powerful pheromone in this species lends credence to this probability.

While we believe that the practical application of the confusion technique, when perfected, will probably be limited to low populations, we must not overlook the possibility of using pheromones to block vital physiological or behavioral processes in insects through fatigue or other mechanisms. Such effects, if they could be achieved and sustained, would probably be independent of density of the natural population.

The confusion method, even if substantial amounts of disparlure are required, should be much less costly to apply than insecticides and, because of the lure's highly selective action, should avoid the hazards often associated with the use of broad-spectrum insecticides. As much as 5 grams of lure per acre at a cost of approximately 30¢ per gram would still be within the cost range of insecticides now required for effective control.

Discussion

Scientists engaged in research on the use of sex pheromones for insect control are fully aware that much more information than is now available is needed before the best techniques for their use can be formulated. However, research—basic, applied, and theoretical—on the synthetic pheromone of the gypsy moth has advanced to the stage that federal and state agencies are prepared to undertake pilot experiments designed to test the feasibility of suppressing or preventing the spread of this pest *(38)*. Because it appears that the effectiveness of disparlure will be governed by the density of the gypsy moth populations, it seems doubtful that disparlure alone will be of benefit where the pest is already causing damage of economic proportions. Yet there are millions of acres of forest and shade trees south and west of the currently infested areas where the pest has not yet become established. There is no reason to assume that the pest cannot and will not spread to forested environments in virtually all parts of the nation and create the same havoc in many other areas as is now experienced in the Northeast unless appropriate countermeasures are applied.

There is good reason to hope that it will be possible and practical to develop a pheromone technology for containing the spread of the gypsy moth. Populations of the insect in advance of the currently infested areas are still generally scattered and are of low densities. The attractant provides an excellent method of early detection, even though the vastness of the area to which the pest might spread makes early detection a difficult and costly program. Isolated infestations can be expected to consist of very few insects, the number of insects in a single incipient infestation being unlikely to exceed 2000 within 4 years of its origin. Infestations of this size, if well delineated, would be well within the range that can be suppressed by pheromone traps or possibly by the confusion system. A critical requirement will be the detection of infestations as soon as possible and the delineation of their exact location by intensive trapping so that the pheromone can be used with maximum effect and efficiency.

If we include the year of initial spread by larval drift, there should be a period of about 3 years during which individual foci of infestations do not consist of more than 200 adult insects. However, in an area that might be 50 miles in advance of well-established populations, there could be hundreds or even thousands of such small incipient populations. Within a year or two the same situation might exist in other areas 25 to 50 miles still further west or south. On the other hand, if pheromone traps or some other system of pheromone utilization were put into operation where natural larval spread had not yet occurred, and if spread by other means was still minimal, the pheromone would have to compete only with scattered moths or with small colonies of perhaps 20 insects each. Under such circumstances as few as 100 traps per square mile might be adequate to prevent the establishment of the pest, although at least 1000 traps per square mile would be contemplated in actual practice. Thus, the advantage of early protective action in the form of a pheromone-treated barrier is readily apparent. In areas closer to the general infestation where small incipient populations might have existed for 1 or 2 years, 500 or more traps per square mile might be adequate, but at least 5000 traps per square mile should be employed. The cost of such a protective barrier should not be prohibitive; present estimates are $100 for 5000 traps and $100 for dispersing them. Thus chances seem good that the immediate spread of the pest into new areas could be largely halted for $200 per square mile. The cost of chemical control would probably be in the order of $2000 per square mile, and it is doubtful that presently available chemicals could halt spread of the moth.

The possibility of traps being used to eliminate incipient populations already existing immediately in advance of the generally infested areas should not be discounted. If the cost of traps and their distribution is as low as anticipated, their use could be less costly than chemical control even if they were distributed at a rate as high as 25,000 per square mile. The cost of this number at 2¢ each would be $500 per square mile, and the cost of application might not be much higher than that of applying fewer traps. Large numbers of traps are likely to overwhelm incipient populations even in the 3rd or 4th year of their becoming established. Furthermore, where localized populations are too high to yield to this suppression, the traps might prevent a general buildup of the insect throughout the area, and populations too high to be controlled in this manner might be delineated and dealt with by other means. Moreover, the existence of a barrier of densely distributed traps should make the more lightly treated protective barrier ahead of it more effective.

The use of the pheromone to halt or delay the unrelenting spread of the moth would give us time to explore other ecologically acceptable ways to cope with the higher populations found in established infestations. There would also be time to perfect techniques for mass rearing and release of sterile male moths to control small isolated infestations, especially in heavily populated urban centers and much-used recreational areas where large numbers of traps might be considered objectionable. The sterile male technique should prove complementary to the use of pheromone traps, which could be particularly effective in suppressing moths in more extensive rural areas. As with the pheromone, population suppression by the release of sterile males would be effective chiefly against low-density moth populations or higher populations that could be reduced to low levels by other means.

The containment of the gypsy moth, if proved feasible by the methods proposed, will be a major undertaking and the cost will be high, probably amounting to several million dollars each year. However, such costs would have to be balanced against the tens of millions of dollars in annual losses of our timber resources that seem inevitable if the pest is allowed to spread unchecked. The aesthetic values of our forests and shade trees must also be taken into account.

There is a tendency today for some ecologists to oppose any pest suppressive measures on the assumption that the measures employed, or even the elimination of the pest itself, will upset the ecology of the area under treatment; often no distinction between native and alien pests is made. In response to such views, we note that should containment be achieved by the use of the pheromone disparlure, supplemented by the use of sterile moths in special areas, implementation of such measures should be without hazard and be completely acceptable from an ecological standpoint. In contrast, even without economic considerations, there is reason for us to have grave concern over the harmful ecological effects of the gypsy moth if this alien pest is left to spread to the limits of its range and become a permanent resident throughout the forest ecosystems of the country.

REFERENCES AND NOTES

1. Comments, *Nature 231*, 142 (1971); N. Wade, *Science 174*, 41 (1971).
2. B. A. Bierl, M. Beroza, C. W. Collier, *Science 170*, 87 (1970).
3. E. H. Forbush and C. H. Fernald, *The Gypsy Moth* (Wright & Potter, Boston, 1896).
4. A. F. Burgess, *J. Econ. Entomol. 33*, 558 (1940).
5. W. V. O'Dell, *ibid. 48*, 170 (1955); *ibid. 50*, 541 (1957).
6. U.S. Forest Service, Southeastern Area, *The Gypsy Moth: A Threat to Southern Forests* (Government Printing Office, Washington, D.C., 1971); D. E. Leonard, *Can. Entomol. 102*, 239 (1970).
7. D. E. Leonard, *J. Econ. Entomol. 64*, 638 (1971).
8. I. Schwinck, *10th Int. Congr. Entomol. Proc. 2*, 577 (1958).
9. *Gypsy Moth Fact Sheet* (U.S. Department of Agriculture, Agricultural Research Service, Washington, D.C., 1970).
10. *The Gypsy Moth* (U.S. Department of Agriculture, Program Aid 910, Washington, D.C., 1970).
11. J. D. Kegg, *J. Forest. 69*, (No. 12) 852 (1971), reports that mortality of the oaks in the Morristown National Historical Park rose from 6 percent in 1967 to 69 percent in the spring of 1970.
12. *New York Times*, 24 June 1970.
13. Several of the many publications on the subject are: L. O. Howard and W. F. Fiske, *U.S. Dep. Agr. Bur. Entomol. Bull. 91*, 1 (1911); A. F. Burgess and S. S. Crossman, *U.S. Dep. Agr. Tech. Bull. No. 86* (1929), p. 1; P. B. Dowden, *U.S. Dep. Agr. Forest Serv. Handb. No. 226* (1962).
14. W. D. Rollinson, F. B. Lewis, W. E. Waters, *J. Invertebr. Pathol. 7*, 515 (1965).
15. C. C. Doane, *ibid. 17*, 303 (1971); F. B. Lewis and D. P. Connola, *U.S. Forest Serv. Res. Pap. NE-50* (1966); C. C. Doane, *J. Econ. Entomol. 59*, 618 (1966).
16. D. E. Leonard, *J. Econ. Entomol. 64*, 640 (1971).
17. C. W. Collins and S. F. Potts, *U.S. Dep. Agr. Tech. Bull. No. 336* (1932).

18. H. L. Haller, F. Acree, Jr., S. F. Potts, *J. Amer. Chem. Soc. 66*, 1659 (1944).
19. E. D. Burgess, *J. Econ. Entomol. 43*, 325 (1950).
20. R. F. Holbrook, M. Beroza, E. D. Burgess, *ibid. 53*, 751 (1960).
21. F. Acree, Jr., M. Beroza, R. F. Holbrook, H. L. Haller, *ibid. 52*, 82 (1959).
22. M. Jacobson, M. Beroza, W. A. Jones, *Science 132*, 1011 (1960); *J. Amer. Chem. Soc. 83*, 4819 (1961).
23. M. Jacobson and W. A. Jones, *J. Org. Chem. 27*, 2523 (1962).
24. M. Jacobson, M. Schwarz, R. M. Waters, *J. Econ. Entomol. 63*, 943 (1970).
25. M. Beroza, B. A. Bierl, J. G. R. Tardif, D. A. Cook, E. C. Paszek, *ibid. 64*, 1499 (1971).
26. E. H. Forbush and C. H. Fernald, *The Gypsy Moth* (Wright & Potter, Boston, 1896), p. 95.
27. Reported at a recent meeting of the Gypsy Moth Advisory Council.
28. E. F. Knipling and J. U. McGuire, Jr., *U.S. Dep. Agr. Inform. Bull. No. 308* (1966).
29. D. D. Hardee, W. H. Cross, E. B. Mitchell, P. M. Huddleston, H. C. Mitchell, M. E. Merkl, T. B. Davich, *J. Econ. Entomol. 62*, 161 (1969); D. D. Hardee, O. H. Lindig, T. B. Davich, *ibid. 64*, 928 (1971).
30. W. L. Roelofs, E. H. Glass, J. Tette, A. Comeau, *ibid. 63*, 1162 (1970).
31. M. Beroza, B. A. Bierl, E. F. Knipling, J. G. R. Tardif, *ibid. 64*, 1527 (1971).
32. Calculations for days 1 and 2 are as follows; those for subsequent days are similar. Fractional values for insects are rounded off to the nearest whole number. On day 1 the number of traps (5000) divided by the number of unmated females (100) equals 50. Thus 1/51 of the 100 males, or 2 males, are attracted to the unmated females and mate. The remaining males (98) are attracted to the traps and destroyed. Since 98 of 100 females remain unmated on day 1, mating suppression is 98 percent. On day 2, traps remain constant at 5000. The competing females consist of 100 newly emerged females plus 75 percent of the 98 unmated females surviving from day 1 for a total of 174 unmated females. The male population consists of 100 newly emerged males plus 75 percent of the 2 males that mated on day 1 for a total of 102. The ratio of traps to unmated females is 28.7 : 1. Thus on day 2, 1/29.7 of the 102 males (or 3 males) mate, and 171 of the females remain unmated. With only 3 females mating on day 2 compared to an expected mating of 100 for an untreated population (for which complete mating of the 100 males and 100 females that emerge each day is assumed), mating is suppressed 97 percent. The following equation summarizes the calculations:

$$M_n = \frac{[m + SM_{n-1}] \cdot [f + S(F_{n-1} - M_{n-1})]}{T + [f + S(F_{n-1} - M_{n-1})]}$$

where M_n is the number of matings on day n, m is the number of males, and f is the number of females emerging each day; F_n is the number of unmated females on day n or $F_n = f + S(F_{n-1} - M_{n-1})]$,

D is the number of days in the mating season, S is the survival rate per day, and T is the number of Traps. Thus $\int_1^D M_n$ gives the total number of matings in the controlled population, and fD is the number of matings in the uncontrolled population.

33. E. F. Knipling, *J. Econ. Entomol. 48*, 459 (1955); *ibid. 53*, 415 (1960); *ibid. 55*, 782 (1962); *Science 139*, 902 (1959); *Sci. Amer. 203* (4), 54 (1960).
34. M. Beroza, *Agr. Chem. 15* (7), 37 (1960).
35. L. K. Gaston, H. H. Shorey, C. A. Saario, *Nature 213*, 1155 (1967); H. H. Shorey, L. K. Gaston, C. A. Saario, *J. Econ. Entomol. 60*, 1541 (1967); J. A. Klun and J. F. Robinson, *ibid. 63*, 1281 (1970).
36. The first test was conducted in April 1971 on an island off the coast of Alabama by A. E. Cameron (Pennsylvania State University) and F. M. Philips (U.S. Department of Agriculture) and the second in Massachusetts about 2 months later by L. Stevens and his staff (U.S. Department of Agriculture; manuscript in preparation). Formulations were prepared by M. Beroza and B. A. Bierl. The Massachusetts test was the more pertinent one because it was conducted in the area where the insect is found and live females as well as traps baited with pheromone were used. Live females were not used in the Alabama test.
37. C. C. Doane, *Ann. Entomol. Soc. Amer. 61*, 768 (1968).
38. Such cooperative programs are under way, the principal effort being made by Pennsylvania State University.

13. *An Idea We Could Live Without—the Naked Ape*

DAVID PILBEAM

". . . fierce aggression and status-seeking are no more 'natural' attributes of man than they are of most monkey and ape societies. The degree to which most behaviors are developed depends very considerably indeed upon cultural values and learning. Territoriality likewise is not a 'natural' feature of human group living; nor is it among most other primates. . . . As to sex roles . . . so much of human role behavior is learned that we could imagine narrowing or widening the differences almost as much or as little as we wish." — Dr. David Pilbeam has written extensively in the fields of physical anthropology and primate paleontology as related to human evolution. Recent books are *Evolution of Man* Funk and Wagnalls, 1969, and *The Ascent of Man: An Introduction of Human Evolution*, Macmillan, 1972. He is associate professor of anthropology and associate curator of the Peabody Museum of Natural History, Yale University.

Last fall CBS Television broadcast a National Geographic Special, in prime time, called "Monkeys, Apes, and Man." This was an attempt to demonstrate how much studies of primates can tell us about our true biological selves. In a recent *Newsweek* magazine article, Stewart Alsop, while discussing problems of war, stated that nations often quarrel over geopolitical real estate when national boundaries are poorly defined: his examples were culled from areas as diverse as the Middle East, Central Europe, and Asia. One of his introductory paragraphs included the following:

"The animal behaviorists—Konrad Lorenz, Robert Ardrey, Desmond Morris—have provided wonderful insights into human behavior. Animals that operate in groups, from fish up to our ancestors among the primates, instinctively establish and defend a territory, or turf. There are two main reasons why fighting erupts between turfs—when the turfs are ill-defined or overlapping; or when one group is so weakened by sickness or other cause as to be unable to defend its turf, thus inviting aggression."

Here Alsop is taking facts (some of them are actually untrue facts) from the field of ethology—which is the science of whole animal behavior as studied in naturalistic environments—and extrapolating directly to man from these ethological facts as though words such as *territoriality, aggression*, and so forth describe the same phenomena in all animal species, including man.

Source: Copyright 1972 Peabody Museum of Natural History, Yale University. Reprinted by permission of the publisher and author from *Discovery* (Peabody Museum of Natural History), 7(2):63-70, Spring 1972.

Both these examples from popular media demonstrate nicely what can be called "naked apery" or the "naked-ape syndrome." When Charles Darwin first published *The Origin of Species* and *The Descent of Man*, over 100 years ago, few people believed in any kind of biological or evolutionary continuity between men and other primates. Gradually the idea of man's physical evolution from ape- or monkey-like ancestors came to be accepted; yet the concept of human behavioral evolution was always treated with scepticism, or even horror. But times have changed. No longer do we discriminate between rational man, whose behavior is almost wholly learned, and all other species, brutish automata governed solely by instincts.

One of the principal achievements of ethologists, particularly those who study primates, has been to demonstrate the extent to which the dichotomy between instinct and learning is totally inadequate in analyzing the behavior of higher vertebrate species—especially primates. Almost all behavior in monkeys and apes involves a mixture of the learned and the innate; almost all behavior is under some genetic control in that its development is channeled—although the amount of channeling varies. Thus, all baboons of one species will grow up producing much the same range of vocalizations; however, the same sound may have subtly different meanings for members of different troops of the same species. In one area, adult male baboons may defend the troop; those of the identical species in a different environment may habitually run from danger. Monkeys in one part of their species range may be sternly territorial; one hundred miles away feeding ranges of adjacent groups may overlap considerably and amicably. These differences are due to learning. Man is the learning animal par excellence. We have more to learn, take longer to do it, learn it in a more complex and yet more efficient way (that is, culturally), and have a unique type of communication system (vocal language) to promote our learning. All this the ethologists have made clear.

Studies of human behavior, at least under naturalistic conditions, have been mostly the preserve of social anthropologists and sociologists. The anthropological achievement has been to document the extraordinary lengths to which human groups will go to behave differently from other groups. The term "culture," a special one for the anthropologist, describes the specifically human type of learned behavior in which arbitrary rules and norms are so important. Thus, whether we have one or two spouses, wear black or white to a funeral, live in societies that have kings or lack chiefs entirely, is a function not of our genes but of learning; the matter depends upon which learned behaviors we deem appropriate—again because of learning. Some behaviors make us feel comfortable, others do not; some behaviors may be correct in one situation and not in another—forming a line outside a cinema as opposed to the middle of the sidewalk, for example, singing rather than whistling in church; talking to domestic animals but not to wild ones. The appropriate or correct behavior varies from culture to culture; exactly which one is appropriate is arbitrary. This sort of behavior is known as "context dependent behavior" and is, in its learned form, pervasively and almost uniquely human. So pervasive is it, indeed, that we are unaware most of the time of the effects on our behavior of context dependence. It is important to realize here that although a great deal of ape and monkey behavior is learned, little of it is context dependent in a cultural, human sense.

In the past ten years there has been a spate of books—the first of the genre was Robert Ardrey's *African Genesis* published in 1961—that claim first to describe man's "real" or "natural" behavior in ethological style, then go on to explain how these behaviors have evolved. In order to do this, primate societies are used as models of earlier stages of human evolution: primates are ourselves, so to speak, unborn. *African Genesis, The Territorial Imperative, The Social Contract,* all by Ardrey, *The Naked Ape* and *The Human Zoo* by Desmond Morris, Konrad Lorenz's *On Aggression,* and, most recently, Antony Jay's *Corporate Man,* approach the bestseller level. All purport to document, often in interminable detail, the supposedly surprising truth that man is an animal. Also they argue that his behavior—particularly his aggressive, status-oriented, territorial and sexual behavior—is somehow out of tune with the needs of the modern world, that these behaviors are under genetic control and are largely determined by our animal heritage, and that there is little we can do but accept our grotesque natures; if we insist on trying to change ourselves, we must realize that we have almost no room for maneuver, for natural man is far more like other animals than he would care to admit. Actually, it is of some anthropological interest to inquire exactly why this naked apery should have caught on. Apart from our obsessive neophilia, and the fact that these ideas are somehow "new," they provide attractive excuses for our unpleasant behavior toward each other.

However, I believe these general arguments to be wrong; they are based upon misinterpretation of ethological studies and a total ignorance of the rich variety of human behavior documented by anthropologists. At a time when so many people wish to reject the past because it has no meaning and can contribute nothing, it is perhaps a little ironic that arguments about man's innate and atavistic depravity should have so much appeal. The world *is* in a mess; people *are* unpleasant to each other; that much is true. I can only suppose that argument about the inevitability of all the nastiness not only absolves people in some way of the responsibility for their actions, but allows us also to sit back and positively enjoy it all. Let me illustrate my argument a little.

Take, for example, one particular set of ethological studies—those on baboons. Baboons are large African monkeys that live today south of the Sahara in habitats ranging from tropical rain forest to desert. They are the animals that have been most frequently used as models of early human behavior; a lot of work has been done on them, and they are easy to study—at least those living in the savannah habitats thought to be typical of the hunting territories of early man. They are appealing to ethologists because of their habitat, because they live in discrete and structured social groups, and because they have satisfied so many previous hypotheses.

Earlier reports of baboon behavior emphasized the following. Baboons are intensely social creatures, living in discrete troops of 30 to 50 animals, their membership rarely changing; they are omnivorous, foraging alone and rarely sharing food. Males are twice as big as females; they are stronger and more aggressive. The functions of male aggression supposedly are for repelling predators, for maintaining group order, and (paradoxically) for fighting among themselves. The adult males are organized into a dominance hierarchy, the most dominant animal being the one that gets his own way as far as food, grooming partners, sex, when to stop and eat, and when and where the troop should move are concerned. He is the most aggressive, wins the most fights, and impregnates the most desirable females. Females, by the way, do little that is exciting in baboondom, but sit around having babies, bickering, and tending to their lords. Adult males are clearly the most important animals—although they cannot have the babies—and they are highly status conscious. On the basis of fighting abilities they form themselves into a dominance hierarchy, the function of which is to reduce aggression by the controlling means of each animal knowing its own place in the hierarchy. When groups meet up, fighting may well ensue. When the troop moves, males walk in front and at the rear; when the group is attacked, adult males remain to fight a rearguard action as females and young animals flee to safety in the trees.

Here then we have in microcosm the views of some men—and it is a very male view—of the way our early ancestors may well have behaved. How better to account for the destructiveness of so much human male aggression, to justify sex differences in behavior, status seeking, and so forth. I exaggerate, of course, but not too much. But what comments can be made?

First, the baboons studied—and these are the groups that are described, reported, and extrapolated from in magazine articles, books, and in CBS TV specials—are probably abnormal. They live in game parks—open country where predators, especially human ones, are present in abundance—and are under a great deal of tension. The same species has been studied elsewhere—in open country and in forest too, away from human contact—with very different results.

Forest groups of baboons are fluid, changing composition regularly (rather than being tightly closed); only adult females and their offspring remain to form the core of a stable group. Food and cover are dispersed, and there is little fighting over either. Aggression in general is very infrequent, and male dominance hierarchies are difficult, if not impossible, to discern. Intertroop encounters are rare, and friendly. When the troop is startled (almost invariably by humans, for baboons are probably too smart, too fast, and too powerful to be seriously troubled by other predators), it flees, and, far from forming a rearguard, the males—being biggest and strongest—are frequently up the trees long before the females (encumbered as they are with their infants).

When the troop moves it is the adult females that determine when and where to; and as it moves adult males are not invariably to be found in front and at the rear. As for sexual differences, in terms of functionally important behaviors, the significant dichotomy seems to be not between males and females but between adults and young. This makes good sense for animals that learn and live a long time.

The English primatologist Thelma Rowell, who studied some of these forest baboons in Uganda, removed a troop of them and placed them in cages where food had to be given a few times a day in competition-inducing clumps. Their population density went up and cover was reduced. The result? More aggression, more fighting, and the emergence of dominance hierarchies. So, those first baboons probably were under stress, in a relatively impoverished environment, pestered by humans of various sorts. The high degree of aggression, the hierarchies, the rigid sex-role differences, were abnormalities. In one respect, troop defense, there is accumulating evidence that male threats directed toward human interlopers occur only after troops become habituated to the observers, and must therefore be treated as learned behavior too.

Studies on undisturbed baboons elsewhere have shown other interesting patterns of adult male behaviors. Thus in one troop an old male baboon with broken canines was the animal that most frequently completed successful matings, that influenced troop movements, and served as a focus for females and infants, even though he was far less aggressive than, and frequently lost fights with, a younger and more vigorous adult male. Here, classical dominance criteria simply do not tie together as they are supposed to.

The concept of dominance is what psychologists call a unitary motivational theory: there are two such theories purporting to explain primate social behavior. These are that the sexual bond ties the group together, and that social dominance structures and orders the troop. The first of these theories has been shown to be wrong. The second we are beginning to realize is too simplistic. In undisturbed species in the wild, dominance hierarchies are hard to discern, if they are present at all; yet workers still persist in trying to find them. For example, Japanese primatologists describe using the "peanut test" to determine "dominance" in wild chimpanzees by seeing which chimp gets the goodies. Yet what relevance does such a test have for real chimp behavior in the wild where the animals have far more important things to do—in an evolutionary or truly biological sense—than fight over peanuts? Such an experimental design implies too the belief that "dominance" is something lurking just beneath the surface, waiting for the appropriate releaser.

Steven Gartlan, an English primatologist working in the Cameroons, has recently suggested a much more sensible way of analyzing behavior, in terms of function. Each troop has to survive and reproduce, and in order to do so it must find food, nurture its mothers, protect and give its young the opportunity to learn adult skills. There are certain tasks that have to be completed if successful survival is to result. For example, the troop must be led, fights might be stopped, lookouts kept, infants fed and protected; some animals must serve as social foci, others might be needed to chase away intruders, and so on. Such an attribute list can be extended indefinitely.

If troop behavior is analyzed in such a functional way like this, it immediately becomes clear that different classes of animals perform different functions. Thus, in undisturbed baboons, adults, particularly males, police the troop; males, especially the subadults and young adults, maintain vigilance; adult females determine the time and direction of movement; younger animals, especially infants, act as centers of attention.

Thus a particular age-sex class performs a certain set of behaviors that go together and that fulfill definite adaptive needs. Such a constellation of behavioral attributes is termed a role. Roles, even in nonhuman primates, are quite variable. (Witness the great differences between male behaviors in normal baboon troops and those under stress.) If dominance can come and go with varying intensities of certain environmental pressures, then it is clearly not innately inevitable, even in baboons. Dominance hierarchies, then, seem to be largely artifacts of abnormal environments.

What is particularly interesting in the newer animal studies is the extent to which aggression, priority of access, and leadership are divorced from each other. Although a baboon may be highly aggressive, what matters most is how other animals react to him; if they ignore him as far as functionally important behaviors such as grooming, mating and feeding are concerned, then his aggression is, in a social or evolutionary sense, irrelevant.

I want to look a little more closely at aggression, again from the functional point of view. What does it do? What is the point of a behavior that can cause so much trouble socially?

The developmental course of aggressive behavior has been traced in a number of species: among primates it is perhaps best documented in rhesus macaques, animals very similar to baboons. There are genetical and hormonal bases to aggressive behavior in macaques; in young animals males are more aggressive, on the average, than females, and this characteristic is apparently related to hormonal influences. If animals are inadequately or abnormally socialized, aggressive behaviors become distorted and exaggerated. Animals that are correctly socialized in normal habitats, or richly stimulating artificial ones, show moderate amounts of aggression, and only in certain circumstances. These would be, for example, when an infant is threatened, when a choice item is disputed, when fights have to be interrupted, under certain circumstances when the troop is threatened, and occasionally when other species are killed for food.

Under normal conditions, aggression plays little part in other aspects of primate social life. The idea that the function of maleness is to be overbearingly aggressive, to fight constantly, and to be dominant, makes little evolutionary sense.

How about extrapolations from primates to man that the "naked-apers" are so fond of? Take, for example, dominance. Everything that I have said about its short-comings as a concept in analyzing baboon social organization applies to man, only more so. Behaviors affecting status-seeking in man are strongly influenced by learning, as we can see by the wide variation in human behavior from one society to another. In certain cultures, status is important, clear-cut, and valued; the emphasis placed on caste in Hindu society is an obvious example. At the opposite extreme, though—among the Bushmen of the Kalahari Desert, for example—it is hard to discern; equality and cooperativeness are highly valued qualities in Bushman society, and hence learned by each new generation.

I've used the term "status-seeking" rather than "dominance" for humans, because it describes much better the kind of hierarchical ordering one finds within human groups. And that points to a general problem in extrapolating from monkey to man, for "status" is a word that one can't easily apply to baboon or chimp society; status involves prestige, and prestige presupposes values—arbitrary rules or norms. That sort of behavior is cultural, human, and practically unique.

As we turn to man, let's consider for a while human groups as they were before the switch to a settled way of life began a mere—in evolutionary terms—10,000 years ago. Before that our ancestors were hunters and gatherers. Evidence for this in the form of stone tool making, living areas with butchered game, camp sites, and so on, begins to turn up almost 3 million years ago, at a time when our ancestors were very different physically from us. For at least 2½ to 3 million years, man and his ancestors have lived as hunters and gatherers. The change from hunting to agricultural-based economies began, as I said, just over 10,000 years ago, a fractional moment on the geological time scale. That famous (and overworked) hypothetical visiting Martian geologist of the 21st century would find remains of hunters represented in hundreds of feet of sediments; the first evidence for agriculture, like the remains of the thermonuclear holocaust, would be jammed, together, in the last few inches. Hunting has been a highly significant event in human history; indeed, it is believed by most of us interested in human evolution to have been an absolutely vital determinant, molding many aspects of human behavior.

There are a number of societies surviving today that still live as hunters; Congo pygmies, Kalahari Bushmen, and Australian Aborigines, are three well-known examples. When comparisons are made of these hunting societies, we can see that certain features are typical of most or all of them, and these features are likely to have been typical of earlier hunters.

In hunting societies, families—frequently monogamous nuclear families—are often grouped together in bands of 20 to 40 individuals; members of these hunting bands are kinsmen, either by blood or marriage. The band hunts and gathers over wide areas, and its foraging range often overlaps those of adjacent groups. Bands are flexible and variable in composition—splitting and reforming with changes in the seasons, game and water availability, and whim.

Far from life being "short, brutish, and nasty" for these peoples, recent studies show that hunters work on the average only 3 or 4 days each week; the rest of their time is leisure. Further, at least 10% of Bushmen, for example, are over 60 years of age, valued and nurtured by their children. Although they lack large numbers of material possessions, one can never describe such peoples as savages, degenerates, or failures.

The men in these societies hunt animals while the women gather plant food. However, women often scout for game, and in some groups may also hunt smaller animals, while a man returning empty-handed from a day's hunting will almost always gather vegetable food on his way. Thus the division of labor between sexes is not distinct and immutable; it seems to be functional, related to mobility: the women with infants to protect and carry simply cannot move far and fast enough to hunt efficiently.

Relations between bands are amicable; that makes economic sense as the most efficient way of utilizing potentially scarce resources, and also because of exogamy—marrying out—for adjacent groups will contain kinsmen and kinsmen will not fight. Within the group, individual relations between adults are cooperative and based upon reciprocity; status disputes are avoided. These behaviors are formalized, part of cultural behavior, in that such actions are positively valued and rewarded. Aggression between individuals is generally maintained at the level of bickering; in cases where violence flares, hunters generally solve the problem by fission: the band divides.

Data on child-rearing practices in hunters are well known only in Bushmen, and we don't yet know to what extent Bushmen are typical of hunters. (This work on Bushman child-rearing, as yet unpublished, has been done by Patricia Draper, a Harvard anthropologist, and I am grateful to her for permitting me to use her data.) Bushman children are almost always in the company of adults; because of the small size of Bushman societies, children rarely play in large groups with others of their own age. Aggression is minimal in the growing child for two principal reasons. First, arguments between youngsters almost inevitably take place in the presence of adults and adults always break these up before fights erupt; so the socialization process gives little opportunity for practicing aggressive behavior. Second, because of the reciprocity and cooperativeness of adults, children have few adult models on which to base the learning of aggressiveness.

Thus the closest we can come to a concept of "natural man" would indicate that our ancestors were, like other primates, capable of being aggressive, but they would have been socialized culturally in such a way as to reduce as far as possible the manifestation of aggression. This control through learning is much more efficient in man than in other primates, because we are cultural creatures—with the ability to attach positive values to aggression-controlling behaviors. Thus Bushmen value and thereby encourage peaceful cooperation. Their culture provides the young with non-violent models.

Other cultures promote the very opposite. Take, for example, the Yanomamö Indians of Venezuela and Brazil; their culture completely reverses our ideals of "good" and "desirable." To quote a student of Yanomamö society: "A high capacity for rage, a quick flash point, and a willingness to use violence to obtain one's ends are considered desirable traits." In order to produce the appropriate adult behaviors, the Yanomamö encourage their children, especially young boys, to argue, fight, and be generally belligerent. These behaviors, I should emphasize, are learned, and depend for their encouragement upon specific cultural values.

Our own culture certainly provides the young with violent, though perhaps less obtrusive, models. These I should emphasize again, are learned and arbitrary, and we *could* change them should we choose to do so.

So far we have seen that fierce aggression and status-seeking are no more "natural" attributes of man than they are of most monkey and ape societies. The degree to which such behaviors are developed depends very considerably indeed upon cultural values and learning. Territoriality likewise is not a "natural" feature of human group living; nor is it among most other primates.

As a parting shot, let me mention one more topic that is of great interest to everyone at the moment—sex roles. Too many of us have in the past treated the male and female stereotypes of our particular culture as fixed and "natural": in our genes so to speak. It may well be true that human male infants play a little more vigorously than females, or that they learn aggressive behaviors somewhat more easily, because of hormonal differences. But simply look around the world at other cultures. In some, "masculinity" and "feminity" are much more marked than they are in our own culture; in others the roles are blurred. As I said earlier, among Bushmen that are still hunters, sex roles are far from rigid, and in childhood the two sexes have a very similar upbringing. However, among those Bushmen that have adopted a sedentary life devoted to herding or agriculture, sex roles are much more rigid. Men devote their energies to one set of tasks, women to another, mutually exclusive set. Little boys learn only "male" tasks, little girls exclusively "female" ones. Maybe the switch to the sedentary life started man on the road toward marked sex role differences. These differences are learned though, almost entirely, and heavily affected by economic factors.

So much of human role behavior is learned that we could imagine narrowing or widening the differences almost as much or as little as we wish.

So, what conclusions can be drawn from all this? It is overly simplistic in the extreme to believe that man behaves in strongly genetically deterministic ways, when we know that apes and monkeys do not. Careful ethological work shows us that the primates closely related to us—chimps and baboons are the best known—get on quite amicably together under natural and undisturbed conditions. Learning plays a very significant part in the acquisition of their behavior. They are not highly aggressive, obsessively dominance-oriented, territorial creatures.

There is no evidence to support the view that early man was a violent status-seeking creature; ethological and anthropological evidence indicates rather that pre-urban men would have used their evolving cultural capacities to channel and control aggression. To be sure, we are not born empty slates upon which anything can be written; but to believe in the "inevitability of beastliness" is to deny our humanity as well as our primate heritage—and, incidentally, does a grave injustice to the "beasts."

ACKNOWLEDGMENTS. I would like to thank the following for helpful discussions and/or assistance in translating the manuscript into American English: Patricia Draper, Zelda Edelson, Fredericka Oakley, and Carol Pilbeam.

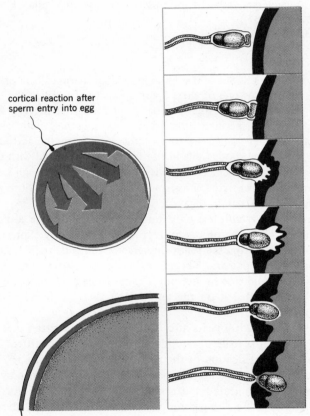

cortical reaction after
sperm entry into egg

fertilization membrane along
inner surface of vitelline membrane

The sequence of events in the union of a sperm with an egg. Upper left shows sperm penetrating the egg and initiating the cortical reaction—the formation of the fertilization membrane (lower left). The fertilization membrane prevents entry of additional sperm. Top right shows sequences of sperm penetration into the egg. When a sperm contacts an egg, the sperm's acrosome releases enzymes that dissolve a pathway into the egg. The sperm nucleus then moves through the egg's cytoplasm and joins its nucleus.

14. *The Sperm Cell*

J. L. HANCOCK

Sperm cells are mini-vehicles bearing the genetic information of the male. During intercourse an ejaculate of semen deposits about 500 million of them in the upper reaches of the vagina. Only a few thousand make it through the uterus and up the Fallopian tube. Several may contact the egg, however, but one is usually involved in fertilization. Dr. James L. Hancock is professor of veterinary anatomy at Royal Veterinary College, University of London. He was senior scientific officer at ARC Animal Breeding Research Organization, Edinburgh, Scotland. His principal research interests have been in studies of fertility and semen fertilizing capacity.

A man normally produces about 500 million individual sperm cells in one ejaculation. In favourable circumstances one of these spermatozoa will fertilize the single egg cell normally shed by a woman at ovulation. Besides stimulating the egg to divide and begin its development, the spermatozoon brings to the egg the genetic contribution of the father which becomes joined to that of the mother in the fertilized egg. The spermatozoa are made from body cells in the testis and are mixed with the secretions of several glands to form the semen which is ejaculated into the vagina of the female during intercourse. The spermatozoa assisted by muscular contractions of the womb reach the Fallopian tubes where the egg is fertilized.

These facts have been established only during the past 100 years, although the idea of the role of the semen in the process of reproduction is clearly expressed in the Old Testament as can be seen from the following verse: "And Judah said unto Onan, go in into thy brother's wife and marry her and raise up seed to thy brother. And Onan knew that the seed should not be his and it came to pass when he went in into his brother's wife that he spilled it on the ground lest that he should give seed to his brother" (*Genesis 38, VIII*).

It was not until some 2500 years after this passage was written that the essential ideas about the role of the male in reproduction were finally clarified. The first real step forward was the identification of the spermatozoa in semen in 1677 by a young Dutch medical student Johan Hamm. His observations were confirmed by Antonie van Leeuwenhoek who communicated the observations to the Royal Society in 1679; these observations were the starting point of a profound controversy about the role of the spermatozoa in reproduction. Two distinct schools of thought developed: "ovists" believed that the sperm merely nourished or stimulated the female seed into development, whereas the "animists" believed that the sperm itself contained the

Source: Reprinted from the June 1970 issue of *Science Journal* by courtesy of Syndication International Ltd., London.

entire embryo in miniature. The animists even produced pictures showing a tiny, but perfectly formed, man (the "homunculus") curled up in the head of the sperm.

The next major step was the discovery of the mammalian egg by K. E. von Baer in 1827, but it was not till 1843 that the penetration of the egg by the spermatozoon was recorded by M. Barry. The essential details in the production of sperm in the testis were first described by R. Kolliker in 1841 and the role of the spermatozoon in fertilization was finally established by E. van Beneden in 1883.

Improvements to the light microscope had meanwhile made possible a continuing growth of knowledge, so that by the beginning of this century classical cytology had furnished a picture of the structure of spermatozoa and of their development. By this time the resolving power of the light microscope had been developed near to its theoretical limits, but in recent years the electron microscope has provided the cytologist with a more powerful instrument and there has been a renewal of interest in sperm structure.

Spermatozoa are produced in the seminiferous tubules of the testis and leave the testis suspended in a fluid produced by specialized cells of the collecting tubules of the rete testis [Fig. 1]. The rete discharges the sperm into the epididymis, which is essentially a single coiled tube leading eventually, via a shorter uncoiled portion, the vas deferens, to the penis. Here, at the time of ejaculation, the contributions of the accessory glands (Cowper's glands, the prostrate and the seminal vesicles) are added. The male sterilization operation known as "vasectomy" is performed by severing the vas deferens.

The head of the spermatozoon carries the deoxyribonucleic acid (DNA) which, at fertilization, forms the genetic contribution of the father [see Fig. 3]. The head is covered in front by a cap, the acrosome. This is the site of certain enzymes which enable the sperm to penetrate the layer of cells covering the fresh ovulated egg. Wide variations in the shape of the nucleus and the acrosome are found in various species of animals. The middle-piece of the sperm contains a sheath of mitochondria which are the site of other enzymes whose activity provides for the transfer of energy, derived from metabolism of sugars, to the tail fibres. The middle-piece is therefore the generator which provides the energy for the active movement of the spermatozoon. The sperm tail articulates with the head by means of what is effectively a ball and socket joint; the capitulum of the connecting piece is the ball and this articulates with the basal plate which lines the concavity at the base of the head of the spermatozoon.

The tail itself is composed of fibres arranged in a basic pattern which consists of a central pair of fibres with an outer ring of nine coarse fibres and an inner ring of nine fine fibres. This same basic pattern of fibres is found in the flagellae of both animal and plant cells.

The spermatozoa of normal fertile males are similar under the microscope and it is now well recognized that gross variations from the normal shape for the species are associated with infertility in both man and domestic animals. Samples of sperm from infertile men may contain spermatozoa with widely different shapes. Abnormalities commonly found are spermatozoa with two heads or two tails and others with

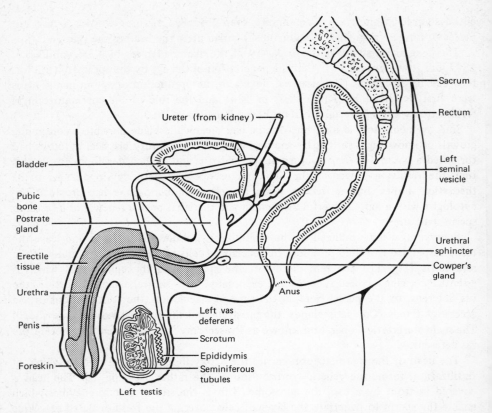

Fig. 1. Spermatozoa are produced in the seminiferous tubules of the testis and then pass into the epididymis and the vas deferens. Secretions from the seminal vesicles and Cowper's glands are mixed with the spermatozoa and this mixture—the semen—is ejaculated during intercourse. Erectile tissue in the penis fills with blood causing it to stiffen during sexual excitement.

misshapen heads or tails. In addition, infertility is often associated with a shortage of sperms in the ejaculate.

Before puberty the seminiferous tubules of the testis are lined by special nurse cells and the germ cells from which the spermatozoa are formed. At puberty the germ cells (spermatogonial) begin to multiply by a series of divisions and eventually differentiate to form a new class of cell—the primary spermatocyte [see Fig. 2].

The production of a continuous supply of spermatozoa from the testis is ensured by the continued existence of germ cells which form a reservoir of stem cells from which future spermatozoa are derived. These stem cells are set aside from the main population of spermatogonia which multiply and differentiate to meet the immediate needs of sperm production. Each primary spermatocyte divides twice to produce two

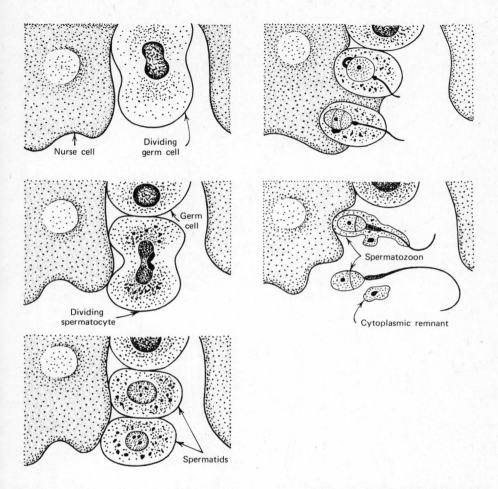

Fig. 2. Spermatozoa are formed from germ cells in the tubules of the testis. The germ cells divide to form spermatocytes which divide again to form spermatids with half the number of chromosomes. The spermatids become embedded in nurse cells and begin to grow tails. Then the cytoplasm contracts and leaves sperm to swim free. Spermatids originating from the same germ cell normally develop at the same rate.

daughter cells and each of these again divides to produce two spermatids. The spermatids differentiate and acquire the specialized form of the spermatozoon. The two divisions which result in the formation of our spermatids are "reduction divisions" in which the chromosome number of the diploid primary spermatocyte is halved to give the haploid number (23 in humans) found in spermatozoon and egg. At the same time an important redistribution of genes between the homologous chromosomes takes place. The whole process of sperm production from the first division of the

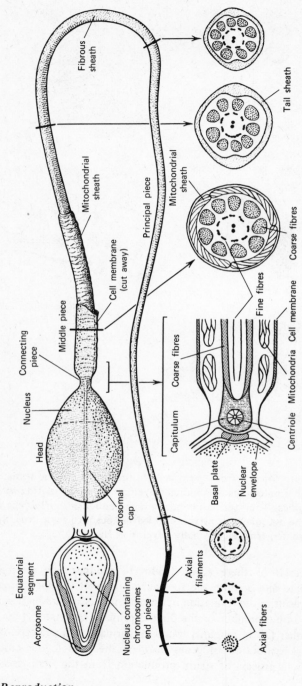

Fig. 3. The spermatozoon differs from any other kind of body cell. The head of the sperm contains the nucleus with its genetic material, covered in front by the acrosome which carries powerful enzymes to assist penetration of the egg. Behind this is the midpiece—the energy generator of the sperm with its battery of mitochondria (shown where the outer membrane is cut away). Finally, there is the tail which consists of both coarse fibres and fine fibres which contract causing the sperm to move.

sperm cells to production of sperm in the ejaculate takes 72 days.

Spermatozoa entering the epididymis from the testis are incapable of fertilization but acquire this ability during the journey through the epididymis. The final part of the epididymis acts as a storage organ for spermatozoa and within it the spermatozoa can survive for several weeks or months, although they will survive for only a few hours within the female tract or maintained at blood heat outside the body. One reason for this seems to be that spermatozoa metabolize more slowly in the epididymis, possibly because of a reduced oxygen supply.

After long periods of sexual rest the epididymis may contain a considerable proportion of dead or senescent spermatozoa but the fate of unejaculated spermatozoa remains uncertain. Australian workers have shown that in sexually rested rams the total sperm production can be accounted for in the numbers of cells recovered from the urine.

Sperm production is under the control of the anterior pituitary gland, removal of which results in the development of infantile testes. Administration of steroid hormones can temporarily halt sperm production. However, some cases of infertility in man are treated with such hormones because when the steroid treatment is stopped there is a rebound phenomenon with a resulting stimulation of the testis by the sudden release of stored pituitary hormone.

In man and in most other mammals normal production of spermatozoa can proceed only if the testis is maintained at a temperature lower than that of the body. The changes in position of the testis which occur in response to changes in surrounding temperature are signs of the operation of these mechanisms. When the surrounding temperature is high, the scrotum relaxes; in cold weather, the scrotum contracts holding the testes closer to the body.

If, in man, the testes fail to descend into the scrotum—as they normally do in the month before birth—no spermatozoa are made. If the testis is then removed surgically to the scrotum normal sperm production usually results. There have been repeated suggestions that the dress and habits of urban civilized man tend to impair the normal mechanisms of temperature control around the testis. Indeed, it has been shown that men treated by exposure of the whole body to environmental temperatures as high as 40-47°C show reduced sperm counts from three to seven weeks after treatment. Similar findings have been recorded after raising testis temperature by local heat treatment for periods of not more than 30 minutes. Indian clinicians have recorded evidence of the damaging effects produced by a particular mode of dress in which the scrotum is strapped in close contact with the body.

In addition to temperature, the effects of altitude are worth mentioning. A 17th century author tells us that for 53 years the Spanish settlers in the mountain city of Potosi in Bolivia failed to produce children. It has been suggested that lack of oxygen rather than temperature may have been the cause.

Other agents which damage spermatozoa and reduce production include irradiation and certain radiomimetic agents which have a similar effect as X-rays. However, the usefulness of such agents for the control of male fertility are limited by undesirable side effects in people who have been treated.

There is a great variation between species in the relative contributions of the different accessory glands to the semen. The differences in glandular development are associated with variations in the volume of the semen is about 3ml—it varies from about 0.5 to 10ml—whereas a boar may ejaculate up to 500ml. Using old fashioned non-metric measures a man might be said to produce about two teaspoonfuls of semen and a boar as much as a pint.

In man the liquid semen coagulates rapidly by the action of a clotting enzyme in the prostrate on a substance produced by the seminal vesicles, but the clot liquefies again with a few minutes by the action of another prostatic enzyme. The formation of a "copulation plug" of coagulated semen is a feature of the ejaculate of many rodents. Indeed, it seems that in the rat matings are only fertile if such a copulation plug is formed inside the vagina. However, the reason why human semen should clot then liquefy is not at all clear.

The components of semen are not necessarily ejaculated as a homogeneous mixture. As long ago as 1786 John Hunter noted that the semen which constitutes the first part of the emission more closely resembled the contents of the vas deferens than did the succeeding fractions. A sperm-rich fraction can also be readily distinguished from the rest of the ejaculate in the dog and boar.

When living spermatozoa are examined in dilute suspension it is difficult to analyze their precise movements but cine pictures show that bull spermatozoa swim by propagating two dimensional waves along their tails, starting at the head end. Apparently the amplitude of the wave does not dimish as it travels along the tail and this is taken to mean that the energy needed is being fed into the tail along its whole length. One explanation of the way in which the detailed features of the sperm tail provide the technical basis for its motility is that the individual fibres contract like muscle fibres in a fixed order in space and time.

Sperm "models" may be prepared by procedures which destroy the cell membrane and remove the soluble organic and inorganic constituents and enzymes. If the extraction medium is replaced with the containing calcium ions and a source of energy such as glucose, the tails bend from side to side with about the same frequency as living ones, but the waves are not propagated along the tail and the spermatozoa do not move forwards. Apparently the extraction procedure destroys an essential part of the propagating system, which results in the ordered contraction and relaxation of the fibres. The essential part which is destroyed may be the centriole, a cell organelle better known for its role in cell division.

Dense suspensions of ram and bull spermatozoa examined by transmitted light show wave-like formations, apparently due to temporary local variations in optical density. The cause of wave motion appears to be associated with changes in orientation of spermatozoa; it may be significant that, where wave motion is conspicuous, as in bull and ram sperm, the heads of the spermatozoa are extremely flattened in the horizontal plane. It is likely that the physical changes responsible for wave motion also produce the changes in electrical impedance, which can be detected when a pair of electrodes is used to pass a small alternating current through a suspension of sperm cells. Human semen does not show changes in electrical impedance, probably because of its relatively low sperm concentration.

The question of the directional movement of spermatozoa is of some interest in considering their transport in the female tract. Spermatozoa become orientated facing upstream when placed in a shallow trough containing a moving solution, but this situation is related to purely physical laws and does not imply the existence of any sensory system in the spermatozoon itself.

Orientation of spermatozoa occurs in the female tract for quite different reasons. Spermatozoa penetrating the mucus at the neck of the womb become orientated in a direction determined by the orientation of long chain molecules of the mucus. By this means, the sperm moves in the appropriate direction when penetrating the neck of the womb. The orientation of the molecules in the mucus is determined by simple physical forces. In fact, the effect can be studied in the laboratory by allowing spermatozoa to penetrate mucus which has been drawn out into a thread; the spermatozoa follow paths parallel to the long axis of the thread. In women, the consistency of the mucus at the neck of the womb varies with the menstrual cycle. It is most fluid at mid cycle—exactly the time when conception is most likely to follow as an immediate result of intercourse.

Immune properties of spermatozoa are of particular interest as it has been suggested that they may affect fertility. Spermatozoa may act as antigens and thus be capable of stimulating the production of antibodies in the blood.

Sperm antigens may be of several kinds: they may belong only to the sperm itself or they may also be found in other tissue cells; in some cases antigens may be acquired by passive transfer from other cells. The spontaneous appearance of antibodies to spermatozoa in men has been found to be associated with infertility. These antibodies are apparently more likely to occur in men where obstruction of the ducts leading from the testes provides conditions for absorption of spermatozoa into the blood stream.

Although spermatozoa do not normally come into contact with antibodies from the blood, it has been suggested that they do meet similar antibodies in the female genital tract, and that the binding of these antibodies to the sperm may have some physiological significance. It seems that motile living spermatozoa tend not to bind antibody, in contrast to immotile spermatozoa which do so. Coating of the moribund spermatozoa with antibody may be a necessary preliminary to the removal of spermatozoa from the female tract by the scavenging white cells of the blood.

The sex determining mechanism of mammals is based upon the existence of two different kinds of spermatozoa. Half of these carry the female determining X chromosomes and half carry the male determining Y chromosome. Chromosomes in the body cells of the adult exist as pairs, one from the father and the other from the mother. The adult body cells of the human male contain 22 identical pairs of homologous chromosomes and an unequal pair—one X and one Y—of so-called sex chromosomes. In the female, there are two X chromosomes which form a homologous pair in addition to the other 22 pairs of chromosomes.

When the germ cells in the testis divide, the number of chromosomes is halved so that each egg of spermatozoon contains 22 individual chromosomes plus a single X or Y chromosome. The sex determining mechanism is based upon the existence of these two populations of spermatozoa. Half carry the X chromosome and can give rise to female offspring, and half carry the Y chromosome and can only give rise to male offspring.

As a result of these and other rearrangements of genetic material which take place during spermatogenesis the spermatozoa have a gene complement which, although derived from that of the parent cells, is only a part of it. Although many of the characteristics of sperm cells are known to be determined by the genetic constitution of the parent cells from which they are derived, there is evidence that the gene complement of the sperm itself can determine some of its characteristics. There is therefore reason to believe that it may be possible to separate X and Y carrying spermatozoa by biological of physical methods.

For instance, it has been suggested that X and Y spermatozoa are likely to differ in specific gravity, and claims have been made for successful separation of the two classes of sperm using methods based on this. However, recent experiments using fractions separated by such procedures have failed to confirm earlier claims for the separation of X and Y carrying spermatozoa.

Workers in the Soviet Union have claimed that X and Y spermatozoa have electric potentials of opposite sign and that they can be separated by applying an electric current to opposite ends of a suspension of sperm in liquid. However, it has since been shown that all spermatozoa have a net charge of the same sign, although the magnitude of the charge on the head may be different from that on the tail. When placed in an electric field the spermatozoa become orientated relative to the electrodes according to the distribution of the net charge between the head and the tail. 'Head anode' spermatozoa move at varying speeds head first towards the anode, while 'tail anode' spermatozoa have their tail orientated towards the anode but move either towards the anode or the cathode according to how fast they can move against the electric current. The relative sizes of the two populations of spermatozoa which can be so distinguished vary with the experimental conditions in a way which makes it difficult to accept earlier claims that X and Y spermatozoa have been separated by electric currents.

A further possibility is that at least some of the antigenic properties of spermatozoa may be determined by the genotype of the spermatozoon itself. For example, it has been suggested that males of blood type AB produce two different types of spermatozoa characterized by the presence of either A or B antigen. However, recent evidence suggests that this is not the case and that whether or not red cell antigens are found on spermatozoa depends on the so-called secretor status of the individual. The AB antigens of secretor men are found on spermatozoa and in the saliva, whereas those of non-secretors are invariably absent from the spermatozoa and saliva. However, the possibility remains that genes on the spermatozoa may determine some of the antigenic properties of spermatozoa and that these could be used to identify and separate the two sperm populations.

Despite many claims of success in separating the male and female spermatozoa, no one has yet succeeded in satisfying scientists as a whole that this has been done.

Insemination of mammals was first demonstrated in 1784 by the Italian biologist Lazero Spallanzani who recorded the birth of three pups to a bitch inseminated 62 days previously. John Hunter is reported to have attempted artificial insemination in man about the same time but the documentation is incomplete.

In recent years improved methods for the preservation of semen outside the body have greatly extended the use of artificial insemination in the breeding of farm animals. Semen may be preserved at normal refrigeration temperature (+4°C) for several days and after freezing to the temperature of liquid nitrogen (−196°C) or of solid carbon dioxide (−79°C) has been kept for at least 10 years. For the preservation of semen at room temperature other methods have to be used to depress metabolic activity; for example, the fertilizing capacity of boar semen can be maintained for several days by gassing the diluting medium with carbon dioxide. If semen is cooled to 4°C or below it is necessary to dilute it with an agent (usually egg yolk) to protect it against the otherwise damaging effects of cooling. If semen is frozen, glycerol or antifreeze agent is necessary to prevent the formation of damaging ice crystals.

About 60 per cent of the 4.5 million adult cattle in the UK are bred by artificial insemination (AI). It is theoretically possible to inseminate 1000 cows from a single ejaculate but in practice bulls in artificial service in the UK are used to inseminate only about 2000 cows per year. However, in the application of AI to other farm species, it has proved impossible to achieve quite the same degree of efficiency of insemination. In pigs, for instance, fertility declines when less than 2000 million spermatozoa are used.

Artificial insemination in man does not usually require the preservation of semen for more than a few hours but insemination with previously frozen semen has resulted in successful conception. Artificial insemination in man is used in two ways. AIH, artificial insemination by husband, is sometimes used where the husband has a very low fertility or is impotent. AID, artificial insemination by donor, is used when AIH is not possible or when the husband and wife prefer it because of hereditary defects carried by the husband. It has also been suggested that AID might be used in severe cases of rhesus incompatibility. Legal problems about the legitimacy of the offspring have prevented this technique from becoming as popular as it might otherwise have done but nevertheless thousands of AID children exist already in Europe and the United States. In the US as many as 7000 children may be conceived by AID annually.

Although at present human semen is seldom stored, human 'semen banks' containing sperm samples from men with many different characteristics are a real possibility. It could yet become possible for a woman to give birth to a child whose characteristics were selected before she underwent insemination.

15. *The Egg and Fertilization*

C. R. AUSTIN

"Eggs are produced at an early stage in embryonic development, but before they can be fertilized in the adult animal they must ripen and mature. Once fertilized by a sperm, an egg rapidly develops a blocking mechanism to prevent further sperms from fertilizing it." — *Science Journal.* Dr. Colin R. Austin, a native of Australia, received his graduate education at the University of Sydney. He is Charles Darwin Professor of Animal Embryology at Cambridge University. His interests are in fertilization, early developmental anomalies, electron microscopy, and research of primate reproduction.

Fertilization involves the fusion of an egg, or female germ cell, and a spermatozoon, or male germ cell. The fertilized egg then undergoes cell division and multiplication, and begins to grow and develop into an embryo. But at an early stage in embryonic development, special primordial germ cells are formed, and it is from these cells that the future eggs and spermatozoa of the adult animal derive.

The feature that characterizes these primordial germ cells in the human embryo is the ability to stain their cytoplasm with dyes that detect alkaline phosphatases; they are seen first in tissue that originates from the fertilized egg but lies outside the true body of the embryo and are thus said to have an extra-embryonic origin. Shortly, in the course of embryonic development, the cells migrate from this site into the body of the embryo and move towards the genital ridges, regions in which the future gonads, the ovary and testis, are due to develop. To begin with there may be no more than half a dozen of the primordial cells, but as they move through the tissues they multiply rapidly so that large numbers come to occupy the genital ridges. From this point on, the history of the germ cells differs according to whether they will form eggs or spermatozoa.

Cells which will form eggs gather just below the surface of the developing ovary, and their multiplication ceases shortly before birth. The germ cells now become progressively involved in a preliminary stage of the process known as meiosis, which results in a halving of the number of chromosomes so that, when an egg and a sperm fuse, the full adult number of chromosomes—and no more—is restored. As meiosis begins in the germinal cells the nuclei assume a distinctive spherical form—the germinal vesicle—and the cells at this stage are called primary oocytes.

At birth all the germ cells in the developing ovary are present as primary oocytes. They stay at this stage until shortly before the time of ovulation (the release of the

Source: Reprinted from the June 1970 issue of *Science Journal* by courtesy of Syndication International Ltd., London.

oocyte from the ovary), which for some oocytes will be reached in 12 or 13 years at puberty, but others will have to wait until later in reproductive life, even as long as 50 or more years. Once the oocyte has entered upon meiosis no further multiplication can take place, and so the human female is born with her full stock of potential eggs. The number of oocytes, nevertheless, appears considerably in excess of requirements—of the order of 500,000 to one million. If one allows for the ovulation of one oocyte in each menstrual cycle, and 12 menstrual cycles a year and 40 years of reproductive life, it is clear that the average woman will ovulate no more than about 500 oocytes. All the remaining oocytes degenerate.

Soon after the germ cells become primary oocytes they are enclosed in a single layer of cells that derive from the surface of the developing ovary and thus form the primordial follicle. No further change then takes place until puberty, after which, at various times, follicles with their contained oocytes undergo a considerable degree of growth. The oocyte swells to several hundred times its original volume, and the surrounding follicle cells multiply. Growth of oocyte and follicle proceeds initially at an equivalent rate, but at a certain point difference in growth rate becomes apparent. This is because the oocyte has reached its maximum size, while the follicle continues to enlarge mostly through the formation of a fluid-filled space—the antrum—between the cells.

The fully grown oocyte is more commonly called an egg or ovum, but there is no clear difference in meaning between these three terms, nor much uniformity in their usage. With the formation of an antrum the follicle is then called a Graafian follicle, after the anatomist de Graaf, who first recognized the follicle as the source of the embryo. The final events in the life of the follicle occur a few hours before ovulation and under the influence of two pituitary hormones acting in succession. Follicle stimulating hormone (FSH) causes an increase in size of the follicle, and luteinizing hormone (LH) brings about changes that result in ovulation.

Shortly before ovulation the events of meiosis, which were discontinued at about the time of birth, are resumed—the chromosomes become visible and the envelope of the germinal vesicle disappears. Soon the chromosomes arrange themselves on a division apparatus—the spindle—and then undergo separation into two groups, so completing the first meiotic division. As its name implies, the spindle is a small spindle-shaped body of cytoplasm in which the chromosomes are embedded. Fine thread-like 'microtubules,' which run to each end of the spindle, form attachments with the chromosomes, and by their contraction produce the movements seen in meiosis. The spindle next rotates through 90°, and at the same time a cleft runs through the immediately adjacent cytoplasm in such a way as to cut off a small portion which becomes the first polar body. The group of chromosomes at one pole of the spindle are included in the polar body. The remaining chromosomes left within the oocyte promptly become involved in the start of the second meiotic division, and they stay in this stage during the process of ovulation and for a number of hours after the arrival of the oocyte in the oviduct. Further progress of the second meiotic division occurs at the start of fertilization. [See Figs. 1 and 2]

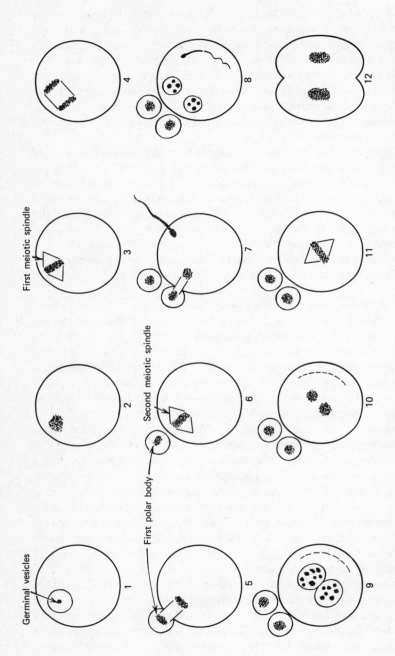

Germinal vesicles

First meiotic spindle

Second meiotic spindle

First polar body

Figs. 1 and 2. Transformation of the primary oocyte (above). The envelope of the germinal vesicule disappears shortly before ovulation, and the chromosomes then arrange themselves on the spindle and separate into two groups, so completing the first meiotic division. The chromosomes remaining in the oocyte promptly become involved in the start of the second meiotic division, which is not completed until fertilization. At this stage, as a result of the two meiotic divisions, both the egg and the sperm contain only half of the normal number of chromosomes (the haploid number), so that when they fuse the full

....
(continued on next page)

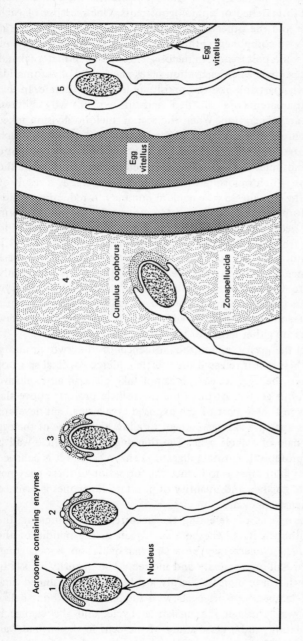

diploid number of chromosomes is restored. The final diagram shows the first division of the fertilized egg. Before a sperm can fertilize an egg, it must undergo a structural change called the acrosome reaction (below). When the sperm leaves the male genital tract it is not capable of penetrating the zona pellucida and gelatinous matrix surrounding the egg. But while it is in the female tract, it undergoes a physiological change—capacitation—by which small perforations are produced in the acrosome wall for release of enzymes which enable the sperm to digest a path into the egg.

When the chromosomes become visible at the beginning of the first meiotic division they are seen to be arranged in matched or homologous pairs. One member of each pair originates from the father and the other from the mother. Each chromosome at this time is clearly made up of two components—the chromatids—so that in fact sets of four chromatids are seen. At some point early in meiosis, chromatids of maternal and paternal origin exchange parts, so that when separation takes place pairs of chromatids of a new composition are drawn apart and pass to opposite poles of the spindle. In this way, the heridary factors of the parents are 'shuffled' and this explains why children differ from their parents in each generation. When the second meiotic division takes place at fertilization, the chromatids of each chromosome are separated, each chromatid being then recognized as a chromosome. As a result of the two meiotic divisions the number of chromosomes in the ovum is reduced to half. Thus, while the normal human tissue cells have 46 chromosomes (the diploid number), as a result of meiosis the oocyte will come to contain 23 (the haploid number). A reduction of this kind is also involved in the formation of the spermatozoon—as the individual sperm cells are called—and is clearly necessary if the diploid number is to be re-established in the individual that derives from the fusion of two germ cells. [See Fig. 3.]

The course of meiosis is not always smooth. Errors sometimes occur, and these usually have special significance for embryonic development. The errors commonly take one of two forms—failure of a complete meiotic division, leading to an embryo with three sets of chromosomes (triploidy) instead of two, or failure of separation of a pair of chromosomes, which results in an embryo with one too many or one too few chromosomes. Both conditions are either lethal, or lead to abnormal embryos.

The egg is released from the ovary under the influence of the two pituitary hormones FSH and LH. Just before it is released the follicle undergoes a final spurt of growth and then opens to free the egg. Release does not take place in an explosive fashion, as once was thought, but gently. At this time the follicle projects above the surface of the ovary, and in a restricted part of the exposed area a small slit develops which slowly widens until the aperture is large enough to allow the escape of the egg together with a surrounding halo of follicle cells. The thinning of the follicle wall is thought to be attributable to proteolytic (protein digesting) enzymes, and it is believed that the action of the pituitary hormones is to induce the formation of these enzymes in the follicle. The appropriate balance and quantity of pituitary hormones is reached about half-way through the menstrual cycle.

As yet we have no reliable means of detecting the occurrence or estimating the precise time of ovulation, although it is believed that a pain in the middle of the menstrual cycle known as *mittelschmerz* may be a sign that ovulation is occurring. Objective signs of ovulation are all approximate and indirect. The quantity of LH in the circulating blood has been shown to increase sharply just before the presumed time of ovulation; the basal body temperature often shows a sudden rise of about half a degree centigrade about half-way through the menstrual cycle, and the change is thought to coincide with ovulation; and a close study of the cells in the vaginal wall shows an increase in the number with deeply staining nuclei at about the same time.

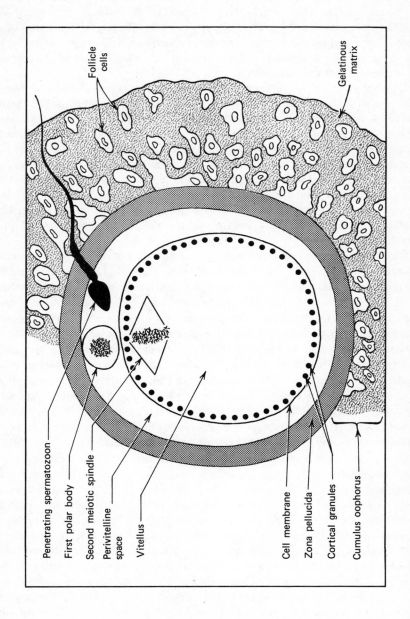

Fig. 3. Egg is about to undergo fertilization. The chromosomes on the second meiotic spindle will separate, and half will be emitted in the second polar body, leaving the ovum with only half the normal number. The sperm also contains this number of chromosomes, so that when germ cells fuse, full diploid number of chromosomes is restored.

Follicle cells

Gelatinous matrix

Penetrating spermatozoon

First polar body

Second meiotic spindle

Perivitelline space

Vitellus

Cell membrane

Zona pellucida

Cortical granules

Cumulus oophorus

When the egg with its surrounding follicle cells emerges from the ruptured follicle it is caught up in a stream produced in the peritoneal fluid by the beating motion of fine hair-like structures—the cilia—which line the inside of the oviduct. The egg is thus wafted into the opening of the oviduct and passes along it largely by the action of the cilia, aided by muscular contractions of the oviduct wall. The egg comes to rest temporarily in the upper one-third of the oviduct, a region known as the ampulla, which represents the site of fertilization.

Accidents may take place during any of the several stages just considered. The egg may fail to leave the follicle, or, on leaving the follicle, it may fail to reach the opening of the oviduct and become transferred instead to some part of the peritoneal cavity. Such oocytes are unlikely to be fertilized, but spermatozoa are able to traverse the oviduct and find their way to the follicle or to other parts of the peritoneal cavity, and so very occasionally they do fertilize eggs in abnormal locations. In this way ectopic pregnancy may originate in the body cavity or in the ovary itself. Alternatively, these conditions may arise through eggs fertilized in the ampulla in the normal way later passing back into the peritoneal cavity. Ectopic pregnancy rarely continues for long, and is often terminated surgically because of the risk of haemorrhage and other complications.

Ovulation can be provoked artificially by the injection of appropriate hormones. Initially extracts of human pituitary gland containing FSH and LH were used, but this was far from convenient and other sources of equivalent hormonal activity were sought. It was found that the urine of women who had passed the menopause contained a hormone (called HMG) having an action similar to FSH, and the urine of pregnant women another hormone (HCG) with an action similar to that of LH. An effective treatment consists of injecting HMG for one or more days in the first half of the menstrual cycle to stimulate follicle growth, followed by one or two injections of HCG with the object of precipitating ovulation. The procedure has in fact proved successful in a number of cases of infertility due to failure of ovulation, but results vary a good deal, and sometimes the response exceeds expectations. It is in this way that some of the cases of quintuplets and sextuplets reported in the newspapers came about. It is now becoming more common first to test the sensitivity of the patient to various doses of hormone by chemical means and grade the treatment accordingly.

At intercourse spermatozoa, suspended in semen, are deposited in the upper reaches of the vagina and appear to make their way more or less under their own power—by movements of their tails—into the neck or cervix of the uterus. Conditions within the cervix are especially favourable to spermatozoa and some people regard this region as a kind of reservoir from which spermatozoa continue to pass up the tract for a number of hours after coitus. Passage of spermatozoa through the uterus and oviduct is brought about chiefly by the action of vigorous contractions in the walls of these organs. The result is a mixing of the contents of the organs, and suspended objects such as spermatozoa are soon carried to all parts of each organ. Additionally, the currents and counter currents produced by the beating movements of the cilia may well aid the spermatozoa in their journey. Thus, some spermatozoa reach the site of fertilization in the ampulla of the oviduct, and judging from observations that have been made in several species other than man the time of transport is remarkably rapid: in many instances they reach their goal in a bare five minutes from the time of deposition in the vagina.

During the passage through the genital tract, the spermatozoa undergo a considerable reduction in number, partly by their being diluted in the tract secretions and partly by the sheer inefficiency, as it were, of the transport mechanisms. From among the hundreds of millions of sperms that are usually deposited in the vagina, there may be only a few thousand that succeed in entering the oviduct and perhaps only a few dozen that actually come within reach of the egg. Perhaps one should indeed consider it a function of the female genital tract to prevent too many spermatozoa reaching the site of fertilization—the advantage of the small numbers being simply that the prospects of penetration of excessive numbers of sperm into the eggs, known as polyspermy, would be kept at a low level.

The egg consists of a spherical mass of cytoplasm which is referred to as the vitellus, and this contains the nucleus and various other important bodies such as the mitochondria, ribosomes, Golgi components and cytomembranes. The vitallus is limited by a cell membrane and surrounded by a moderately thick, transparent coat called the zona pellucida, which is separated from it by the fluid-filled perivitelline space. About the zona pellucida lie follicle cells which accompany the egg from the follicle, embedded in a gelatinous matrix which binds the cells to the egg. The mass of follicle cells and matrix are collectively referred to as the cumulus oophorus.

In order, therefore, to bring about fertilization the spermatozoon must penetrate through the cumulus as well as the zona pellucida before it can reach the egg surface, and it has long been a matter of debate as to how it achieves this feat. An early clue was provided by the fact that an enzyme capable of digesting the cumulus could be extracted from spermatozoa. Later several enzymes were identified and there was evidence that they were carried in a small sac-like container—the acrosome—covering part of the sperm nucleus. But it was not clear how spermatozoa were able to release the enzymes at the appropriate moment. The explanation that has emerged is in fact quite a complex one. From an extensive range of experiments it became known that the spermatozoon on leaving the male tract is not immediately capable of fertilizing an egg: it must first undergo a physiological change called capacitation, which normally takes place while the spermatozoon is in the female tract. Capacitation seems to involve a kind of destabilization, achieved perhaps by removal of a protective coating. The spermatozoon is then able to undergo a structural change, the acrosome reaction, by which small perforations are produced in the acrosome wall permitting escape of the enzymes. Thus, the spermatozoon is now in a position to digest a path for itself through the cumulus and the zona pellucida surrounding the egg.

Passage of the spermatozoon through the zona pellucida is peculiar in that it always follows a curved path. The reasons are unknown, but the result is that the spermatozoon projects into the perivitelline space and comes to rest with its head flat on the surface of the vitellus. Generally at this stage the tail is still protruding through the slit that is left in the zona pellucida, and it may still be actively waving. Soon after the spermatozoon head has reached the vitellus it becomes firmly attached to the surface and then a process of fusion follows. This involves the union of the cell membranes of spermatozoon and egg in such a way that two cells come to be enclosed within the same envelope.

Thus, a single cell is formed from the two gametes, although at this point they still look as if they are independent bodies. Shortly, however, the spermatozoon nucleus moves farther into the egg and enlarges to form a distinctive structure known as the male pronucleus. Entry of the sperm tail into the egg progressively ensues, but this is unlikely to play any important role in the economy of the embryo. Although the motility of the spermatozoon seems unlikely to be responsible for its passage through the female tract it probably aids penetration of the egg.

The egg reacts quite rapidly to sperm contact. Among the earliest events are changes in the zona pellucida and the surface of the vitellus which tend to prevent the entry of more than one spermatozoon. Neither change is perfect, however, so that sometimes two or more spermatozoa do participate in fertilization. When such polyspermy occurs it dooms the embryo to an early death. The mechanism whereby the two changes which block polyspermy take place is as yet quite unknown, but may possibly be related to another structural change stimulated by sperm penetration—the evacuation of a number of small bodies known as cortical granules which lie just under the surface of the vitellus. The precise function of the cortical granules remains a mystery; it has been suggested that the material they release passes across the perivitelline space and is responsible for inducing the zona reaction. A fourth change involves the resumption of the second meiotic division, which ends with the formation of the second polar body. The group of chromosomes now remaining within the egg develop into the female pronucleus.

The most obvious events during fertilization involve formation and growth of the male and female pronuclei. They develop synchronously. Nuclear envelopes become evident about the swollen sperm head and the egg chromosomes, and with further enlargement small spheres—the nucleoli—make their appearance within each pronucleus. Pronuclei grow considerably in size and, as they do so, the number of size of the nucleoli also increase, being the only structure now visible within each pronucleus. When they have reached their full dimensions the two pronuclei come together into intimate contact in the centre of the egg, and a kind of reversal of events occurs: the pronuclei shrink, the number of nucleoli diminishes, the nuclear envelopes gradually fade out and finally the chromosomes reappear. The chromosomes then arrange themselves at the centre of another spindle—the first cleavage spindle.

At this point fertilization can be regarded as complete. It has taken about 12 hours from the time of sperm entry, and the hereditary material donated by the spermatozoon and the egg have now become united. The egg now is often referred to as the zygote. The next step will involve division of the zygote into two equal parts each containing an equal share of the combined heriditary material. Thus embryonic development begins with the formation of the two cell embryo. After a few hours of existence the first two cells divide so that a four cell embryo is produced. Further division leads eventually to the formation of a hollow sphere—the blastocyst.

Soon after the blastocyst stage has been reached, the inner cell mass (a mass of larger cells situated at one pole of the blastocyst) begins to assume the form of the foetus which will eventually become the child itself, and it is at about the start of this phase that one sometimes sees an unexpected change. Instead of developing into a single new individual, the inner cell mass or early foetus splits into two distinct halves, and each of these then reorganizes itself and grows into a foetus of normal size and form. In this way true or 'identical' twins come into being. The process is quite different from the mode of origin of unlike or fraternal twins, which involves the simultaneous fertilization of two eggs and development of two quite separate embryos. Similarly, identical triplets or quadruplets arise through the subdivision of the early foetus into three or four portions, while the fraternal counterparts derive from a corresponding number of eggs. So-called 'Siamese' twins born joined together represent the result of incomplete separation of the portions of the original dividing cell mass. Identical twins are necessarily of the same sex and resemble each other very closely indeed, whereas fraternal twins can be of opposite sex and can differ just as much in other ways (except age) as do brothers and sisters born at different times. Twinning is rather uncommon, occurring in one out of 88 births, but fraternal twins are more likely than identical twins which are encountered in only one out of about 300 births. The figures vary somewhat with race—in Japan, for example, the frequency of twins of both kinds is about one in 155 births. The frequency of triplets of either kind in white races is of the order of one in 88 x 88 (about one in 8000) births, and larger 'litters' are correspondingly rarer. One would expect identical quintuplets, such as the Dionne girls, once in roughly 8000 million births.

Twinning shows a slight tendency to run in families, and so can be considered heritable to a small degree. This seems to apply to both kinds of twins, and there is also a greater likelihood of triplets occurring in families with a history of twins than in those without. Fraternal twinning also becomes a little more common in mothers approaching the late 30s, after which it is less frequent.

At fertilization equal portions of genetic material—the deoxyribonucleic acid (DNA)—are contributed by egg and spermatozoon, but in each case the amount provided by these haploid cells is equivalent only to half that of the normal (diploid) body cell. And so, if there were no further change, the zygote would come to contain only the same amount of DNA as a normal body cell and, when it divided, the two cells of the early embryo would each receive only half the amount of DNA. This state of affairs is avoided by the synthesis of DNA during the course of fertilization—an important function of the pronuclei. The amount of DNA is in fact doubled, so that each cell of the two cell embryo receives the same amount of DNA as in a normal body cell.

The fertile life of the egg is comparatively brief. Precise estimates are difficult, especially in the human subject where the time of ovulation is uncertain. There are indications that in monkeys the eggs remain capable of undergoing fertilization for nearly 24 hours, and this may be true also for the human egg. However, there is another point to be considered here, namely the consequence of egg ageing, since a good deal of evidence in experimental animals indicates that the egg deteriorates progressively, and though remaining fertilizable loses the capacity to give rise to a normal embryo. Anomalous embryos usually die during the course of pregnancy, but if surviving to birth they could develop as physically or mentally abnormal children. It has been suggested that the increased frequency of abnormal children born to older mothers may be in part caused by the fertilization of old stale eggs which undergo meiosis abnormally with loss of chromosomes.

The eggs of several mammalian species, including man, can be fertilized outside the body and under controlled conditions in the laboratory. The advantage of this manoeuvre is considerable, for the whole process of fertilization can be watched through a microscope, and effects of various experimental agents can readily be tested. To obtain the fertilization of mammalian eggs *in vitro* requires rather special conditions, but these have now been defined for man as well as for the rabbit, mouse, hamster and cow. Furthermore, it is possible also to recover human oocytes from ovarian follicles at a stage before maturation has begun, and place these under appropriate laboratory conditions; after some hours maturation proceeds. In addition, it has proved practicable, in mouse and man, to obtain some early stages of embryonic development under laboratory conditions.

These accomplishments are encouraging, but we have of course a very long way to go before anything resembling the production of 'test-tube babies' can be achieved. For one thing the process of implantation of the embryo in the uterine tissues presents a considerable difficulty, and as yet we really have no idea how this can be mimicked under laboratory conditions. There are also several problems yet to be tidied up at earlier stages. Thus, although we can successfully persuade eggs to undergo maturation in the laboratory, experimental results suggest that, after fertilization, these eggs are not capable in many instances of giving rise to normal embryos. It would seem that, though the conditions are appropriate for the successive steps of meiosis and polar body extrusion, some abnormality is inadvertently introduced, and much more research is needed before this kind of work can be applied in man.

Fertilization may go awry in several ways and this appears to be due mainly to ageing of the egg before sperm penetration. One of the commonest anomalies associated with the fertilization of ageing eggs is the occurrence of polyspermy. In polyspermy two male pronuclei are formed (nearly always two spermatozoa rather than a larger number are involved) and both develop in an apparently normal way, so that at the end of fertilization three groups of chromosomes, the third being those of the egg, gather together on the first cleavage spindle. Thus polyspermy results in triploidy—a common cause of miscarriage during early pregnancy.

Another anomaly involves rather a curious modification. The second meiotic spindle moves inwards from the surface of the egg so as to take up a position more or less in the centre. When the spindle divides, the whole egg cleaves into two more or less equal parts instead of forming the second polar body. In a sense a polar body is formed, but it has the same size and appearance as the rest of the egg. Both products of division can be fertilized, and if a spermatozoon enters each half extensive development is quite possible. Indeed there seems to be no reason why such an embryo should not continue development in a normal way, though it would be distinguished by the fact that it would consist of two equal populations of cells with different sets of chromosomes, and therefore different hereditary characters. The condition is known as mosaicism. The embryo could develop to birth and the new individual possibly pass unnoticed, unless the two sets of hereditary features clearly differed from each other. Of course, if one of the fertilizing spermatozoa carried an X chromosome and the other a Y, the product would be a mixture of both sexes, a gynandromorph. Just how frequently events follow the course just described is hard to say, but there are records in the medical literature of patients showing mosaicism, including forms of gynandromorphism.

Usually, of course, the egg fails to be fertilized at all. In experimental animals, under certain circumstances, unfertilized eggs have been observed occasionally to take it upon themselves to begin development—a process known as parthenogenesis. It can be induced in rabbits by quite simple means, such as subjecting the eggs to a cold shock, and parthenogenetic rabbit embryos may develop to the stage of implantation, but as yet there is no sound evidence that any go as far as birth. Lack of full development is thought to be attributable to the fact that in the comparatively large complement of genetic factors that exists in mammaliam chromosomes there are almost inevitably several lethal genes. When development follows fertilization, the lethal genes from one parent can be dominated by normal genes from the other parent. In parthenogenesis, however, the lethal gene is dominant and consequently its lethal effect is expressed. Nevertheless, there is a remote possibility that the genotype of an egg could be lacking in lethal genes, and therefore that parthenogenesis could go to full development. The human population, by virtue of its enormous size, could indeed contain a few parthenogenones. These would be very difficult to distinguish from normal individuals: they would of course be female and they would resemble their mothers very closely but otherwise they need show no distinctive features.

16. The Obsolescent Mother (A Scenario)

EDWARD GROSSMAN

> " 'Did you know that a woman can now have children without a man?'
> " 'But what on earth for?'
> " 'You can apply ice to a woman's ovaries, for instance. She can have a child. Men are no longer necessary for humanity.'
> "At once Ella laughs, and with confidence. 'But what woman in her senses would want ice applied to her ovaries instead of a man?' "
>
> Doris Lessing, *The Golden Notebook*

In a German woodcut from the sixteenth century, a woman is giving birth. She squats on a birth-stool, something that looks like a portable three-legged toilet, various models of which turn up in mosaics, drawings, and paintings all the way from Rome until the early 1800s in America. This German woman (or rather girl: the flesh of her placid face is rounder and firmer than the saggy jowls on the midwives who attend her) grasps with her left hand the bottom of the stool while with the right she presses down on the swelling under her clothes. She is fully dressed, like the others in the picture, one midwife (perhaps the assistant) standing behind her, holding her up by the armpits, while the senior hag sits on a very low chair facing the girl's spread legs, her arms thrust up the girl's voluminous skirts, giving the artist a chance to show off his technique with light and shadow on folds, and exposing the girl's feet.

These feet—they are bare, muscular, contorted. They are in contrast with the rest of the picture, which is stylized, with little except formal lines of tension. They are just about the only sign that in real life the scene was not serene, but hectic, noisy, painful, with sweat and juices running, screams, moans, and curses, and in the air a possibility of death. If this girl died in labor, or her child from infection in the first few days of life, it was as unremarkable as the father of her child being slaughtered in the wars of religion.

Other animals don't seem to feel pain or run much risk in giving birth: their bodies take on a casual attitude, the mother doesn't need help. The human mother is an exceptional case that it took the whole progress of evolution to produce. For almost as long as Man has been human, birth has been a big event, a disturbance, and more or less a shock, so much so that at least one mythology has accounted for the pain of it as punishment for an original sin against Almighty God. Considering how long Man has been human—defined biologically or mythologically—it is just yesterday that childbirth began to be mitigated. For thousands of years the experience of women was

Source: Reprinted by permission of International Famous Agency. Copyright © 1971 by Edward Grossman. From the *Atlantic*, May 1971.

pretty much that of the German girl. It was against the law to dissect, and no one understood the reproductive organs. Obstetrical tools were scarce and clumsy, and midwives relied on the cunning of their (dirty) hands. There was no pain-killer, so women went through the thing wide-awake, unless they were lucky enough to faint. An ordeal, and one which thoughtful men must have been happy to be spared.

Yet the transformation of childbirth from painful and dangerous event to safe and efficient routine was thanks to men, to the energy of males. The beginning of this transformation dates only from a little before the time when drawings and paintings start to show men, instead of women, working around the swollen body of the woman in labor. What part of this male energy, channeled into ingeniousness, derived from a curiosity separate from feeling, what part from compassion or guilt? Leonardo's drawings of the fetus may well be just an expression of his need to find nature out and record it; yet without such anatomical study, childbirth could not have changed. Peter Chamberlen, inventor of the forceps in the year Shakespeare published *Love's Labour's Lost,* kept the details of his simple and great invention secret, haggling for a price with his colleague-rivals while women suffered. In contrast to him, there have probably been any number of doctors of whom we could think that their sympathy for women was at least as active as their desire to get rich or be famous. Maybe it is true that the motives of most of the doctors were mixed, and that only the humble practitioners were pure.

In 1842, James Young Simpson, professor of midwifery at the University of Edinburgh, at the age of thirty became the first to use anesthesia to help women in labor. Before he got any honor for it, he had to contend with the gloomy abuse and doubt of the people who ran the medical and religious Establishments—that is, other men. Ministers, for example, objected to Simpson because:

"Pain during operations is, in the majority of cases, even desirable! Its prevention or annihilation is, for the most part, hazardous to the patient. In the lying-in chamber, nothing is more true than this: pain is the mother's safety, its absence her destruction. Yet, there are those bold enough to administer vapor of ether, even at this critical juncture, forgetting it has been ordered, that 'in sorrow shall she bring forth.' "

This combination of bad moralizing and bogus medicine was not considered stupid or cruel; it sounded convincing to most people, even women. Some of the letters Dr. Simpson got from his colleagues were better informed, but even agnostic medical men were apt to believe that (a woman's) pain was somehow necessary, without bothering to prove it. While women couldn't have begun to be liberated from the special pains and dangers of their biology without the help of certain men, there have always been, at each "critical juncture," other men opposing this liberation, or at least the technical means of achieving it.

Simpson's answers to the reactionaries and the skeptics radiate good sense and also concern for the suffering of women. He has to remind a Dr. Meigs of Philadelphia, one of his critics, that the fact is "the contractions of the uterus, and not pain, is the essential to the progress of labor." But aside from the medical facts, he is obliged to go in for some moralizing, too:

"Like other physicians you deem it, I doubt not, your duty to wield the powers of your art, in order to free those that submit themselves to your medical care, from these and from other similar sufferings. But if it is right for you to relieve and remove these pains, why is is not right for you also to relieve and remove the pains accompanying the act of parturition? I cannot see on what principle of philosophy, or morality, or humanity, a physician should consider it his duty to alleviate and abolish, when possible, the many minor pains to which his patients are subject, and yet should consider it improper to alleviate and abolish, when possible, pains of so aggravated a character, that, in your own language, they are 'absolutely indescribable and comparable to no other pains,' 'Pains for which there is no other name but Agony.' "

Even Simpson was not entirely altruistic, however. During the time it took for pain-killers to be accepted by the medical Establishment and acquiesced in by the Church (about twelve years, until Queen Victoria's much publicized labor under chloroform), he fretted, not only about the women who continued to go through childbirth cold-sober but also about his professional reputation and whether he would get the glory coming to him. Innovators, great benefactors of women like Simpson, have generally had complicated feelings, their sympathy being sharpened by ambition. Only women, once they have realized the meaning for themselves of these innovations, have displayed uncomplicated enthusiasm, which finally proves more effective than anything else in overcoming tradition. The first woman on whom Simpson tried chloroform, in 1848 (until that time he had been making do with ether), was so grateful, and perhaps unconsciously clear about the improved future of her half of the human race, that she had her girl-child christened "Anesthesia."

Other men made further advances: antisepsis in surgery, antibiotics, refinements in diagnosis and delivery. Because of this accumulation of knowledge, to have a child in an industrial country today is no longer dangerous. The process may be bothersome and uncomfortable for many, but very few women die in childbirth anymore. However, despite this, it is still not possible to say that we take birth as casually as animals do. A tinge of mystery remains; the thing still seems formidable, perhaps least to obstetricians, more to women, and most to men.

As long as men have had the artifice to represent, they have made the image of the beatific Madonna and Child, an image that speaks to some profound need, love, guilt, fear, or reverence, or all of these together subsumed under awe. Today, medical knowledge has grown, and conception and childbearing are apparently less mysterious than they once were. It is understood that the egg when it bursts from its follicle in the ovary is about as big as the period at the end of this sentence; that if it happens to be fertilized during its journey down the fallopian tube (in Latin, *oviduct*), the genetic material from it and the sperm combine and rearrange, and the new organism (an embryo) divides, and divides again, and again, until it arrives in the uterus looking something like a segmented soccer ball, and implants itself into the mucus lining of the uterus, where about two weeks after fertilization there begins to arise between the embryo and its mother one of the most complicated structures in nautre: the placenta. This is an entire environment, a universe for the embryo alone, separating it from and connecting it to its mother by membranes so fine that blood cells pass through undamaged, bringing the embryo nourishment and carrying away poisons. For the next eight months, so long as she eats, breathes, and excretes for herself, the mother eats, breathes, and excretes for the embryo, then fetus, which develops from a plasm with creases, to a mollusk, to a fish, to a pig, and attains the morphology (Greek, *morphe*, "form") of a human being, with two clear and open jelly eyes which rapidly move in movements associated with dreaming. Then, in a normal pregnancy, about nine months after conception, the gigantically stretched uterus begins to contract and expel, and birth takes place.

All this is known today, yet the process of childbearing, with its final event as if something were coming to inexorable term, still has about it a sense of prehistory, savage and elemental, even though it is surrounded by rubber gloves and stainless steel. It is a spectacle that impresses the civilized no less than the savage mind as awesome, and together with the other striking biological events associated with a woman's body, may lead a man like Sigmund Freud to write, "Anatomy is destiny." However, this epigram, with its numerous social, sexual, economic, and political implications, has had its portion of incontrovertible truth reduced in the years since it was written, and again, this has been thanks to men. The female's circular, periodic, excitable, "destined" biology has been brought closer to the linear biology of the male as a result of new knowledge of the chemistry of sexual differentiation and functioning, and the technology that this knowledge has made possible—above all, the Pill.

But a stubborn remnant of biological fact and cultural myth, that men and women alike are affected by, persists. So long as we reproduce ourselves, we also reproduce the spectacle of a woman withdrawing into herself, becoming huge, and in blood and tumult bringing forth the succeeding generation.

This is the stuff myths are made of, customs (such as the French custom of kissing a lady's hand, which originated not as a compliment to her but a symbolic gesture of gratitude to all women together for what they endure in childbirth) and practice (such as the practice of the mother, rather than the father, caring for children, because for nine months it was impossible to decide whether she and the child constituted two organisms or one).

Even the mother may experience this mystic awesomeness, so strange and somehow at odds with the present and the future. Yet it is only the remnant of a myth, and technology, which has gone part of the way toward destroying it, may yet destroy the rest.

II

According to the Nobel Prize Committee, the great advances in the science of biology in the years from the middle fifties to the middle sixties involved advances in knowledge of the genetic code. The most famous hypothesis put forward and confirmed was Crick and Watson's model of DNA, but also of importance was later work, identifying, under tremendous magnification, individual genes. Biologists say that with the cracking of the genetic code and the visualization of the smallest unit in genetics, a period of intense exploration and significant discovery has come to an end; the work remaining is not speculative, but by way of filling in and reconfirming. There is a sense, among younger men, that the "excitement" has gone out. Many have shifted their attention to the closely related field of embryology, which is rich in tantalizing problems. For example, the knowledge of how the basic genetic material reproduces itself has been helpful to the embryologist, but he still does not know exactly how the genes themselves dictate the orderly development of a fertilized egg into a complicated and highly differentiated multicelled organism: what makes an eye an eye, and an arm not an eye, and what puts together limbs and organs so as to make an individual creature? Not only are such questions of "pure" interest, but since they go to the center of the reproductive process, the answers to them promise to have practical application to problems of genetic defect and birth control, two worldly matters much on the minds of scientists and nonscientists alike.

The research of the last few years in embryology, or molecular and reproductive biology, has already yielded a number of interesting hypotheses, as well as laboratory experiments, such as the one called "cloning." Theoretically, since every cell of an animal carries a load of genetic information unique to that individual, it ought to be possible, by destroying (with radiation) the female complement of chromosomes in the nucleus of an unfertilized frog cell, say, and implanting in its place an entire nucleus lifted out from another cell of a second frog, to fertilize the first cell and fabricate an embryo which would grow, be born, and mature into an adult frog that would not merely resemble the donor frog but would be its absolute replica. In cloning, this theory has living proof. A whole crowd (or *clone*; Greek, "a throng") of identical creatures has been bred in the laboratory, literally and predictably identical in every way to the donor frog except that some of the copies are older than others. Among other things, this shows that an individual creature, previously unique and mortal, may be rid of its uniqueness and have immortality conferred on it. The experiment is dazzling, and some of the knowledge gained ("spinoff") may eventually have practical human application, but so far the major interest has been "pure": no one suggests that humans can be immortal, or that there is any good reason why they should be. Much the same thing can be said of experiments in parthenogenesis, or the asexual fertilization, by mechanical means, of mammalian eggs. The first reports of such "immaculate conception" in lower animals in the laboratory came many years ago, when the French biologist Jacques Loeb caused "traumatic parthenogenesis" by touching a sea urchin egg with dry ice, but recently the accuracy of the method has been improved, and it has been made to work on mammals such as rabbits. Again, however, the direct human application is not apparent for many reasons, among them that human eggs fertilized by parthenogenesis always deteriorate and die; that because of the chromosome business, only females are conceived; and that, taking a long view, parthenogenesis is undesirable because the machinery of adaptive evolution requires sexual reproduction.

There are many embryologists at work in many laboratories, and much going on besides cloning and parthenogenesis. Some of it focuses right on the human egg, sperm, and embryo. This work has closer applicability to human affairs; indeed, the money that pays for it comes mostly from funds allotted not for "pure" research, but for applied—specifically, research to be applied to real and actual problems of contraception, fertility, sterility, and birth defects. To gain knowledge which would lead to new and better technology and social programs in these areas of daily concern, embryologists have needed to observe, measure, and experiment with the first phases of human reproduction, which they have long suspected are crucial in many respects (such as genetic defect), but which they have had mostly to speculate about, since the events were hidden in the female body. Now, these first phases, including culturing of the human egg, its fertilization by sperm, and development as an embryo, have been carried out under the microscope, or in the scientific terminology, *in vitro,* "under glass."

There were an unusual number of obstacles that had to be overcome before this was accomplished. The first and by no means least was that human eggs suitable for experimentation were hard to get. While there has never been a problem getting sperm (the current rate of remuneration for a sample by masturbation is $25), eggs could only be obtained, until very recently, either from fresh cadavers or from ovarian tissue cut out during gynecological surgery and given by the surgeon to the embryologist as a favor. Both methods were unsatisfactory, yielding few and often damaged, stale, or immature eggs. It was much easier to get eggs from mice, rabbits, and hamsters, which could simply be killed at the right time in the egg-maturing cycle, and their eggs picked out of the follicles of the ovary or flushed from the oviduct. The first experiment of this kind was conducted by the German Schenk, who in 1878 put some rabbit eggs in a culture dish and added sperm; nothing happened, however. For about fifty years a small number of biologists in scattered laboratories fitfully persisted in trying to fertilize mammalian eggs outside the body. In 1934 an American, Dr. Gregory Pincus, later one of the developers of the Pill, published findings that suggested he might have succeeded in fertilizing rabbit eggs. In 1940, Dr. John Rock, another American, who was to play an important role in legalizing and propagandizing birth control, said that he had put some human eggs, which he had managed to get, into the presence of sperm, and that a small number of the eggs had been fertilized and had actually divided. Ten years later, Landrum Shettles, an obstetrician at Columbia-Presbyterian Hospital in New York, also claimed to have gotten fertilization and growth of the human egg *in vitro*, this time to a stage where the embryo was a solid mass of cells.

There was considerable skepticism, however, about the validity of Pincus', Rock's, and Shettles' results. One of their fellow scientists who was most skeptical, and offered the most persuasive criticism, was M.C. Chang of the Worcester Foundation for Experimental Biology in Shrewsbury, Massachusetts. Chang said that if you take an egg out of the body and just leave it alone in a culture dish, as likely as not it will show signs of "dividing"—signs that will inevitably prove misleading, for in fact, the egg will be deteriorating and dying. The events described by Pincus, Rock, and Shettles, Chang more than implied, could just as well have taken place without the presence of sperm, and therefore, the case for fertilization was unconvincing. Furthermore, Chang reminded biologists that the very definition of fertilization had still not been agreed on (this was in the mid-fifties).He advanced the view, which has since been accepted by everyone, that fertilization, rather than being a single event which happens instantaneously (the mythical "moment of conception"), is a process which takes place over several hours. Fertilization, in other words, is not the simple penetration of the sperm into the outermost layer of the egg; rather, it includes a whole series of events and reactions which probably cannot be said to be complete until the division (or cleavage) of the egg into two cells, each carrying a load of maternal and paternal chromosomes in its nucleus.

Chang now set out to make a thorough study of the process of fertilization in mammals, with the goal of eventually devising and carrying out an experiment whose meaning no one could dispute. His diligent work during the fifties rescued this sector of embryology from the realm of the exotic. One of Chang's first and most important discoveries while working with rabbits was that the sperm had to have something happen to it while it was in the female reproductive tract before it would penetrate an egg cell: the sperm had to be "capacitated." The precise mechanism of this physiological change in sperm is even now not completely understood, but exploiting what he had learned, and after extensive investigation in timing and mapping egg development, Chang was finally able, in 1959, to fertilize a mammalian egg outside the body, using "capacitated" sperm recovered from the uterus of a female rabbit killed soon after coitus. Chang's evidence was beyond dispute, because he took the egg he had fertilized and implanted it in the uterus of another rabbit, certified not pregnant, which was segregated from males and in due course gave birth.

In the next few years Chang's experiment was successfully repeated by at least two other research teams. The reports quickened activity in *in vitro* fertilization in several countries. In 1961 a physiologist at the University of Bologna, Dr. Daniele Petrucci, said that he had fertilized a human egg *in vitro*, cultured the embryo for twenty-nine days ("a heartbeat was discernible"), and then destroyed it because "it became deformed and enlarged—a monstrosity." He said that as a womb substitute he had used a silicone container filled with amniotic fluid (liquid material that separates the growing embryo from the innermost membrane of the natural placenta), which he extracted from pregnant women. Petrucci, who professed himself a good Catholic, told the Italian newspapermen that his aim was just to find a way to culture organs that would resist the rejection phenomenon when transplanted. A couple of days after the story was printed, the Vatican's *L'Osservatore Romano* ran an editorial which read in part, "God surrounded the act of creation of a human being with the most supreme assistances of love, nature, and conscience. It would be monstrous to violate these conditions," which suggested to some that certain pressures had been put on Petrucci to stop. Another kind of reaction came in an editorial in *Jenmin Jin Pao*, the paper of

"These are achievements of extreme importance, which have opened up bright perspectives for similar research. . . . Nine months of pregnancy is no light or easy burden and such diseases as poisoning due to pregnancy are detrimental to health. If children can be had without being borne, working mothers need not be affected by childbirth. This is happy news for women."

The Russians were also impressed. Petrucci was invited to Moscow and spent two months at the Institute of Experimental Biology, returning to Bologna with a Soviet medal. Presumably, he gave the Russians the benefit of his expertise, and since then, while he has desisted from further work, there have been rumors from time to time that professors Anokhin and Maiscki in Moscow have followed up on Petrucci, "and have got even further than he did."

But rumors only. In another context, writing about "The Ethical Basis of Science," Bentley Glass, professor of biology at the State University of New York and former president of the American Association for the Advancement of Science, said:

"A full and true report is the hallmark of the scientist, a report as accurate and faithful as he can make it in every detail. The process of verification depends upon the ability of another scientist, of any other scientist who wishes to, to repeat the procedure and to confirm the observation."

Dr. Petrucci offered no photographs; he never even published a report in a scientific journal to describe how he had far surpassed anything that had ever been done before. Because of this, among biologists in the West who understand science roughly as Glass does, Petrucci's "experiment" is said to be "incompletely documented," which seems to be a polite formula for saying, not that there is an honest difference of opinion as there was between Shettles and Chang, but that Petrucci is a fraud.

Meanwhile, more serious men, many quite young, were initiating projects whose findings appeared in the *Journal of Embryology and Experimental Morphology*, *Science*, the *International Journal of Fertility*, and other sober and reputable publications. Their work would converge, toward the end of the sixties, in successful *in vitro* fertilization and growth of the human egg. Three of the most important projects were run by Americans:

Joseph C. Daniel, Jr., professor of biology at the University of Colorado, used rabbits, ferrets, and mink to investigate the development of the embryo just before it implants in the uterus. He found that certain proteins and other compounds are crucial at this phase. Professor Daniel was supported by grants from the National Institutes of Health and the Atomic Energy Commission.

Ralph Brinster, professor of veterinary medicine at the University of Pennsylvania, perfected a combination of nutritive substances (the "medium") in which eggs can be cultured best. He also described the changing biochemistry of the embryo during the time from fertilization to implantation, and devised an efficient incubator and culturing chamber for *in vitro* work, hooked up to CO_2 gas and maintained at 37° centigrade. Professor Brinster's work was also sponsored by the NIH.

Dr. Wesley Whitten of the Jackson Laboratory at Bar Harbor and Dr. John Biggers of Johns Hopkins cultured mice eggs from fertilization to just before implantation. For the first time, they observed the whole development of the embryo in this crucial stage when it is "free," not yet attached to the mother, and can be manipulated *in vitro* and reimplanted without apparent harm to the offspring. The work of Whitten and Biggers was supported by the National Institute for Child Health (in addition, Biggers' work was supported by the Population Council).

With the knowledge and experience gained from these and other experiments available to him, Dr. Robert G. Edwards of Cambridge University was ready in 1969 to try to achieve convincing *in vitro* fertilization of the human egg. He had been making various preparations for this for ten years. Among other things, he and his group had plotted out the stages a human egg must go through on its own between the time it leaves the ovarian follicle and the time it is ready to be fertilized. Edwards would not have been able to do this preliminary research, let alone achieve fertilization, had he not exploited a new method of obtaining large numbers of usable human eggs.

Women volunteers were injected with a hormone, gonadotrophin, which caused them to "superovulate" many eggs (usually only one egg matures per menstrual cycle). Thirty hours after the injection, the volunteer would undergo an operation called "Laparoscopy" (Greek, *lapara*, "flank" or "abdomen"), performed by a surgeon in Edwards' group, Patrick Steptoe. Two small punctures were made in the woman's side, into the ovary. Through one of these openings Dr. Steptoe introduced a slender hollow suction tube, and through the other an ingenious miniature optical device with a tiny flashlight which allowed him to look into the ovary. While he held the optical image steady in his left hand, with his right he maneuvered the suction probe from follicle to follicle, sucking out the eggs. In this way the Edwards group has been able to collect the basic, unsubstitutable, natural material they needed, which no other investigators had ever had in such quantity and quality.

In 1969, the eggs were incubated and washed, and three hours after collection, each one in its separate dish was put in the presence of sperm, which the volunteer woman's husband had contributed. The ratio of nutrients in the dish, the regulation of temperature, gas, and acidity followed closely the work of the Americans on mice, rabbits, and hamsters; Edwards counted, as it turned out only partly correctly, on the uniformity in timing and chemistry among eggs, sperm, and embryos of all mammals. As for "capacitation" of the sperm, Edwards was relieved to find that it would not be necessary to ask the volunteers to have intercourse, and then to operate to recover sperm from the uterus; the masturbated sperm could be capacitated simply by adding to the medium of the culturing dish some serum from the blood of a lamb fetus.

Having put eggs and sperm together, Edwards and his colleagues watched through microscopes to see what would happen. "Suddenly," Edwards was later to recall, "to our unbounded delight, the sperm started penetrating the eggs." In a minority of the dishes, there was to be definite proof of the first stage in fertilization, defined by the formation of "pronuclei" and the expulsion from the center of the egg of something called the "second polar body." Edwards was entitled to his emotion, for this was a considerable success, but there was still no evidence that he had been able to push the process of fertilization to its acknowledged completion: the cleavage of the egg into two cells each having its own nucleus and full number of chromosomes. Possible reasons for this failure suggested themselves, and the Edwards group prepared another series of experiments under somewhat different conditions. It was decided to take the eggs from the volunteers before they were ovulated, and to mature them *in vitro*, so that the timing in placing them with sperm could be exact.

In 1970, the laparoscopies were repeated, the eggs and sperm placed together under new conditions, and by thirty-eight hours after this *in vitro* insemination, many eggs had undergone a cleavage into two cells; by forty-six hours some had divided again into four cells; by sixty-two hours, some had divided again into eight cells; by eighty-five hours a few had divided again into sixteen cells. According to the report of the Edwards group published in the English journal *Nature*, none of the embryos matured past sixteen cells. When no cleavages had been seen to occur for two days, the embryos were removed from their dishes.

The photographs published in *Nature* show a clear, jellylike mass, not much like a soccer ball, more like a bunch of grapes.

These photographs and other data are substantial evidence that true fertilization and development took place *in vitro*. The evidence is not, however, indisputable. A determinedly skeptical embryologist might still insist, even in the face of the symmetry and regularity of the cleavages, that there is a chance the eggs were simply deteriorating all the time; he would not be convinced unless, as in Chang's experiment with rabbits, an egg was put back into the uterus of one of the volunteer women, she was then segregated from her husband, and eight and three quarters months later she gave birth. It should be noted that no embryologist has expressed such doubts formally; Edwards' colleagues in the field seem to be convinced by the evidence available. Yet this is not the end of it.

"One or more embryos [Edwards wrote in *Nature*] have been produced from twenty-nine of the forty-nine patients under treatment in this work. The normality of embryonic development and the efficiency of embryo transfer cannot yet be assessed, although conditions for implantation in the treated patients should be favorable."

Edwards thus indicates that when it is taken, the next step for his group will in effect duplicate Chang's absolute test with rabbits. The question is, When will that step be attempted?

Indeed, the question may be asked, Why have Edwards and his group been at their work with human eggs, sperm, and embryos at all? Has it been just for the satisfaction of research and discovery? Edwards' answer is a definite "no." His answer, in interviews and statements, is that he wishes to relieve the suffering of the women who come to his laboratory as volunteers. These women are sterile; they have tried, and failed, to have babies. Their husbands have adequate sperm, but there has been no conception. They and their husbands are very unhappy to be childless, and much preferring to follow any hope of conceiving their own child rather than adopt one, they have been referred by obstetricians and gynecologists to the lab in Cambridge. Edwards says, "We tell these women, 'Your only hope is to help us.' " Because of their motives, his success in fertilizing and culturing eggs cannot be the end of it; rather it is a first step, which makes them eager for the obvious next one.

It is known that roughly a fourth of sterile women fail to conceive because their oviducts are either blocked or nonexistent, thus preventing sperm from reaching the egg. In the technique used by Chang, and by now routine in lower mammals, the egg is fertilized *in vitro*; it completes *in vitro* the divisions it would normally undergo in the oviduct; then, the embryo is implanted in the uterus, where its presence provokes the growth of a placenta, within which it matures into a fetus that is eventually born and is normal. The infertility that most of Edwards' volunteers are suffering from is attributable to blocked or absent oviducts. They come to Edwards for one reason only, to be cured of their barrenness. And the money which pays for Edwards' work (American money, Ford Foundation) is specifically awarded for research in fertility. This is why the next step is obvious.

And yet the decision to take it is not easy. What makes him hesitate, Edwards says, is that much testing remains to be done on the *in vitro* embryos to make sure that the manipulation of eggs and sperm does not damage the chromosomes, which would show up in more or less serious birth defects. "The last thing we want is abnormal babies." In order to do this minute checking up on the genetic material in the nuclei of the cells, a procedure called "karyotyping" (Greek, *karyo*, "nucleus"), the embryo has to be removed from its dish, which endangers its survival.

According to the published reports, all the genetic work has been done on embryos which spontaneously stopped growing. In fact, officially, the problem of deciding to take the next step is postponed and somewhat eased by the report that none of the embryos has survived *in vitro* past the sixteen-cell stage, while only an embryo developed well past that stage can successfully implant itself or be implanted into the uterus. However, when the Edwards group does succeed in culturing embryos to the implantation stage, the decision to implant a given embryo will have to be based on statistical evidence, and on hope—it will not be possible to karyotype the embryo itself. As Edwards' colleague, Dr. Steptoe, says, it will call for a "brave decision."

Toward the other end of the process of childbearing, or gestation (Latin, *gestare*, "to bear," "to carry"), other researchers, quite independent of the sort of *in vitro* embryological work being done by Edwards, have been devising ways to save babies when a woman's natural machinery fails and the fetus is born too soon. This new branch of medicine is called "fetology." The fetus, its umbilical cord to the natural placenta having been cut, is placed in an incubator which supplies heat and oxygen. There it is fed intravenously, and its breathing is forcibly assisted by an iron lung. By such means, doctors are now able to save most seven-month-old premature babies (average weight two pounds), some six-and-a-quarter- to seven-month-old "premies," and a very few under six and a quarter months. Incubators, which have been in use for a long time, substitute for many of the functions of the womb or placenta in order to permit the premature baby to gain size and weight. However, the baby is indeed a baby, and not a fetus anymore, because the umbilicus has been cut and the lungs are working. Some fetuses are expelled from the mother's body even earlier in gestation, and they die in an incubator. The challenge is to build an environment that duplicates the ordinary environment of a fetus, in which it will not have to do things for which its body is unready. Fetologists are trying various approaches. Dr. Robert Goodlin at Stanford has put fetuses born less than six months after conception into a thick steel chamber where a saline solution saturated with oxygen is kept under a pressure of 200 pounds per square inch, roughly the same experienced by a deep-sea diver at a depth of 450 feet. This immense pressure drives the oxygen through the skin of the fetus, sparing its lungs the need to work. But Goodlin has not solved the problem of carbon dioxide and other poisonous wastes, and no fetus has lasted more than forty-eight hours. Dr. Geoffrey Chamberlain, a British scientist then in Washington, D.C., on a year's research scholarship, has kept alive some much younger fetuses, weighing only 300 to 980 grams, which were obtained during Cesarean section for therapeutic abortions. Chamberlain's method is to save the umbilical cord and connect it to a combination heart-lung-kidney machine. In a report in the *Ob-Gyn Observer* on his most successful experiment, Dr. Chamberlain wrote, "A brisk spontaneous flow [of blood] was noted 22 minutes postpartum; the fetus was kept on the circuit for 5 hours and 8 minutes. Only when the cannula slipped out by accident and could not be reintroduced was the experiment halted." Earlier during gestation (when the organism is still actually an embryo), Dr. D.A.T. New of the Strangeways Laboratory in England has cut out from their mothers mice only 2 millimeters in length and cultured them on drops of blood plasma and nutrient solution. The embryos have rudimentary hearts and nervous systems. During the time they continue to grow in Dr. New's experiment, they quadruple in length, their hearts begin to beat, their brain, spinal cord, eyes, ears, guts, and kidneys develop, and their limbs begin to bud. However, as in Goodlin's experiments with human fetuses, the lack of a placenta of placenta-substitute to draw off poisonous wastes has been fatal to the mice within forty-eight hours. Perhaps the most promising of all the approaches so far is that of Drs. Warren Zapol and Theodor Kolobow at the National Heart Institute, Bethesda, Maryland. Zapol and Kolobow separate a lamb fetus from its mother at a gestation age of 125 days. The complete gestation cycle in sheep is 147 days, so the equivalent human fetus would be seven and a half months, an age at which the ordinary incubator is pretty adequate. However, instead of incubating the premature lamb, causing its lungs to function, Zapol and Kolobow keep it alive by what they call "extracorporeal perfusion." This involves placing the fetus in an unpressurized bath of solution resembling amniotic fluid (the same stuff used by the Italian Petrucci in his "experiment"). The lamb's umbilical cord is attached by a catheter (segmented polyurethane Lycra, Du Pont Corporation) to a circuit of machinery including a pump, a bag of adult-sheep blood, a silicone membrane lung (Medical Division, Dow Corning Corporation), and a bottle of antibiotics and nutrients. As the fetus is removed from its mother, the system is attached and the lung and pump are put to work to take over. The significant thing about this method is that it seems to achieve both nutrition and carrying-off of waste products. The levels of various chemicals in the lamb's body seem to stay fairly constant: in technical language, " . . . the fetus remains in a metabolically stable state lasting several days." Nor is death, after several days, caused by poisoning: "During perfusion, the fetus rested quietly in the artificial amniotic bath. About once each hour it moved its head or legs spontaneously. It exhibited a strong sucking reflex as well as a withdrawal reflex when pinched. After 55 hours of perfusion, the fetus abruptly underwent cardiac arrest." Apparently Zapol and Kolobow have problems remaining to contend with. But perhaps these are not insuperable.

The editorialist of the *New Scientist* has said, ". . . the development of the 'perfect' artificial placenta can only be a matter of time."

III

Early in 1970, the National Academy of Sciences published the results of a study undertaken by a panel of distinguished scientists, at the request of the federal government, on "The Consequences of Technology." The scientists were asked to estimate when various technological developments could be expected to take place. One of these developments was described as the "capability of fertilizing a human ovum *in vitro* and implanting it in a surrogate mother." (Surrogate motherhood will mean either a kind of space-age wet-nursing, or prenatal adoption, the principle in both cases being that an embryo from the egg of one woman may be implanted in the uterus of another woman at the right moment of her menstrual cycle.) At the time this panel of experts was asked for its estimate, the Edwards group had already succeeded in getting the first stages of *in vitro* fertilization, and was on the verge of getting cleavage to sixteen cells and perhaps beyond. Yet the average estimate of the panel for the date of achievement of this technology was 1995.

It was not the first time that those who are relied on for advice and prediction have failed to see how imminent the future is. The reasons for this failure are not clear, but perhaps the general bewildering multiplicity of events and reports has something to do with it. Also, a panel entrusted with technological forecasting may be under strains of advocacy and opposition that are eased by pushing the future safely into the next century, or to the very end of this century. However, in a democracy it is not only the experts who prepare the way for the future.

Two familiar phenomena of democratic societies are the *respected opinion-maker* and the *public opinion poll*. Anyone who is interested in the social setting of embryological research should consult the evidence from respected opinion and public opinion; while the amount of it is not substantial yet, it does invite some working conclusions to be drawn.

The New York *Times*, under the headline, "Test Tube Babies Ahead?" published an editorial about the Edwards experiments praising the hope they hold out for childless couples. Then the *Times* said:

"Ultimately the prospect looms of human babies engendered by fertilization and development completely outside any woman's body—test tube babies, in the most literal sense. . . . Abuses are easy to envisage, but it is encouraging that so far at least there is no evidence of such abuses in the use of artificial insemination to help women conceive. The real question even now is whether—and how—people can develop the sense of social responsibility that will be required if, by the year 2000 or earlier, women are able to have children without any of the morning sickness, special diets and other discomforts and dangers pregnancy now entails."

The *Times* mentions artificial insemination. At least 20,000 babies are conceived by mechanical means in the United States each year, and there may be a million Americans now alive who were so conceived. This indicates a widespread acceptance of the technique, though a recent Harris poll on "New Methods of Reproduction" had only 5 percent of the sample knowing what artificial insemination is. However, once it was explained to them, 49 percent of men and 62 percent of women approved insemination with the husband's sperm in cases of infertility; 24 percent of men and 28 percent of women approved insemination with anonymous donor sperm. The poll-takers also explained other techniques still not in existence, and got opinions on these. Thirty-two percent of men, 39 percent of women would approve of embryo implants of the sort planned by Edwards—37 percent of men, 48 percent of women said they "would feel love" for a baby of their own conceived in this way. The poll-takers then asked about "test-tube babies," babies who at no time would be inside the mother's body. Thirty percent of men, 35 percent of women were of the opinion that "this would be justified if wife might die or be crippled from childbirth." However, if "a woman just wanted to skip pregnancy and have a baby too," more than 90 percent of men and women would disapprove. Forty-seven percent of men, 53 percent of women said they "would feel love" toward such a "test-tube baby" of their own (for some reason, the percentage here, for both men and women, was higher than in the case of the embryo implant). Fifty-five percent of men, 61 percent of women said they believed a "test-tube baby would feel love for [its] family." A striking aspect of the results of this poll is that women invariably display a greater readiness to consider "new methods of reproduction" than men. This readiness is enhanced when the responses are broken down by age group: for example, of the women under thirty, fully 57 percent approved embryo implants. And yet, there was confusion too. Many of the men and women who approved of the new methods, including "test-tube babies," said they saw in them a way to bolster the ideal of monogamous marriage by ensuring that no couple need be childless; yet it was admitted that the new methods might have exactly the opposite effect—that is, of undermining further the ideal of the family.

The *Times* editorial and the Harris poll seem to show that there is important "public" enthusiasm for the goals, both official and possible, of the embryological research now under way, even if this enthusiasm is qualified by some doubts and fears. Indeed, it may be debatable how long the vociferous reaction in the case of a deformed baby from embryo-implant would persist, stifling grants and research. And if exquisite care is taken choosing the first embryo for implant, and the baby is born apparently sound, the excitement and enthusiasm will probably overwhelm doubts and criticism. The whole idea of "new methods" will be given a boost, and methods which had seemed fit only for science fiction will undergo a strange metamorphosis: "test-tube babies," for example.

Leaving aside for the moment the question whether such a method is desirable—is it feasible? The answer would have to do with technology, with whether ways could be devised to transfer the embryo fertilized *in vitro* to an artificial placenta which would duplicate for eight and a half months the environment of the natural placenta. The complexity of this invention of nature has already been hinted at. It has taken scientists five years just to get to understand the physiological process involved in the diffusion of oxygen and carbon dioxide across the placental membrane of the pig. How other substances—amino acids, vitamins, sugars, proteins—are passed is still a mystery. In addition, it is known that just a bit too much of a substance, or too little, too late or too early, can cause peculiar things to happen to embryos, whether they are mammalian and placental, or lower down on the phylogenetic scale. For example, if lithium chloride or magnesium chloride is added to fish eggs, the fish that are hatched will be Cyclopean. So it would be, at the least, a delicate, painstaking, and drawn-out task of plotting the career of a human embryo and fetus in the placenta from minute to minute, and then fabricating the machinery to duplicate the placenta and a computer to monitor and direct it and oversee the piping-in of nutrients and carrying-off of wastes. The technological problems here are formidable, as fetologists working on the margins already know; but are they more formidable than those involved in, say, Apollo 11? Probably not, and using Apollo as a hackneyed but serviceable example, it might be said that for the United States, there is no technological project that is not assured of success provided the decision is made to invest whatever talent and money are necessary; provided also that there is a strong enough sense of national priority so that any misfortune (such as the death by fire of three astronauts on the pad at Cape Kennedy) does not endanger the life of the project itself. It would be too much to expect an artificial womb to "work" the first time, and people would have to be ready to accept the death of a fetus, even though, in contrast with Gus Grissom, the fetus never volunteered.

What reason would there be to make the development of an artificial womb a national priority? Once they are compiled, the specific and predictable benefits of an efficient artificial womb make an impressive list:

1. Fetal medicine would be much improved. By being able to monitor growth and development continuously, fetologists would be able to catch, and perhaps treat, sickness that occurs in the natural womb but does not show up until after birth.

2. Likewise, fetologists would be able to immunize a child for the diseases it would be likely to contract in the world, but while it is still in the sterile safety of the womb.

3. Tissue samples could be taken from the fetus, cultured, and frozen for storage,

which would resist the rejection phenomenon should the human born ever require organ transplants.

4. An efficient artificial womb, far from increasing the incidence of birth defects, would reduce them by keeping the fetus in an absolutely safe and regular environment; safe, for example, from infection by German measles or drugs taken by the mother. There are now thousands of babies born in the United States each year with defects, ranging from relatively minor ones like harelip, to deformed limbs and congenital diseases of the nervous system. Whatever the magnitude of the defect, it is disastrous: doctors say that the immediate and overwhelming response of the parents is not love or pity, but anger; they are angry at the doctor, and angry at their deformed child for choosing them as its parents. This behavior is evident on the part of both parents alike: there is no special redemptive motherlove. Some parents will reject the child, or, after a guilty reaction, some will gird themselves for the job of lifelong sacrifice, of being "noble."

5. The same new conditions that would allow fetologists to prevent birth defects would allow geneticists eventually to be able to program a fetus' development for some superior trait on which society could agree: larger brain capacity, for example. This would seem to be the direction that is being taken anyway now, with genetic counseling. The artificial womb would lift such work out of the realm of the haphazard.

6. An artificial womb would make "sexing" (choosing the sex of the embryo) a simple matter.

7. That part of the population which would use the artificial womb would not have to worry about illegitimacy or doubtful paternity. For the first time it will be possible to prove beyond a shadow of a doubt that a man is the father of his children.

8. Women who are prone to miscarry, or who because of body structure or constitution run a danger of injury in childbirth, would be spared the unhappiness, disappointment, and danger. Other women would be spared the discomfort.

9. Women who decided to have children by the artificial womb might choose to undergo the operation in which the fallopian tubes are tied. This would not affect fertility, but it would be an instant, guaranteed, and permanent barrier to conception from sexual intercourse. No other "birth control" would have to be exercised, and the Pill, together with its harmful and unknown side effects, could be dispensed with. Of course, these women would never have to have abortions, either.

It would seem that from the development and use of an artificial womb, all of society would benefit, but women would stand to gain the most. The artificial womb would set about breaking to pieces the stubborn remnant of biological fact and cultural myth that makes all women pay. The invidious question whether women are different from men in some ultimate and irreducible metaphysical way, whether as a result they should be set and should set for themselves different goals and different styles of life, would be removed from the context of biological difference, which has so far complicated its resolution with gratuitous factors, and would be set in a context of biological equity. Culturally, if the artificial womb "catches on," it will mean that the awefulness associated with pregnancy and childbirth will have nothing to feed on, and motherhood, if it continues to excite any awe at all, will not do so more than fatherhood. This will have its inevitable effect on the relation of women to men, women to their children, and the society or state to children. Once a woman has no more difficult or lengthy role in reproduction than a man (or not much more difficult or lengthy: she will still have to undergo laparoscopy once, when several dozen eggs will be collected and put into cold storage), she will find that society does not expect her to have a special relation to her offspring that takes up years of her life, and also she will not expect it of herself. Too, a society that can grow fetuses in a laboratory will be more disposed to have meaningful day- and night-care centers and communal nurseries on a large scale, for the state, being a third parent, will wish to provide for the maintenance and upbringing of its children.

Natural pregnancy may become an anachronism. The two tiny laparoscopy scars, exposed by a bikini on the beach, will be as ordinary as our smallpox vaccination, but women will no longer have lost their figures in childbearing. The uterus will become appendixlike, though the ovaries will be as crucial as before. At the age of twenty, each girl will be able to choose to be superovulated and her eggs collected and frozen, as it is known that babies conceived by young women are less likely to suffer from mongolism and other birth defects. If there are advances in prenatal care, it may not be necessary to prohibit natural childbearing in the interest of public health and eugenics. In that case, the women who wish to put up with the old style and all that it implies will be free to do so. But it will be a throwback and increasingly rare as the manifest advantages of the artificial womb make it likely to win the competition.

Most, if not all, of its *disadvantages* might be more apparent to us than real to the next generations. We bear it in mind that a man-made mutation like this, finishing what the Pill started, unprecedented in evolution perhaps since sea creatures grew lungs and came out on land or apes developed the ability to touch thumb to forefinger, must have its effect on the body and mind of everyone in society, men not much less than women. Might not everyone, and particularly women, also *suffer* from the artificial womb? The myth of the beatific Madonna has, after all, among its various sources the fact that some women do experience unusual well-being when they have a baby. A more recent myth is that women on the Pill for a long time, who have much sex but never have a baby, suffer the opposite of the bodily and psychic happiness of the Madonna. Does this mean the body has its own wisdom and that for women to be given access to an artificial womb would be to go against the deepest instincts provided by nature?

Again, the question may be invidious. In the first place, as a matter of fact, for every beatific mother in our society there is at least another with "post-partum blues." To propose a "fundamental nature" for women (or men) to which it is immoral or unwise to offer an alternative may be to support a fallacy which is really old-fashioned. The well-being of the Madonna, her rosy complexion, may have as simple an explanation as that during pregnancy and lactation, her body's production of estrogen has shot up: maybe a woman having a baby by the artificial womb might take estrogen orally.

But won't women be "alienated," as we say, from their children, causing further distance to be put between all of us from the crucial beginning, which is not what we need? Again, perhaps an invidious question. In the 1840s, opponents of Dr. Simpson asked whether (actually they claimed that) anesthesia during labor would make children "strangers" to their mothers. Has this proven true? Maternal love does not seem to be connected with the pain of childbirth, or even with childbirth; we know that some women beat the children they have borne, while others love the children they have adopted.

However, by "creating life" won't we be raising ultimate questions that we are not prepared to answer, such as "What is a human being?" The effect may well be to raise such questions. As for "creating life," that is to misunderstand what the artificial womb will do. It will not "create" life, for the materials which contain all the factors for differentiation, growth, and genetic coding—the egg and sperm—will not be created or fabricated: they will only be given another environment in which to work out their process. Sexual reproduction will be preserved; only intercourse and reproduction will be separated, once and for all.

The "ultimate" questions will be harder to ignore, perhaps. But this has been predicted. Jean Rostand, the French biologist and Nobel Prize winner, has considered what life will be like and what questions people will have to face up to when the artificial womb and other "inevitable" technologies become a reality:

"People will live for two hundred years, or even more. There will be no more failure, no more fear, no more tragedy. Life will be safer, easier, longer. But will it still be worth living? . . . How shall we contrive to exercise the formidable powers allotted to us . . . ? How . . . shall we avoid finding ourselves on the perilous slope and yielding to the abuses of a Promethean intoxication?"

Well, Rostand says, "our task will be to improvise the solution, taking account of the collective mentality, of the social and moral situation [and remembering Bacon's warning], 'Knowledge, if it be taken without the true corrective [charity] hath in it some nature of venom or malignity.' "

Rostand assumes the artificial womb and other such innovations are "inevitable": that seems a peculiar idea. He also predicts that when the womb arrives it must have a universal effect. Here he is evidently right. If there is a single prototype artificial womb made and successfully tested, it is unlikely that it will turn out—as the Apollo missions might—an exorbitant stunt without consequence for "the man in the street." Because it will literally be down to earth, an artificial womb will have the potential to change the life of every person. But is the artificial womb "inevitable," as Rostand says? "Inevitable" seems to imply that something will come to pass without our doing anything or despite our intervention, which in the case of an artificial womb is nonsense. And yet there is a meaning of "inevitable" which, in this context, is not ridiculous. This is the meaning which in effect asks to what extent human beings exercise free will, and to what extent they are determined by forces forever beyond them. A very ancient question. Yet is it conceivable that if Dr. Simpson, in 1842, had not decided to give anesthesia to a woman in labor, anesthesia would not be routinely used in labor today? It is not conceivable. If he had not done it, some other doctor would, driven by the combination of curiosity, sympathy, and ambition which many men, not just a single indispensable man, are endowed with. The picture of the solitary scientist breaking ground may be excessively romantic. Without denying the medical scientist in particular his glory, his Nobel Prize, and the gratitude of the people whose suffering he has eased, it may be said that the scientist is far from being on his own, that he is, as we all are, an agent of something, determined by a force, a momentum which blurs distinctions between "it has become possible to do it," "it should be done," "it must be done," and is resolved in the inevitable: "it will be done."

Looked at in this light, it would not seem to make much difference (except, obviously, to the volunteer childless couples at Cambridge) whether Dr. Edwards decides to implant an embryo: the operation is going to take place pretty soon whatever he decides or does. Likewise the question whether what is about to be done in embryology, and what is about to be done in fetology, will ever come together in its logical consummation. If it becomes a national priority, it will be achieved sooner; if unlimited money and support are not forthcoming, it will be achieved later. But it is hard to imagine it not being achieved at all.

Certainly it would take more exertion, over the long run, to prevent it than to achieve it, and why prevent it? Who is to say that Monsignor Vallainc, the Vatican press officer who branded the Edwards experiments "immoral acts and absolutely illicit," is not the hapless spokesman of cruelty and stupidity, our contemporary version of the ministers who damned Simpson? In any case, it is not the thirteenth century anymore, and the centers of research happen not to be in Russia or in Roman Catholic theocracies. There is no forbidding most things and no arresting Dr. Edwards and charging him with murder. More than this: the research is conducted quietly, indoors; it does not require the vast hardware of an Apollo project, and there are no thunderous blasts and clouds of smoke. The expunging of perhaps our foremost Myth, with its ancient, numberless effects of inspiration and practice, habit and suffering, may be accomplished both inevitable and quietly—which leads to the ironic part, that whether anyone, or any movement, comes out for the artificial womb, or not, will make little difference in the end. The only difference it might make—and perhaps this is no small thing for the race—is that at least we will be able to say that our liberation did not catch us by surprise.

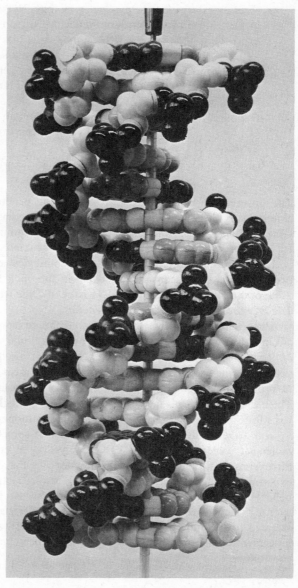

A model of a portion of the DNA molecule. Courtesy John Oldenkamp, Psychology Today.

17. *Molecular Genetics—A Survey of Highlights*

GORDON EDLIN

As is indicated by the title, this paper reviews some of the major events that have recently taken place in molecular biology. Notable advances have been obtained in isolating protein repressor molecules responsible for the regulation of gene expression; identifying a RNA tumor virus that causes the synthesis of DNA (a valuable clue for cancer researchers), and the isolation and synthesis of lactose genes from *E. coli.* The dream of correcting genetic defects is closer to reality, having been accomplished in both mouse and human cells in laboratory cultures. The successful fusion of mouse and human cells has provided a valuable new technique for locating linked genes on human chromosomes. The author, Dr. Gordon Edlin, is a member of the Department of Genetics, University of California at Davis.

The upsurge of public interest in cellular and genetic research is evidenced by the almost daily headlines pertaining to the latest developments in curing cancer or schizophrenia or the common cold and to the imminent development of techniques for genetic engineering which will allow mankind to attain a state of biological perfection hithertofore unimagined. The biological revolution that is occurring has been documented in considerable detail in the book by G. R. Taylor, *The Biological Time Bomb* and in the April 19, 1971 issue of *Time* which devoted an entire section to *The Promise and Peril of the New Genetics.*

The scientific discipline which has contributed in a large measure to the remarkable advance in our understanding of biological, cellular and genetic processes is termed *Molecular Biology* or *Molecular Genetics.* This discipline came into being some 20 years ago when scientists, largely trained in the areas of physics and chemistry, began to probe the molecular and atomic nature of the reactions carried out by living cells. These studies have advanced enormously our biological understanding and form the basis for the extrapolations and speculations mentioned above.

This article is devoted to a factual review of some basic and exciting discoveries in molecular genetics over the past few years. A comprehensive survey of recent discoveries is not attempted nor intended. In the discussion of selected areas and experiments I have sought to focus on those aspects of molecular genetics which I believe have contributed much in the past and are likely to be in the vanguard of exciting discoveries to come in the months and years ahead. Thus, omissions should be ascribed to the limitations of my own particular view and should not be construed as a judgement of any experiments which have not been included.

Source: Reprinted with permission of the publisher and author from *Bioscience, 22(2):*77-81, February 1972.

DNA Replication

The elucidation of the molecular structure of DNA by Watson and Crick in 1953 (Watson and Crick, 1953a, b) was an epochal discovery comparable to Darwin's theory of natural selection in that both discoveries irrevocably altered our view of living systems. Knowing the structure of DNA opened the way to a fundamental understanding of the chemical nature of the hereditary material and to the posing of such basic genetic questions as the mechanisms governing chromosome replication, genetic recombination and mutation. Compelling evidence for DNA being the genetic material had been presented in 1944 by Avery and his collaborators (Avery et al., 1944) who had equated the bacterial "transforming principle" with DNA. In these experiments they showed that a nonpathogenic strain of pneumococcus could be transformed into a pathogenic strain after being treated with DNA extracted from the pathogenic strain. Because it was believed that their DNA preparations were contaminated with other cellular components such as proteins, few scientists were prepared to accept the conclusion that DNA was the genetic material. However, with the publication of the Watson and Crick model, DNA was accepted universally as the genetic material. Genes, which hitherto had been abstract units of inheritance, now became defined sequences of purine and pyrimidine bases along the DNA chain. The unique achievement of the Watson-Crick model was that it was correct in all respects; the α-helical structure of the molecule, the number of strands and the nature of the base pairing, and the hydrogen bonds which hold the two strands together.

Although the structure of DNA and a model for replication were set forth in 1953, it has become apparent in the past several years that we still understand painfully little about the molecular basis of DNA replication in bacteria, not to mention higher organisms. The enzyme DNA polymerase, isolated from *E. coli* by Kornberg and his collaborators was presumed until very recently to be the enzyme responsible for DNA replication in the bacterium *E. coli* (Kornberg, 1969; Englund et al., 1968). Indeed the crowning achievement of years of isolation and purification of the DNA polymerase was the utilization of the enzyme to replicate in vitro the DNA extracted from a small bacterial virus, $\phi\chi174$ (Goulian and Kornberg, 1967; Goulian et al., 1967). Moreover, the newly synthesized virus DNA which had been made in vitro, i.e., outside of a living cell, was shown to be biologically functional. When this newly synthesized DNA was added to bacteria which were made permeable to DNA molecules by partial removal of their cell walls, synthesis of complete new virus particles which were themselves infectious was observed. This was a reconstruction of the natural process which occurs when a bacterium is infected by a virus. In nature the virus attaches to a bacterium and injects its DNA. This DNA is replicated and directs the synthesis of new viruses which can repeat the cycle all over again.

Thus, these experiments in which functional virus DNA was synthesized by the bacterial DNA, polymerase seemed to provide irrefutable evidence that DNA polymerase was indeed the enzyme responsible for DNA (chromosome) replication in *E. coli*. However, a number of properties of the polymerase left some doubts as to whether the Kornberg enzyme was really the enzyme responsible for DNA replication in vivo. One of the observations that gave rise to such doubts was that, among the hundreds of bacterial mutants which were defective in DNA replication under certain restrictive growth conditions, none had a mutation which affected the bacterial DNA polymerase. Thus, there was considerable excitement when J. Cairns reported the isolation of a mutant of *E. coli* which grew normally and yet lacked any measurable DNA polymerase activity (DeLucia and Cairns, 1969). The finding of such a mutant meant that the DNA polymerase was not the enzyme responsible for replication and must perform another and relatively non-essential function in the bacterium. It is now known that the function of this polymerase is to repair DNA molecules which have been damaged by radiation or chemicals.

Subsequent work has identified yet another DNA polymerase (DNA polymerase II) in *E. coli* which has many of the properties expected of the true replicase (the term applied to the enzyme or enzymes actually responsible for DNA replication) (Knippers, 1970; Kornberg and Gefter, 1970; Moses and Richardson, 1970). However, some evidence has been presented which suggests that polymerase II is not the replicase either and so the search for the true enzyme continues (Brazill et al., 1971).

Still another puzzling aspect of the replication problem is how both strands of the double helix can be replicated simultaneously and in the same direction as all the experiments seem to indicate. Both polymerase I and polymerase II function chemically in one direction only, i.e., they can replicate one of the DNA strands by moving from left to right along the DNA template but they cannot function when moving from right to left. Since the two strands of the DNA double helix are joined in a chemically antiparallel sense it is not easy to envisage how both strands can be replicated simultaneously by the same enzyme moving in one direction (Fig. 1). One solution to this problem has been to propose that one strand is replicated continuously in one direction and the other strand is replicated discontinuously in the opposite direction by production of small fragments which are subsequently joined together

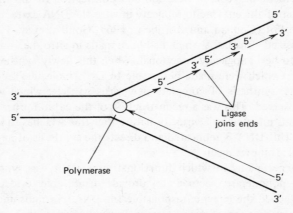

Fig. 1. Schematic model of DNA replication. The dark strands are parental DNA and the light strands are newly synthesized DNA.

end to end by the enzyme DNA ligase (Gellert, 1967; Olivera and Lehman, 1967; Gellert et al., 1968). Such fragments were shown to occur thereby giving some support to such a model (Okazaki et al., 1968). However, it is still possible that these fragments are artifacts of the techniques used to isolate and analyze the DNA so that, as yet, there is no completely convincing model for DNA replication. The problems involved in DNA replication have been comprehensively reviewed recently (Richardson, 1969; Gross, 1971) so I shall not belabor the point here. Suffice to say that 20 years after the structure of DNA was solved, the mechanism of its replication is still very much a mystery. Problems pertaining to DNA replication are being intensively studied in many laboratories, for the question of DNA replication is central to the understanding of cell division and the faithful transfer of genetic information from generation to generation.

Expression of Genes

One of the most satisfying discoveries in recent years was the isolation and purification of two different protein molecules responsible for the regulation of specific gene expression. The existence of such repressor molecules had been postulated in 1961 by Jacob and Monod in the presentation of their, now famous, operon model (Jacob and Monod, 1961). This model, based almost entirely on data from bacterial genetic experiments, explained how the genes which code for the synthesis of the lactose fermenting enzymes are regulated at the level of DNA. To understand this model one must realize that many bacterial genes involved in related functions are situated adjacent to one another in the DNA and such a group of genes together with certain regulatory genes is called an operon. Because all the products whose information is encoded in these genes need to be synthesized in a coordinated fashion, a single messenger RNA molecule is transcribed from this block of genes. This single messenger RNA is translated into a number of different enzymes whose activities are related. In the case of the lactose operon three enzymes are necessary for lactose fermentation and the genetic elements for this operon are shown in Fig. 2.

There are three structural genes which encode the information for the synthesis of three enzymes involved in lactose utilization; z, beta-galactosidase; y, permease and a, acetylase. The synthesis of the messenger RNA for these three enzymes is regulated by the o or operator site which, in turn, is regulated by the repressor protein which is the product of the i gene. Finally, the p site is responsible for the binding of the RNA polymerase to the DNA and is the site where messenger RNA transcription begins. The regulation is effected in the following manner. When lactose is not present in the medium, the i-gene in each bacterium produces an active repressor protein which binds tightly to the operator (o-site) thus preventing any messenger RNA from being synthesized. If lactose is added to the cells, it binds to the repressor protein and greatly reduces the affinity of the repressor for the operator. Now the RNA polymerase can attach to the p-gene and synthesize a messenger RNA molecule which is subsequently translated into the enzyme molecules.

Thus, the operon model posited the existence of a repressor protein which would

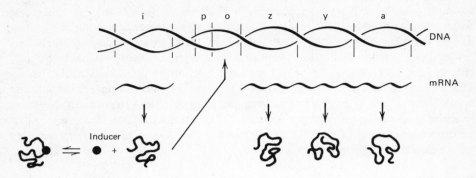

Fig. 2. The lactose operon. The letters refer to the following genes: *i*, repressor; *p*, promotor site where the RNA polymerase binds to the DNA and initiates transcription of messenger RNA; *o*, operator site where the repressor protein binds to the DNA; *z*, *y*, *a*, genes for the lactose fermenting enzymes.

bind to a specific site on the DNA. In 1969 Gilbert and his collaborators successfully isolated and purified the lactose repressor protein and showed that it had the predicted properties (Gilbert and Muller-Hill, 1968). Just prior to the isolation of the lactose repressor, Ptashne isolated and characterized another repressor molecule, the lambda repressor (Ptashne, 1967). Lambda is a virus which is frequently found associated with the bacterium *E. coli*. It does not kill the bacterium because the virus DNA has become integrated into the bacterial DNA (a process termed lysogeny) and the virus genes are unexpressed. The reason the virus DNA remains unexpressed is due to the presence of the lambda repressor which prevents the virus DNA from being transcribed into messenger RNA and, hence, no virus proteins are produced. If this lambda repressor is destroyed or inactivated the virus genes are expressed, virus particles are produced and in a short while the bacterium is killed. Thus, the viral repressor protein is vital to the continued viability of the bacterium and its loss results in the death of the host bacterium.

The enormous importance of this negative control mechanism, i.e., the prevention of gene expression by repressor proteins, derives from the possibility that aberrant gene regulation in animal cells may be involved in the formation of malignant cells. Evidence is increasing which implicates the cellular presence of viruses with certain forms of cancer. It is possible that many animal cells normally harbor virus DNA which remains unexpressed due to various repressor molecules. If these repressor molecules were inactivated then virus particles might be produced. Instead of killing the cell as in the case of the bacterium, these viruses might cause the animal cell to grow rapidly and uncontrollably—what is generally referred to as cancerous growth.

The whole question of regulation of gene expression and cancer received new impetus with the discovery of an RNA-dependent DNA polymerase in animal cells infected by RNA tumor viruses (Temin and Mizutani, 1970; Baltimore, 1970). It had been known for some time that RNA was transcribed from DNA but the reverse synthesis had not been demonstrated. Actually, some years earlier, Temin had suggested that DNA was transcribed from RNA in animal cells infected with RNA viruses and more than ten years ago Crick had suggested the possibility of such a synthesis on theoretical grounds when he formulated the "central dogma" hypothesis (Crick, 1958, 1970).

Infection of animal cells by a wide variety of RNA tumor viruses is known to result in the transformation of normal cells, i.e., the cells acquire malignant characteristics. The existence of an enzyme capable of transcribing information from RNA to DNA led to the speculation that the information in the virus RNA is transferred to the DNA of the host cell and in this way alters the normal expression of the host cells' genes. Since the initial discovery, much attention has focused on this enzyme but, as yet, its function in the cell is not clearly understood.

Another area in which repressors may play a vital role is in the process of cellular differentiation. When an animal ovum is fertilized, cells begin to divide and proliferate and gradually these cells differentiate into the special tissues and organs that make up the mature animal. It is conceded that every cell in our body contains a complete complement of genetic information so there must exist mechanisms whereby each special group of cells expresses only those genes necessary for the survival and function of the specialized cells. For example, heart cells must express different genes from lung or liver cells and it is likely that repressor molecules are involved in regulating the different cellular functions. Understanding of the mechanism of repressor action and identification of repressors or their counterparts in animal cells is one of the central problems of current biological research.

Isolation and Synthesis of Genes

The detailed study of any molecule depends on its being available in a pure form and tangible amount. The study of genes is no exception. Thus, the widely publicized isolation of the lactose genes signaled an important step toward understanding the molecular nature of genes and how they function (Shapiro et al., 1969). It should be noted, however, that DNA containing ribosomal RNA genes or parts thereof was actually isolated prior to the isolation of the lactose genes (Kohne, 1968). The DNA corresponding to ribosomal RNA genes was isolated by the technique of RNA-DNA hybridization. Ribosomal RNA was chosen because it can be extracted from bacteria and is readily purified. Purified bacterial ribosomal RNA was hybridized (joined to the DNA by complementary base pairing) to DNA and the RNA-DNA hybrid molecule was exposed to an enzyme that degrades single-stranded DNA. DNA is digested in this process everywhere except where it has bound RNA. In this way the specific DNA segment complementary to ribosomal RNA could be isolated. However, since the DNA is first fragmented and subsequently digested by the nuclease, it is probable that the DNA isolated does not represent intact ribosomal RNA genes but only fragments.

The lactose genes were isolated from *E. coli* by a series of complicated genetic manipulations. Two bacterial viruses λ and φ80 are capable of having their DNA integrated into the bacterial chromosome. Bacterial genes which are adjacent to the viral DNA can be recovered with the viral DNA when the viruses are induced to replicate and grow. Such viruses, called transducing viruses or transducing phage contain bacterial genes as well as virus genes in virus particles. Thus, the lactose genes were inserted into the DNA of both λ and φ80 viruses but in opposite orientation, i.e., in one case the genes were inserted in the order o-z-y-a and in the other the order was a-y-z-o. The DNA's from the viruses were extracted and the individual strands of the double helix separated. Finally, one strand from the λ virus was reannealed with the complementary strand of the φ80 virus. Since the only region the two strands had in common was the lactose genes, the strands would hybridize to form a double stranded helix in that region but in no other. The non-homologous DNA strands could be digested away with a single-stranded nuclease leaving pure DNA containing the lactose genes. The particular importance of this isolation lies in the fact that the expression and regulation of the lactose genes can be studied in vitro. Considerable success in this regard has already been achieved (Crombrugghe et al., 1971).

The complementary approach to the isolation of genes is their synthesis. To attempt this, however, it is necessary to know the sequence of nucleotides in the DNA specific for a given gene. The most expeditious way to determine the sequence is by isolating and sequencing the nucleotides in the gene product, i.e., the RNA transcribed from a specific gene. The first RNA product of a gene to be isolated and sequenced was the transfer RNA for the amino acid alanine (Holley et al., 1965). This fact led Khorana and his co-workers to undertake the chemical synthesis of the gene which codes for alanine transfer RNA. The complete synthesis was achieved last year and with the techniques developed the way is clear for the synthesis of other genes as soon as the correct nucleotide sequences become known (Agarwal et al., 1970). These two dramatic accomplishments, the isolation of genes and their synthesis have fueled the predictions of "genetic engineering" and "genes made to order" for future generations of man. While it is premature to predict what practical applications will be served by these discoveries, it is certain that our understanding of gene regulation will be markedly advanced.

Somatic Cell Hybrids

Approximately ten years ago the successful hybridization of two different lines of malignant mouse cells was demonstrated (Barski et al., 1960). This hybridization which constitutes the fusion of two disparate animal cell types into a single cell should not be confused with the hybridization referred to earlier, which was the specific joining together of two segments of nucleic acid. The mouse hybrid cells isolated were shown to have properties characteristic of both parents and to contain chromosomes derived from both parental cell lines.

With improved techniques it became possible to construct hybrid cells derived from very dissimilar animals such as mouse/hamster or mouse/human (Davidson et al., 1966; Weiss and Green, 1967). An especially interesting aspect of the mouse/human hybrid cells is that some of the hybrid cell lines lose most of the human chromosomes upon subsequent cell growth. This fact makes it possible to begin to associate specific human enzyme activities with specific human chromosomes and to establish genetic linkage between different human genes. For example, in one mouse/human hybrid cell line it was demonstrated that the genes for a human enzyme, lactate dehydrogenase, are linked to the genes for the enzyme peptidase. (Ruddle et al., 1970).

The hybrid cell lines also provide an experimental system for study of gene regulation in animal cells. Since cell lines which differ in the expression of a number of enzymatic activities can be joined to give a hybrid cell line, it is possible to ask which of the parental functions are expressed in the hybrid cell. In a hybrid constructed from a mouse fibroblast cell and a rat hepatoma cell it was possible to show that the enzyme, tyrosine aminotransferase, which is inducible in the hepatoma cells and uninducible in the fibroblast cells becomes uninducible in the hybrid cell (Schneider and Weiss, 1971). One interpretation of these experiments is that some product of the mouse fibroblast cell represses the expression of the tyrosine aminotransferase genes on the rat hepatoma chromosome. Again, the importance of the bacterial repressor studies mentioned earlier is evident.

Recently, a first step has been made in correcting a genetic defect in a mammalian cell (Schwartz et al., 1971). Mouse fibroblast cells which lack the enzyme, inosinic acid pyrophosphorylase, were cultured in vitro. To these cells were added chick erythrocyte nuclei and the cells were allowed to fuse. The chick erythrocyte cells are able to synthesize the pyrophosphorylase enzyme. The hope was that the DNA which codes for the enzyme would be transferred from the chick nuclei to the mouse cells and become intergrated into the mouse nuclei. Indeed such cells were selected and were shown to have all the characteristics of the mouse cells except for having gained the capacity to synthesize inosinic acid pyrophosphorylase. Moreover, electrophoretic analysis of the enzyme showed that it was the chick enzyme so it could be inferred that the mouse cells had acquired the chick gene for pyrophosphorylase.

An even more dramatic alteration has been claimed for human fibroblast cells cultured in vitro (Merril et al., 1971). There is a congenital disease in humans called galactosemia and persons suffering from this disease lack an enzyme α-D-galactose-1-phosphate uridyl transferase which is essential for the metabolism of galactose, a common sugar. E. coli bacteria also posess the gene for this enzyme and it was possible to produce a stock of the bacterial virus λ which had integrated the bacterial transferase gene into its own genome. When the defective human fibroblast cells were treated with these bacterial viruses, a few of the human cells regained the capacity to synthesize the transferase enzyme. It appears that a few of the human cells were able to incorporate the bacterial gene and utilize it to synthesize the enzyme that previously they were incapable of synthesizing.

The experiments cited above are a brief sample of the exciting reports that are emerging from studies on somatic hybrid cell lines. As has frequently happened in the past in other fields, we may find outselves able to manipulate mammalian genetic material to achieve certain goals long before the mechanisms are understood or even before the mechanisms are understood or even before the consequences of such manipulation can be appreciated. Readers interested in the prospects for genetic intervention in man may wish to read a review of the feasibility of human genetic engineering by B. D. Davis (1970).

Genetics and Neurobiology

Possibly the foremost biological problem remaining to be attacked is the question of the molecular basis of thought and memory. Many of the molecular biologists who have contributed to the remarkable advances of the past two decades have turned their attention to neurobiological problems. As in any uncharted waters, the question is where to begin and where to go. The most publicized work to date has involved measuring changes in RNA content in brain cells following various learning programs but other approaches are being tried also. One interesting approach was selected by S. Brenner working in England a few years ago. He chose to work with a relatively simple animal, a nematode, in the hope that this organism could be analyzed in sufficient detail to provide an understanding of basic neurological processes. It was precisely this approach using microorganisms and viruses that led to the remarkable successes of molecular biology. A brief report of these experiments appeared in the April 30, 1971 issue of *Nature*.

The nematode being studied, *Caenorhabditis elegans*, has a nervous system containing 200-300 nerve cells and the entire animal has about a thousand cells. It is hermaphroditic, has a generation time of three and one-half days and produces a couple of hundred progeny each generation.

In addition to the ease with which it can be grown, the nematode is genetically rather uncomplicated having only six chromosomes. It has a number of behavioral characteristics such as a rhythmic swimming motion, mechanoreceptors which allow it to respond to touch, and probably chemoreceptors. Behavioral mutants have been isolated and mapped genetically. Next comes the arduous task of correlating the mutant phenotype with a change in a particular nerve cell or cluster of cells. This is being attempted by first doing an electron microscopic examination of all of the nerve cells in the animal to determine the normal pattern of nerve connections. The process then has to be repeated for the mutant animal in the hope of discovering an alteration in the nerve connections.

Studies such as these are just beginning in many laboratories. The goal is formidable—to understand the mechanisms and to identify the molecules that enable us to think and to remember. With that knowledge, of course, arises the possibility not only of understanding the mechanism of our mental processes but also of their manipulation and alteration. Since the knowledge is bound to come in any case, it is encumbent on all of us to see that it is used constructively and wisely. Therefore, it is encouraging to find many eminent geneticists and biochemists who already are concerned with the application of our new biological knowledge and who are discussing ways to ensure that mankind benefits from our increased understanding of biological processes. One such meeting was held a few months ago under the auspices of the Weitzmann Institute and was reported in the July 2, 1971 issue of *Nature*.

REFERENCES

Agarwal, K. L. et al. 1970. Total synthesis of the gene for an alanine transfer ribonucleic acid from yeast. *Nature, 227*: 27-34.

Avery, O. T., C. M. Macleod and M. McCarty. 1944. Studies on the chemical nature of the substance inducing transformation of pneumococcal types. *J. Exp. Med., 79*: (2)137-158.

Baltimore, D. 1970. RNA-dependent DNA polymerase in virions of RNA tumor viruses. *Nature, 226*: 1209-1211.

Barski, G., S. Sorieul and F. Cornefert. 1960. *C. Rend. Acad. Sci., 251*: 1825-1827.

Bazill, G. W., R. Hall and J. D. Gross. 1971. DNA synthesis in lysates of RecB⁻ and Rec+ *E. coli* cells. *Nature, 233*: 281-283.

Crick, F. H. C. 1958. On protein synthesis. *Symp. Soc. Exp. Biol., 12*: 138.

————. 1970. Central dogma of molecular biology. *Nature, 227*: 561-563.

Crombrugghe, B. de, et al. 1971. Lac DNA, RNA polymerase and cyclic AMP receptor protein, cyclic AMP, Lac repressor and inducer are the essential elements for controlled Lac transcription. *Nature, 231*: 139-142.

Davidson, R. L., B. Ephrussi and K. Yamamoto. 1966. Regulation of pigment synthesis in mammalian cells, as studied by somatic hybridization. *Proc. Nat. Acad. Sci. 56*: 1437-1440.

Davis, B. D. 1970. Prospects for genetic intervention in man. *Science, 170*: 1279-1283.

De Lucia, P. and J. Cairns, 1969. Isolation of an *E. coli* strain with a mutation affecting DNA polymerase. *Nature, 224*: 1164-1166.

Englund, P. T., M. P. Deutscher, T. M. Jovin, R. B. Kelly, N. R. Cozzarelli and A. Kornberg. 1968. Structural and functional properties of DNA polymerase. *Cold Spring Harbor Symp. Quant. Biol., 33*: 1-9.

Gellert, M. 1967. Formation of covalent circles of λDNA by *E. coli* extracts. *Proc. Nat. Acad. Sci. 57*: 148-155.

Gellert, M., J. W. Little, C. K. Oshinsky and S. B. Zimmerman. 1968. Joining of DNA strands by DNA ligase of *E. coli. Cold Spring Harbor Symp. Quant. Biol., 33*: 21-26.

Goulian, M. and A. Kornberg. 1967. Enzymatic synthesis of DNA. XXIII. Synthesis of circular replicative form of phage φχ174 DNA. *Proc. Nat. Acad. Sci., 58*: 1723-1730.

Goulian, M., A. Kornberg, and R. L. Sinsheimer. 1967. Enzymatic synthesis of DNA. XXIV. Synthesis of infectious phage φχ174 DNA. *Proc. Nat. Acad. Sci., 58*: 2321-2328.

Gross, J. D. 1971. DNA replication in bacteria. In *Current Topics in Microbiology and Immunology*. Springer Verlag, Berlin, Heidelberg and New York.

Holley, R. W. et al. 1965. Structure of a ribonucleic acid. *Science, 147*: 1462-1465.

Jacob, F. and J. Monod. 1961. Genetic regulatory mechanisms in the synthesis of proteins. *J. Mol. Biol., 3*: 318-356.

Knippers, R. 1970. DNA polymerase II. *Nature, 228*: 1050-1053.

Kohne, D. E. 1968. Isolation and characterization of bacterial ribosomal RNA cistrons. *Biophys. J., 8:* 1104-1118.

Kornberg, A. 1969. Active center of DNA polymerase. *Science, 163*: 1410-1418.

Kornberg, T. and M. Gefter. 1970. DNA synthesis in cell-free extracts of a DNA polymerase-defective mutant. *Biochem. Biophys. Res. Commun., 40*: 1348-1355.

Merril, C. R., M. R. Geier and J. C. Petricciani. 1971. Bacterial virus gene expression in human cells. *Nature, 233*: 398-400.

Moses, R. E. and C. C. Richardson. 1970. A new DNA polymerase activity of *E. coli*. I. Purification and properties of the activity present in *E. coli* pol Al. *Biochem. Biophys. Res. Commun., 41*: 1557-1564.

Olivera, B. M. and I. R. Lehman. 1967. Linkage of polynucleotides through phosphodiester bonds by an enzyme from *E. coli. Proc. Nat. Acad. Sci., 57*: 1426-1433.

Okazaki, R., T. Okazaki, K. Sakabe, K. Sugimoto, and A. Sugino. 1968. Mechanism of DNA chain growth. I. Possible discontinuity and unusual secondary structure of newly synthesized chains. *Proc. Nat. Acad. Sci., 59*: 598-605.

Ptashne, M. 1967. Isolation of the λ repressor. *Proc. Nat. Acad. Sci. 57*: 306-313.

Richardson, C. C. 1969. Enzymes in DNA metabolism. *Ann. Rev. Biochem., 38*: 795-840.

Ruddle, F. H., V. M. Chapman, T. R. Chen, and R. J. Klebe. 1970. Genetic analysis with man-mouse somatic cell hybrids. *Nature, 227*: 248-257.

Shapiro, J., L. Machatti, L. Eron, G. Ihler, K. Ippen and J. Beckwith. 1969. Isolation of pure *lac* operon DNA. *Nature, 224*: 768-774.

Schneider, J. A. and M. C. Weiss. 1971. Expression of differentiated functions in hepatoma cell hybrids, tyrosine aminotransferase in hepatoma-fibroblast hybrids. *Proc. Nat. Acad. Sci. 68*: 127-131.

Schwartz, A. G., P. R. Cook and H. Harris. 1971. Correction of a genetic defect in a mammalian cell. *Nature, 230*: 5-8.

Temin, H. M. and S. Mizutani. 1970. RNA-dependent DNA polymerase in virions of Rons Sarcoma Virus. *Nature, 226*: 1211-1213.

Watson, J. D. and F. H. C. Crick. 1953a. A structure for desoxyribose nucleic acids. *Nature, 171*: 737-738.

————. 1953b. The structure of DNA. *Cold Spring Harbor Symp. Quant. Biol., 18*: 123-131.

Weiss, M. C. and H. Green. 1967. Human-mouse hybrid cell lines containing partial complements of human chromosomes and functioning human genes. *Proc. Nat. Acad. Sci., 58*: 1104-1111.

18. *Prospects for Genetic Intervention in Man*

BERNARD D. DAVIS

How close are we to correcting man's faulty genes? Some writers say, "very soon," and others are quite pessimistic. The author of this article calls attention to the genetic principles involved, and reviews the techniques presently available for manipulating an organism's genes. Some of the techniques sound promising, but none have been attempted on humans. Should such success be attained, new problems may arise, as this article points out. Who defines what a defective gene is? Toward what goals should man's genes be altered? Despite these problems, genetic research should not be curtailed because it holds the promise of considerably more good than harm for mankind. The author is Adele Lehman Professor of Bacterial Physiology at Harvard Medical School. He is an M.D. with special interests in the metabolism and genetics of microorganisms.

Extrapolating from the spectacular successes of molecular genetics, a number of essays and symposia (*1*) have considered the feasibility of various forms of genetic intervention (*2*) in man. Some of these statements, and many articles in the popular press, have tended toward exuberant, Promethean predictions of unlimited control and have led the public to expect the blue-printing of human personalities. Most geneticists, however, have had more restrained second thoughts.

Nevertheless, recent alarms about this problem have caused wide public concern, and understandably so. With nuclear energy threatening global catastrophe and with so many other technological advances visibly damaging the quality of life, who would wish to have scientists tampering with man's inner nature? Indeed, fear of such manipulation may arouse even more anxiety than fear of death. The mass media have accordingly welcomed sensational pronouncements about the dangers.

While such dangers clearly exist, it also seems clear that some scientists have dramatized them (*3*) in order to help persuade the public of the need for radical changes in our form of government (*4*). But however laudable the desire to improve our social structure, and however urgent the need to improve our protection against harmful uses of science and technology, exaggeration of the dangers from genetics will inevitably contribute to an already distorted public view, which incresingly blames science for our problems and ignores its contributions to our welfare. Indeed, irresponsible hyperbole on the genetic issue has already influenced the funding of research (*5*). It therefore seems important to try to assess objectively the prospects for modifying the pattern of genes of a human being by various means. But let us first note two genetic principles that must be taken into account.

Source: Copyright 1970 American Association for the Advancement of Science. Reprinted with permission of the publisher and author from *Science, 170*:1279-1283, 18 December 1970.

Relevant Genetic Principles

Polygenic Traits and Behavioral Genetics. The recognition of a gene, in classical genetics, depends on following the distribution of two alternative forms (alleles) from parents to progeny. In the early years of genetics, after the rediscovery of Mendel's laws in 1900, this analysis was possible only for those genes that exerted an all-or-none control over a corresponding monogenic trait—for example, flower color, eye color, or a hereditary disease such as hemophilia. The study of such genes has continued to dominate genetics. However, monogenic traits constitute a small, special class. Most traits are polygenic: that is, they depend on multiple genes, and so they vary continuously rather than in an all-or-none manner. Moreover, each gene itself is polymorphic—that is, it is capable of existing, as a result of mutation, in a variety of different forms (alleles); and though the protein products of these alleles differ only slightly in structure, they often differ markedly in activity.

For our purpose it is especially pertinent that the most interesting human traits—relating to intelligence, temperament, and physical structure—are highly polygenic. Indeed, man undoubtedly has hundreds of thousands of genes for polygenic traits, compared with a few hundred recognizable through their control over monogenic traits. However, the study of polygenic inheritance is still primitive; and the difference from monogenic inheritance has received little public attention. Education on the distinction between monogenic and polygenic inheritance is clearly important if the public is to distinguish between realistic and wild projections for future developments in genetic intervention in man.

Interaction of Heredity and Environment. The study of polygenic inheritance is difficult in part because it requires statistical analysis of the consequences of reassortment, among the progeny, of many interacting genes. In addition, even a full set of relevant genes does not fixedly determine the corresponding trait. Rather, most genes contribute to determining a *range of potential* for a given trait in an individual, while his past and present environments determine his phenotype (that is, his actual state) within that range. At a molecular level the explanation is now clear: the structure of a gene determines the structure of a corresponding protein, while the interaction of the gene with subtle regulatory mechanisms, which respond to stimuli from the environment, determines the amount of protein made. Hence, the ancient formulation of the question of heredity versus environment (nature versus nurture) in qualitative terms has presented a false dichotomy, which has led only to sterile arguments.

Possibilities in Genetic Manipulation

Somatic Cell Alteration. Bacterial genes can already be isolated (6) and synthesized (7); and while the isolation of human genes still appears to be a formidable task, it may also be accomplished quite soon. We would then be able to synthesize and to modify human genes in the test tube. However, the incorporation of externally supplied genes into human cells is another matter. For while small blocs of genes can be introduced in bacteria, either as naked DNA (transformation) or as part of a nonlethal virus (transduction), we have no basis for estimating how hard it will be to overcome the obstacles to applying these methods to human cells. And if it does become possible to incorporate a desired gene into some cells, in the intact body, incorporation into all the cells that could profit thereby may well remain difficult. It thus seems possible that diseases depending on deficiency of an extracellular product, such as insulin, may be curable long before the bulk of hereditary diseases, where an externally supplied gene can benefit only those defective cells that have incorporated it and can then make the missing cell component.

Such a one-shot cure of a hereditary disease, if possible, would clearly be a major improvement over the current practice of continually supplying a missing gene product, such as insulin. (It could be argued that improving the soma in this way, without altering the germ cells, would help perpetuate hereditary defectives; but so does conventional medical therapy.) The danger of undesired side effects, of course, would have to be evaluated, and the day-to-day medical use of such material would have to be regulated: but these problems do not seem to differ significantly from those encountered with any novel therapeutic agent.

Germ Cell Alteration. Germ cells may prove more amenable than somatic cells to the introduction of DNA, since they could be exposed in the test tube and therefore in a more uniform and controllable manner. Another conceivable approach might be that of *directed mutagenesis*: the use of agents that would bring about a specific desired alteration in the DNA, such as reversal of a mutation that had made a gene defective. So far, however, efforts to find such directive agents have not been successful: all known mutagenic agents cause virtually random mutations, of which the vast majority are harmful rather than helpful. Indeed, before a mutagen could be directed to a particular site it would probably have to be attached first to a molecule that could selectively recognize a particular stretch of DNA (*8*); hence a highly selective mutagen would have to be at least as complex as the material required for selective genetic recombination.

If predictable genetic alteration of germ cells should become possible it would be even more useful than somatic cure of monogenic diseases, for it could allow an individual with a defective gene to generate his own progeny without condemning them to inherit that gene. Moreover, there would be a long-term evolutionary advantage, since not only the immediate product of the correction but also subsequent generations would be free of the disease.

Genetic Modification of Behavior. In contrast to the cure of specific monogenic diseases, improvement of the highly polygenic behavioral traits would almost certainly require the replacement, in germ cells, of a large but specific complement of DNA. Since I find such replacement, in a controlled manner, very hard to imagine, I suspect that such modifications will remain indefinitely in the realm of science fiction, like the currently popular extrapolation from the transplantation of a kidney or a heart, with a few tubular connections, to that of a brain, with hundreds of thousands of specific neural connections. However, this consideration would not apply to the possibility of impairing cerebral function by genetic transfer, since certain monogenic diseases are known to cause such impairment.

Copying by Asexual Reproduction (Cloning). We now know that all the differentiated somatic cells of an animal (those from muscle, skin, and the like) contain, in their nuclei, the same complete set of genes. Every somatic cell thus contains all the genetic information required for copying the whole organism. In different cells different subsets of genes are active, while the remainder are inactive. Accordingly, if it should become possible to reverse the regulatory mechanism responsible for this differentiation any cell could be used to start an embryo. The individual could then be developed in the uterus of a foster mother, or eventually in a glorified test tube, and would be an exact genetic copy of its single parent. Such asexual reproduction could thus be used to produce individuals of strictly predictable genetic endowment; and there would be no theoretical limit to the size of the resulting clone (that is, the set of identical individuals derivable from a single parent and from successive generations of copies).

Though differentiation is completley reversible in the cells of plants (as in the transfer of cuttings), it is ordinarily quite irreversible in the cells of higher animals. This stability, however, depends on the interaction of the nucleus with the surrounding cytoplasm; and it is now possible to transfer a nucleus, by microsurgery or cell fusion, into the cytoplasm of a different kind of cell. Indeed, in frogs differentiation has been completely reversed in this way: when the nucleus of an egg cell is replaced by a nucleus from an intestinal cell embryonic development of the hybrid cell can produce a genetic replica of the donor of the nucleus (9). This result will probably also be accomplished, and perhaps quite soon, with cells from mammals. Indeed, there is considerable economic incentive to achieve this goal, since the copying of champion livestock could substantially increase food production.

Another type of cloning can already be accomplished in mammals: when the relatively undifferentiated cells of an early mouse embryo are gently separated each can be used to start a new embryo (10). A large set of identical twins can thus be produced. However, they would be copies of an embryo of undetermined genetic structure, rather than of an already known adult. This procedure therefore does not seem tempting in man, unless the production of identical twins (or of greater multiplets) should develop special social values, such as those suggested by Aldous Huxley in *Brave New World.*

Predetermination of Sex. Though no one has yet succeeded in directly controlling sex by separating XX and XY sperm cells, this technical problem should be soluble. Moreover, in principle it is already possible to achieve the same objective indirectly by aborting embryos of the undesired sex: for the sex of the embryo can be diagnosed by tapping the amniotic fluid (amniocentesis) and examining the cells released into that fluid by the embryo.

Wide use of either method might cause a marked imbalance in the sex ratio in the population, which could lead to changes in our present family structure (and might even be welcomed in a world suffering from overpopulation). Alternatively, new social or legal pressures might be developed to avert a threatened imbalance (11). But though there would obviously be novel social problems, I do not think they would strain our powers of social adaptation nearly as much as some urgent present problems.

Selective Reproduction. A discussion of the prospects for molecular and cellular intervention in human heredity would be incomplete without noting that any society wishing to direct the evolution of its gene pool already has available an alternative approach: selective breeding. This application of classical, transmission genetics has been used empirically since Neolithic times, not only in animal husbandry, but also, in various ways (for example, polygamy, *droit de seigneur,* caste system), in certain human cultures. Declaring a moratorium on genetic research, in order to forestall possible future control of our gene pool, would therefore be locking the barn after the horse was stolen.

Having reviewed various technical possibilities, I would now like to comment on the dangers that might be presented by their fulfillment and to compare these with the consequences of efforts to prevent this development.

Evaluation of the Dangers

Gene Transfer. I have presented the view that if we eventually develop the ability to incorporate genes into human germ cells, and thus to repair monogenic defects, we would still be far from specifying highly polygenic behavioral traits. And with somatic cells such an influence seems altogether excluded. For though genes undoubtedly direct in considerable detail the pattern of development of the brain, with its network of connections of 10 billion or more nerve cells, the introduction of new DNA following this development clearly could not redirect the already formed network; neither could we expect it to modify the effect of learning on brain function.

To be sure, since we as yet have little firm knowledge of behavioral genetics we cannot exclude the possibility that a few key genes might play an especially large role in determining various intellectual or artistic potentials or emotional patterns. But even if it should turn out to be technically possible to tailor the psyche significantly by the exchange of a small number of genes in germ cells, it seems extremely improbable that this procedure would be put to practical use. For it will always be much easier, as Lederberg *(12)* has emphasized, to obtain almost any desired genetic pattern by copying from the enormous store already displayed in nature's catalog.

While the improvement of cerebral function by polygenic transfer thus seems extremely unlikely, one cannot so readily exclude the technical possibility of impairing this function by transfer of a monogenic defect. And having seen genocide in Germany and massive defoliation in Vietnam, we can hardly assume that a high level of civilization provides a guarantee against such an evil use of science. However, several considerations argue against the likelihood that such a future technical possibility would be converted into reality. The most important is that monogenic diseases, involving hormonal imbalance or enzymatic deficiencies, produce gross behavioral defects, whose usefulness to a tyrant is hard to imagine. Moreover, even if gene transfer is achieved in cooperating individuals, an enormous social effort would still be required to extend it, for political or military purposes, to mass populations. Finally, in contrast to the development of nuclear energy, which arose as an extension of already accepted military practices, the potential medical value of gene transfer is much more evident than its military value; hence a "genetic bomb" could hardly be sprung on the public as a secret weapon. Accordingly, we are under no moral obligation to sacrifice genetic advances now in order to forestall such remote dangers: if and when gene transfer in man becomes a reality there would still be time to assert the cultural and medical traditions that would promote its beneficial use and oppose its abuse.

This last obstacle would be eliminated if it should prove possible to develop a virus that could be used to infect a population secretly with specific genes, and it is the prospect of this ultimate horror that seems to cause most concern. However, for reasons that I have presented above the technical possibility of producing useful modifications of personality by infections of germ cells seems extremely remote, and the possibility of doing so by infecting somatic cells in an already developed individual seems altogether excluded. These fears thus do not seem realistic enough to help guide present policy. Nevertheless, the problem cannot be entirely ignored: in a country that has recently been embarrassed by its accumulation of rockets containing nerve gas even the remote possibility of handing viral toys to Dr. Strangelove will require vigilance.

Genetic Copies. If the cloning of mammals becomes technically feasible its extension to man will undoubtedly be very tempting, on the grounds that enrichment for proved talent by this means might enormously enhance our culture, while the risk of harm seemed small. Since society may be faced with the need to make decisions in this area quite soon, I would like to offer a few comments in the hope of encouraging public discussion.

On the one hand, in fields such as mathematics or music, where major achievements are restricted to a few especially gifted people, an increase in their number might be enormously beneficial—either as a continuous supply from one generation to another or as an expanded supply within a generation. On the other hand, a succession of identical geniuses might exert an excessively conservative influence, depriving society of the richness that comes from our inexhaustible supply of new combinations of genes. Or genius might fail to flower, if its drive depended heavily on parental influence or on cultural climate. And in the literary, social, and political areas the cultural climate surely plays so large a role that there may be little basis for expecting outstanding achievement to be continued by a scion. The world might thus be quite disappointed by the contributions of another Tolstoy, Churchill, or Martin Luther King, or even another Newton or Mozart. Moreover, though experience with monozygotic twins is somewhat reassuring, persons produced by copying might suffer from a novel kind of "identity crisis."

Though our system of values clearly places us under moral obligation to do everything possible to cure disease, there is no comparable basis for using cloning to advance culture. The responsibility for initiating such a radical departure in human reproduction would be grave, and surely many will feel that we should not do so. But I suspect that it would be impossible to enforce any such prohibition completely: the potential gain seems too large, and the procedure would require the cooperation of only a very small group of people. Hence whatever the initial social concensus, I suspect that a stable attitude would not emerge until after some early tests, whether legal or illegal, had demonstrated the magnitude of the problems and the gains.

A much greater threat, I believe, would be the use of cloning for the large-scale amplification of a few selected individuals. Who would wish to send a child to a school with a large set of identical twins as his classmates? Moreover, the success of a species depends not only on its adaptation to its present environment but also on its possession of sufficient genetic variety to include some individuals who could survive in any future environment. Hence if cloning were extended to the point of markedly homogenizing the population, it could create an evolutionary danger. However, we have already lived for a long time with a similar possibility: any male can provide a virtually limitless supply of germ cells, which can be used in artificial insemination; yet genetic homogenization by this means has not become the slightest threat. Since cloning is unlikely to become nearly so easy it is difficult to see a rational basis for the fear that its technical possibility would increase the threat.

Implications for Genetic Research. Though the dangers from genetics seem to me very small compared with the immense potential benefits, they do exist: its applications could conceivably be used unwisely and even malevolently. But such potential abuses cannot be prevented by curtailing genetic research. For one thing, we already have on hand a powerful tool (selective breeding) that could be used to influence the human gene pool, and this technique could be used as wisely or unwisely as any future additional techniques. Moreover, since the greatest fear is that some tyrant might use genetic tools to regulate behavior, and especially to depress human potential, it is important to note that we already have on hand pharmacological, surgical, nutritional, and psychological methods that could generate parallel problems much sooner. Clearly, we shall have to struggle, in a crowded and unsettled world, to prevent such a horrifying misuse of science and to preserve and promote the ideal of universal human dignity. If we succeed in developing suitable controls we can expect to apply them to any later developments in genetics. If we fail—as we may—limitations on the progress of genetics will not help.

If, in panic, our society should curtail fundamental genetic research, we would pay a huge price. We would slow our current progress in recognizing defective genes and preventing their spread; and we would block the possibility of learning to repair genetic defects. The sacrifice would be even greater in the field of cancer: for we are on the threshold of a revolutionary improvement in the control of these malignant heriditary changes in somatic cells, and this achievement will depend on the same fundamental research that also contributes toward the possibility of cloning and of gene transfer in man. Finally, it is hardly necessary to note the long and continuing record of nonmedical benefits from genetics, including increased production and improved quality of livestock and crops, steadier production based on resistance to infections, vastly increased yields in antibiotic and other industrial fermentations, and, far from least, the pride that mankind can feel in one of its most imaginative and creative cultural achievements: understanding of some of the most fundamental aspects of our own physical nature and that of the living world around us.

While specific curtailment of genetic research thus seems impossible to justify, we should also consider briefly the broader proposal (see, for example, 8) that we may have to limit the rate of progress of science in general, if we wish to prevent new powers from developing faster than an inadequate institutional framework can be adjusted to handle them. While one can hardly deny that this argument may be valid in the abstract, its application to our present situation seems to me dangerous. No basis is yet in sight for calculating an optimal rate of scientific advance. Moreover, only recently have we become generally aware of the need to assess and control the true social and environmental costs of various uses of technology. Recognition of a problem is the first step toward its solution, and now that we have taken this step it would seem reasonable to assume, until proved otherwise, that further scientific advance can contribute to the solutions faster than it will expand the problems.

Another consideration is that we cannot destroy the knowledge we already have, despite its potential for abuse. Nor can we unlearn the scientific method, which is available for all who wish to wrest secrets from nature. So if we should choose to curtail research in various fundamental areas, out of fear of possible long-range application, we must recognize that other societies may make a different choice. Knowledge is power, and power can be used for good or for evil; and, since the genie that brings new knowledge is already out of the bottle, we must learn to direct the use of the resulting power rather than curse the genie or try to confine him.

We cannot see how far the use of science as a scapegoat for many of our social problems will extend. But the gravity of the threat may be underscored by recalling that another politically based attack on science, Lysenkoism, utterly destroyed genetics in the Soviet Union and seriously crippled agriculture, from 1935 to 1965 *(13)*. [This development illustrates ironically the unstable relation between political and scientific ideas: for Karl Marx had unsuccessfully requested permission to dedicate the second volume of *Das Kapital* to Charles Darwin *(14)*!] Moreover, the current attacks on genetics from the New Left can build on, and have no doubt contributed to, widespread public anxiety concerning gene technology. Thus while a recent report prepared for the American Friends Service Committee *(15)* presents an open and thoughtful view on such questions as contraception, abortion, and prolongation of the period of dying, it is altogether opposed to any attempted genetic intervention, including the cure of hereditary disease.

Genetics will surely survive the current attacks, just as it survived attacks from the Communist Party in Moscow and from fundamentalists in Tennessee. But meanwhile if we wish to avert the danger of some degree of Lysenkoism in our country we may have to defend vigorously the value of objective and verifiable knowledge, especially when it comes into conflict with political, theological, or sociological dogmas.

REFERENCES AND NOTES

1. P.B. Medawar, *The Future of Man* (Basic Books, New York, 1960); Symposium on "Evolution and Man's Progress," *Daedalus* (Summer, 1961); G. Wolstenholme, Ed., *Man and His Future* (Little, Brown, Boston, 1963); J. Lederberg, *Nature 198*, 428 (1963); J.S. Huxley, *Essays of a Humanist* (Harper and Row, New York, 1964); T.M. Sonneborn, Ed., *The Control of Human Heredity and Evolution* (Macmillan, New York, 1965); R. D. Hotchkiss, *J. Hered. 56*, 197 (1965); J.D. Roslansky, Ed., *Genetics and the Future of Man* (Appleton-Century-Crofts, New York, 1966); N.H. Horowitz, *Perspect. Biol. Med. 9*, 349 (1966).
2. The term "genetic engineering" seemed at first to be a convenient designation for applied molecular and cellular genetics. However, I agree with J. Lederberg [*The New York Times*, Letters to the editor, 26 September (1970)] that the overtones of this phrase are undesirable.
3. Editorials, *Nature 224*, 834, 1241 (1969); J. Shapiro, L. Eron, J. Beckwith, *ibid.*, p. 1337.
4. J. Beckwith, *Bacteriol. Rev. 34*, 222 (1970).
5. P. Handler, *Fed. Proc. 29*, 1089 (1970).
6. J. Shapiro, L. MacHattie, L. Eron, G. Ihler, K. Ippen, J. Beckwith, *Nature 224*, 768 (1969).
7. K.L. Agarwal, and 12 others, *ibid. 227*, 27 (1970).
8. S.E. Luria, in *The Control of Human Heredity and Evoluation*, T.M. Sonneborn, Ed. (Macmillan, New York, 1965), p. 1.
9. R. Briggs and T.J. King, in *The Cell*, J. Brachet and A.E. Mirsky, Eds. (Academic Press, New York, 1959), vol. 1; J.B. Gurdon and H.R. Woodward, *Biol. Rev. 43*, 244 (1968).
10. B. Mintz, *J. Exp. Zool. 157*, 85, 273 (1964).
11. A. Etzioni, *Science 161*, 1107 (1968).
12. J. Lederberg, *Amer. Natur. 100*, 519 (1966).
13. Z.A. Medvedev; *The Rise and Fall of T.D. Lysenko* (Columbia Univ. Press, New York, 1969).
14. T. Dobzhansky, *Mankind Evolving* (Yale Univ. Press, 1962), p. 132.
15. *Who Shall Live?* Report prepared for the American Friends Service Committee (Hill and Wang, New York, 1970).

19. *The Genetic Implications of Population Control*

CARL JAY BAJEMA

The author of this paper examines genetic effects of three different population policies: (1) favoring continued population growth, (2) achieving zero population growth by voluntary means, and (3) achieving zero population growth by compulsory means. Each policy could have different genetic outcomes. At the moment, data indicate that Americans have evidently chosen the second. On the other hand, many nations of the world have yet to choose their policy. Much planning is necessary, since any choice will bring about great changes in future generations. Dr. Carl Bajema is research associate in population studies at the Harvard Center for Population Studies, Cambridge, Massachusetts, and an associate professor of biology at Grand Valley State College, Allendale, Michigan.

Each generation of mankind is faced with the awesome responsibility of having to make decisions concerning the quantity and quality (both genetic and cultural) of future generations. Because of its concern for its increasing population in relation to natural resources and quality of life, America appears to be on the verge of discarding its policy favoring continued population growth and adopting a policy aimed at achieving a zero rate of population growth by voluntary means. The policy that a society adopts with respect to population size will have genetic as well as environmental consequences.

Human populations adapt to their environments genetically as well as culturally (Bajema 1971). These environments have been and are changing very rapidly, with most of the changes being brought about by man himself. Mankind has, by creating a highly technological society, produced a society in which a significant proportion of its citizens cannot contribute to its growth or maintenance because of the limitations (both genetic and environmental) of their intellect. The modern technological societies of democratic nations offer their citizens a wide variety of opportunities for self-fulfillment but find that many of their citizens are incapable of taking advantage of the opportunities open to them.

American society, if it takes its responsibility to future generations seriously, will have to do more than control the size of its population in relation to the environment. American society will have to take steps to insure that individuals yet unborn will have the best genetic and environmental heritage possible to enable them to meet the challenges of the environment and to be able to take advantage of the opportunities for self-fulfillment made available by society.

Source: Reprinted with permission of the publisher and author from *Bioscience, 21 (2):*71–75, 15 January 1971.

The question of genetic quality cannot be ignored for very long by American society because it, like all other human societies, has to cope with two perpetual problems as it attempts to adapt to its environment. First, American society has to cope with a continual input of harmful genes into its population via mutation (it has been estimated that approximately one out of every five newly fertilized human eggs is carrying a newly mutated gene that was not present in either of the two parents). The genetic status quo can be maintained in a human population only if the number of new mutant genes added to the population is counterbalanced by an equal number of the mutant genes not being passed on due to the nonreproduction or decreased reproduction of individuals carrying these mutant genes. Otherwise, the proportion of harmful genes in the population will increase. Second, American society has to adapt to a rapidly changing environment. For instance, the technologically based, sociocultural, computer age environment being created by American society has placed a premium on the possession of high intelligence and creativity. Our society requires individuals with high intelligence and creativity to help it make the appropriate social and technological adjustments in order to culturally adapt to its rapidly changing environment. On the other hand, individuals require high intelligence and creativity in order that they, as individuals, can cope with the challenges of the environment and take advantage of the opportunities for self-fulfillment present in our society.

The proportion of the American population that already is genetically handicapped—that suffers a restriction of liberty or competence because of the genes they are carrying—is not small. Therefore, the genetic component of the human population-environment equation must be taken into account as we attempt to establish an environment that has a high degree of ecological stability and that maximizes the number of opportunities for self-fulfillment available to each individual human being.

Genetic Consequences of American Life Styles in Sex and Reproduction

American life styles with respect to sex and reproduction are currently in a tremendous state of flux and are changing rapidly. This makes it very difficult to accurately predict the genetic consequences of these life styles. Yet, because these life styles determine the genetic make-up of future generations of Americans, it is necessary that we evaluate the genetic consequences of past and present trends and speculate concerning the probable genetic consequences of projecting these trends into the future. Only then will we be able to determine the severity of the problem and to determine what steps, if any, need to be taken to maintain and improve the genetic heritage of future generations.

American society has developed modern medical techniques which enable many individuals with severe genetic defects to survive to adulthood. Many of these individuals can and do reproduce, thereby passing their harmful genes on to the next generation and increasing the frequency of these genes in the population. At present, there is no indication that heredity counseling decreases the probability that these individuals will have children. The life styles of these individuals with respect to reproduction is creating a larger genetic burden for future generations of Americans to bear.

The effect of American life styles in sex and reproduction on such behavioral patterns as intelligence and personality is much less clear. For instance, during most of man's evolution natural selection has favored the genes for intelligence. The genes for higher mental ability conferred an advantage to their carriers in the competition for survival and reproduction both within and between populations. Thus the more intelligent members of the human species passed more genes on to the next generation than did the less intelligent members, with the result that the genes for higher intelligence increased in frequency. As Western societies shifted from high birth and death rates toward low birth and death rates, however, a breakdown in the relation of natural selection to achievement or "success" took place.

The practice of family planning spread more rapidly among the better educated strata of society resulting in negative fertility differentials. At the period of extreme differences, which in the United States came during the great depression, the couples who were poorly educated were having about twice as many children as the more educated couples. The continued observation of a negative relationship between fertility and such characteristics as education, occupation, and income during the first part of this century led many scientists to believe that this pattern of births was a concomitant of the industrial welfare state society and must make for the genetic deterioration of the human race. This situation was, in part, temporary. The fertility differentials have declined dramatically since World War II so that by the 1960s some women college graduates were having 90% as many children as the U.S. average (Kirk, 1969). A number of recent studies of American life styles in reproduction, when taken collectively, seem to indicate that, as the proportion of the urban population raised in a farm environment decreases, as the educational attainment of the population increases, and as women gain complete control over childbearing (via contraception and induced abortion), the relationship between fertility and such characteristics as income, occupation, and educational attainment will become less negative and may even become positive (Goldberg, 1959, 1960, 1965; Freedman and Slesinger, 1961; Duncan, 1965).

The only two American studies which have related the intellectual ability (as measured by IQ) of an individual to his subsequent completed fertility have found the relationship to be essentially zero or slightly positive (Bajema, 1963; Waller, 1969). Further analyses of the data of these two studies indicate that it is difficult to infer the relationship between intelligence (as measured by IQ) and fertility from group differences in fertility with respect to income, occupation, and educational attainment because there is so much variation within these groups with respect to intelligence (Bajema, 1966, 1968; Waller, 1969).

The overall net effect of current American life styles in reproduction appears to be slightly dysgenic—to be favoring an increase in harmful genes which will genetically handicap a larger proportion of the next generation of Americans. American life styles in reproduction are, in part, a function of the population policy of the United States.

What will be the long-range genetic implications of controlling or not controlling population size in an industrialized welfare state democracy such as America?

Genetic Implications of Policies Favoring Continued Population Growth

Most contemporary human societies are organized in such a way that they encourage population growth. How is the genetic make-up of future generations affected by the size of the population? What will be the ultimate genetic consequences given a society that is growing in numbers in relation to its environment? One possible consequence is military aggression coupled with genocide to attain additional living space. This would result in genetic change insofar as the population eradicated or displaced differs genetically from the population that is aggressively expanding the size of its environment. The displacement of the American Indians by West Europeans is an example of this approach to the problem of population size in relation to the environment (Hulse, 1961). Throughout man's evolution such competition between different populations of human beings has led to an increase in the cultural and genetic supports for aggressive behavior in the human species. Violence as a form of aggressive behavior to solve disagreements among populations appears to have become maladaptive in the nuclear age. It will probably take a nuclear war to prove this contention.

If one assumes that military aggression plus genocide to attain additional living space is not an option open to a society with a population policy encouraging growth in numbers then a different type of genetic change will probably take place. Most scientists who have attempted to ascertain the probable effect that overcrowding in a welfare state will have on man's genetic make-up have concluded that natural selection would favor those behavior patterns that most people consider least desirable. For instance, Rene Dubos (1965), in discussing the effect of man's future environment on the direction and intensity of natural selection in relation to human personality patterns, states that:

"Most disturbing perhaps are the behavioral consequences likely to ensure from overpopulation. The ever-increasing complexity of the social structure will make some form of regimentation unavoidable; freedom and privacy may come to constitute antisocial luxuries and their attainment to involve real hardships. In consequence, there may emerge by selection a stock of human beings suited to accept as a matter of course a regimented and sheltered way of life in a teeming and polluted world, from which all wilderness and fantasy of nature will have disappeared. The domesticated farm animal and the laboratory rodent in a controlled environment will then become true models for the study of man."

The genetic and cultural undesirability of either of these two alternative outcomes for mankind makes it imperative that societies move quickly to adopt policies aimed at achieving and maintaining an optimum population size that maximizes the dignity and individual worth of a human being rather than maximizing the number of human beings in relation to the environment.

Genetic Implications of Policies Favoring Control of Population Size by Voluntary Means

There is strong evidence that contemporary societies can achieve control of their population size by voluntary means, at least in the short run.

What will be the distribution of births in societies that have achieved a zero population growth rate? In a society where population size is constant—where each generation produces only enough offspring to replace itself—there will still be variation among couples with respect to the number of children they will have. Some individuals will be childless or have only one child for a variety of reasons—biological (genetically or environmentally caused sterility), psychological (inability to attract a mate, desire to remain childless), etc. Some individuals will have to have at least three children to compensate for those individuals who have less than two children. The resulting differential fertility—variation in the number of children couples have—provides an opportunity for natural selection to operate and would bring about genetic change if the differences in fertility among individuals are correlated with differences (physical, physiological, or behavioral) among individuals.

The United States is developing into a social welfare state democracy. This should result in an environment that will evoke the optimal response from the variety of genotypes (specific combinations of genes that individuals carry) present in the population. It is questionable, however, as to whether a social welfare state democracy creates the type of environment that will automatically bring about a eugenic distribution of births resulting in the maintenance of enhancement of man's genetic heritage. It is also questionable as to whether a social welfare state democracy (or any society for that matter) will be able to achieve and maintain a zero population growth rate—a constant population size—by voluntary means.

Both Charles Darwin (1958) and Garrett Hardin (1968) have argued that universal compulsion will be necessary to achieve and maintain zero population growth. They argue that appeals to individual conscience as the means by which couples are to restrain themselves from having more than two children will not work because those individuals or groups who refused to restrain themselves would increase their numbers in relation to the rest, with the result that these individuals or groups with their cultural and/or biological supports for high fertility would constitute a larger and larger proportion of the population of future generations and *Homo contracipiens* would be replaced by *Homo progenetivis*.

Hardin (1968) raises this problem in his classic paper, The Tragedy of the Commons, when he states:

"If each human family were dependent only on its own resources; if the children of improvident parents starved to death; if, thus, overbreeding brought its own 'punishment' to the germ line—then there would be no public interest in controlling the breeding of families. But our society is deeply committed to the welfare state, and hence confronted with another aspect of the tragedy of the commons.

"In a welfare state, how shall we deal with the family, the religion, the race, or the class (or indeed any distinguishable and cohesive group) that adopts overbreeding as a policy to secure its own aggrandizement? To couple the concept of freedom to breed with the belief that everyone born has an equal right to the commons is to lock the world into a tragic course of action."

The only way out of this dilemma according to Hardin is for society to create reproductive responsibility via social arrangements that produce coercion of some sort. The kind of coercion Hardin talks about is mutual coercion, mutually agreed upon by the majority of the people affected. Compulsory taxes are an example of mutual coercion. Democratic societies frequently have to resort to mutual coercion to escape destruction of the society by the irresponsible. Mutual coercion appears to be the only solution to the problem of pollution. If Hardin is right, it may also be the only solution for any society that is attempting to control the size and/or the genetic make-up of its population.

Hardin's thesis has been questioned on the basis that children are no longer the economic assets they once were in agrarian societies. Rufous Miles has argued that, given today's postindustrial economy, children are expensive pleasures; they are economic liabilities rather than assets. Miles (1970) points out that:

"There is no conflict, therefore, between the economic self-interest of married couples to have small families and the collective need of society of preserve 'the commons.' It is in both their interests to limit procreation to not more than a replacement level. Unfortunately, couples do not seek their self-interest in economic terms alone, but in terms of total satisfactions. They are 'buying' children and paying dearly for them. The problem, therefore, is compounded of how to persuade couples to act more in their own economic self-interest and that of their children; how to assist them in obtaining more psychological satisfactions from sources other than large families; and how to replace the outworn and now inimical tradition of the large family with a new 'instant tradition' of smaller families."

As pointed out earlier in this paper, there is some evidence to support the contention that as American society becomes more urbanized, achieves higher levels of educational attainment, and allows its citizens to exercise complete control over their fertility, reproductive patterns will develop which will lead to a zero or negative population growth rate and a eugenic distribution of births. If this prediction is correct, then there will be no need for the adoption of mutual coercion—compulsory methods of population control—by American society in order to control the size and/or genetic quality of its population. If, on the other hand, these reproductive patterns do not develop or are transitory, it may very well be that reproduction will have to become a privilege rather than a right in social welfare state democracies in order to insure that these societies and their citizens do not have to suffer the environmental and genetic consequences of irresponsible reproduction.

What might the genetic consequences be if a society had to resort to mutual coercion—had to employ compulsory methods of population control—to control its numbers?

Genetic Implications of Compulsory Population Control

There are a number of methods by which compulsory population control can be achieved (Berelson, 1969). Mutual coercion could be institutionalized by a democratic society to ensure that couples who would otherwise be reproductively irresponsible are restricted to having only two children. Compulsory abortion and/or sterilization could be employed to guarantee that no woman bears more children than she has a right to under the rules set up by society.

A democratic society forced to employ mutual coercion to achieve zero population growth will probably assign everyone the right to have exactly two children. Because of the fact that some individuals will have only one child or will not reproduce at all, it will be necessary to assign these births needed to achieve replacement level to other individuals in that population. The assignment of these births could be made at random via a national lottery system. The result would probably be genetic deterioration. While those individuals who have less than two children would constitute a sample of the population with above average frequencies of various genetic defects, the selective removal of their genes would probably not be sufficient to counterbalance the continual input of mutations. Thus the result would probably be genetic deterioration even if the environment remained constant. If the environment were changing (this is about the one thing we can always count on—a constantly changing environment), the population would become even more genetically ill-adapted because those individuals in the society that are best adapted to changing environments and to the new environments would not be passing more genes on to the next generation a per person basis than those individuals less well adapted.

What kinds of eugenics programs could be designed for a democratic society where mutual coercion is institutionalized to ensure that couples who would otherwise be irresponsible are restricted to having two children?

One compulsory population control program designed to operate in a democratic society that has eugenic implications is the granting of marketable licenses to have children to women in whatever number necessary to ensure replacement of the population (say 2.2 children per couple). The unit certificate might be the deci-child or 1/10 a child and the accumulation of ten of these units, by purchase or inheritance or gift, would permit a woman in maturity to have one child. If equality of opportunity were the norm in such a society, those individuals with genetic make-ups that enable them to succeed (high intelligence, personality, etc.) would be successful in reaching the upper echelons of society and would be in the position of being able to purchase certificates from the individuals who were less successful because of their genetic limitations. The marketable baby license approach to compulsory population control, first discussed by Kenneth Boulding (1964) in his book *The Meaning of the Twentieth Century*, relies on the environment, especially the sociocultural environment, to do the selecting automatically, based on economics. The marketable baby license approach would probably bring about a better genetic adaptation between a population and its environment. Remember, the direction and rate of genetic change is to a great extent, a function of the social structure of the human population. The marketable baby license approach ensures that those people selected in society are those who are most successful economically. To ensure genetic improvement society would have to make sure that achievement and financial reward are much more highly correlated than they are at the present.

Another compulsory population program that a democracy might adopt would be to grant each individual the right to have two children and to assign the child-bearing rights of those individuals unable or unwilling to have two children to other individuals based on their performance in one or more contests (competition involving mental ability, personality, sports, music, arts, literature, business, etc.) The number of births assigned to the winners of various contests would be equal to the deficit of births created by individuals having less than two children. Society would then determine to a great extent the direction of its future genetic (and cultural) evolution by determining the types of contests that would be employed and what proportion of the winners (the top 1% or 5%) would be rewarded with the right to have an additional child above the two children granted to all members of society.

A society might even go further and employ a simple eugenic test—the examination of the first two children in order to assure that neither one was physically or mentally below average—which a couple must pass before being eligible to have additional children (Glass, 1967). The assignment of additional births to those individuals who passed the eugenic tests then could be on the basis of a lottery, marketable baby licenses, or contests, with the number of licenses equaling the deficit of births created by individuals who, at the end of their reproductive years (or at time of death if they died before reaching the end of their reproductive years), did not have any children or who only had one child.

The programs designed to bring about a eugenic distribution of births that have been discussed so far may prove to be incapable of doing much more than counteracting the input of harmful mutations. In order to significantly reduce the proportion of the human population that is genetically handicapped, a society may have to require that each couple pass certain eugenic tests before being allowed to become the genetic parents of *any* children. If one or both of the prospective genetic parents fail the eugenic tests, the couple could still be allowed to have children via artificial insemination and/or artificial inovulation utilizing human sperm and eggs selected on the basis of genetic quality. Such an approach would enable society to maintain the right of couples to have at least two children while improving the genetic birthright of future generations at the same time.

Successful control of the size and/or genetic quality of human populations by society may require restrictions on the right of individual human beings to reproduce. The right of individuals to have as many children as they desire must be considered in relation to the right of individuals yet unborn to be free from genetic handicaps and to be able to live in a high quality environment. The short-term gain in individual freedom attained in a society that grants everyone the right to reproduce and to have as many children as they want can be more than offset by the long-term loss in individual freedom by individuals yet unborn who, as a consequence, are genetically handicapped and/or are forced to live in an environment that has deteriorated due to the pressure of human numbers.

Conclusion

Each generation of mankind faces anew the awesome responsibility of making decisions which will affect the quantity and genetic quality of the next generation. A society, if it takes its responsibility to future generations seriously, will take steps to ensure that individuals yet unborn will have the best genetic and cultural heritage possible to enable them to meet the challenges of the environment and to take advantage of the opportunities for self-fulfillment present in that society.

The way in which a society is organized will determine, to a great extent, the direction and intensity of natural selection especially with respect to behavioral patterns: The genetic make-up of future generations is also a function of the size of the population and how population size is regulated by society. The genetic implications of the following three basic types of population policies were explored in this paper:

1. policies favoring continued population growth;
2. policies aimed at achieving zero population growth by voluntary means; and
3. policies aimed at achieving zero population growth by compulsory measures (mutual coercion mutually agreed upon in a democratic society).

If societies adopt compulsory population control measures, it will be for the control of population size and not for the control of the genetic make-up of the population. However, it is but a short step to compulsory control of genetic quality once compulsory programs aimed at controlling population size have been adopted. The author personally hopes that mankind will be able to solve both the quantitative and qualitative problems of population by voluntary means. Yet one must be realistic and consider the alternatives. This is what the author has attempted to do in this paper by reviewing the genetic implications of various population control programs.

REFERENCES

Bajema, C. 1963. Estimation of the direction and intensity of natural selection in relation to intelligence by means of the intrinsic rate of natural increase. *Eugen. Quart., 10:*175-187.

_____1966. Relation of fertility to educational attainment in a Kalamazoo public school population: A follow-up study. *Eugen. Quart. 13:*306—315.

_____ Relation of fertility to occupational status, IQ, educational attainment and size of family of origin: a follow-up study of a male Kalamazoo public school population. *Eugen. Quart., 15:*198–203.

Bajema, C. (ed.). 1971. *Natural Selection in Human Populations: The Measurement of Ongoing Genetic Evolution in Contemporary Human Societies.* John Wiley & Sons, New York. (In press.)

Berelson, B. 1968. Beyond family planning. *Science, 163:*533–543.

Boulding, K. 1964. *The Meaning of the Twentieth Century: The Great Transition.* Harper & Row, New York.

Darwin, C. 1958. *The Problems of World Population.* Cambridge University Press, Cambridge, England, 42 p.

Dubos, R. 1965. *Man Adapting.* Yale University Press, New Haven, Conn.

Duncan, O. 1965. Farm background and differential fertility. *Demography, 2:*240-249.

Freedman, R., and D. Slesinger 1961. Fertility differentials for the indigenous non-farm population of the United States. *Population Stud., 15:*161–173.

Glass, B. 1967. What Man Can Be. Paper presented at the American Association of School Administrators Convention, Atlantic City, N.J. 23p.

Goldberg, D. 1959. The fertility of two-generation urbanites. *Population Stud., 12:*214–222.

Goldberg, D. 1960. Another look at the Indianapolis fertility data. *Milbank Fund Quart. 38:*23–36.

Goldberg, D. 1965. Fertility and fertility differentials: Some observations on recent changes in the United States. In: *Public Health and Population Change,* M. Sheps and J. Ridley (eds.), p. 119-142. University of Pittsburgh Press, Pittsburgh, Pa.

Hardin, G. 1968. The tragedy of the commons. *Science, 162:*1243-1248.

Hulse, F. 1961. Warfare, demography and genetics. *Eugen. Quart., 8:*185–197.

Kirk, D. 1969. The genetic implications of family planning. *J. Med. Educ., 44* (Suppl 2):80–83.

Miles, R. 1970. Whose baby is the population problem? *Population Bull., 16:*3–36.

Waller, J. 1969. The relationship of fertility, generation length, and social mobility to intelligence test scores, socioeconomic status and educational attainment. Doctoral Thesis, University of Minnesota, Minneapolis. 100 p.

Fossil of a palm leaf. Courtesy the American Museum of Natural History.

20. *Chemical Origins of Life*

J. LAWRENCE FOX

Every so often, research efforts into the origin of life seem to falter and then suddenly new data inject excitement into the field. Such a finding was the isolation of amino acids from stony meteorites and the identification of small organic molecule precursors in deep space. This in no way proves the existence of extraterrestrial life, but it does say that simple organic materials are formed elsewhere in the universe. This article reviews previous research efforts on chemical evolution as well as recent progress. Dr. Fox does an unusually good job of clearly presenting a rather difficult, technical field of study. He is an assistant professor of biophysics at the University of Texas, Austin. This article was the basis of a seminar he presented at the annual meeting of the National Science Teachers Association in 1971.

Any discussion of the origin of life must necessarily be based upon a definition of life. Fig. 1 shows what at first appearance seem to be a mushroom, *Acetabularia* (a delicate genus of green algae), and the well-known mitotic division process of karyokinesis. However, these pictures are the clever work of a former Frenchman, S. LeDuc. They are formed from dyes and blotter paper and inorganic salts in solution. They are purely inorganic artifacts. Thus, a further investigation belies their initial appearance.

We might define a living thing as having the ability to divide and form copies of itself. Under the proper conditions, oil droplets may be made to divide, but they are not considered living cells. Another definition which we might adopt includes the ability to selectively assimilate chemicals and to bring about chemical transformations to produce needed biochemicals. But any respectably sized chemical or drug plant would fit this description. Thus, a loose definition would be insufficient.

We see that a detailed definition of life is essential to eliminate nonliving artifacts. Single criteria are insufficient; witness how easily LeDuc's artifacts deceived us. Note, too, that our exploration of space is partially concerned with the search for the presence of life elsewhere in the universe. We must, therefore, not restrict our definition of life to the point that it excludes forms of life unfamiliar to earth. Because we all generally have feelings for what is living and what is not, the burden of obtaining a suitable definition will be left up to the reader.

Since earliest recorded history man has sought an answer to the question of where he came from. The ancient Egyptians felt that various forms of amphibians, reptiles, and even mammals could arise spontaneously out of the muddy banks of the Nile in response to sunlight. The ancient Greeks believed that life could arise out of the basic

Source: Reprinted with permission of the publisher and author from *The Science Teacher, 39 (1)*:29-34. January 1972.

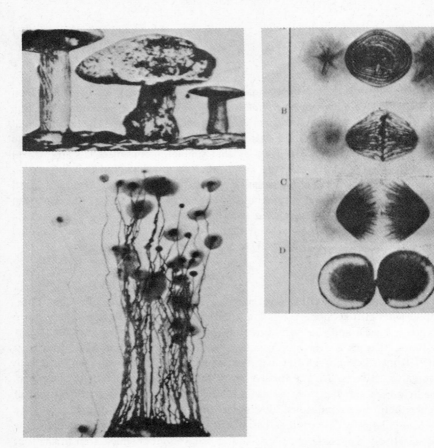

Fig. 1. Defining or even recognizing life presents real problems. These three photos, which seem to depict mushrooms, *Acetabularia*, and mitotic division, are actually inorganic artifacts formed from dyes, inorganic salts, and blotter paper. (From: Mechanism of Life by S. Le Duc. William Heinemann, Publisher, London. 1911.)

elements: sun, fire, earth, and water. These views continued well into the eighteenth century. They can best be described as the *spontaneous generation* of living forms. Inherent in the notion of spontaneous generation was a belief that something other than the elements, a *vital force,* was necessary. These vitalistic theories gained support and were essentially the only prevalent theories at the end of the Middle Ages.

As a result of the Renaissance, however, experimentation came into vogue, and attempts were made to confirm or deny the spontaneous generation of living organisms on an experimental basis. Early in the seventeenth century the famous Brussels physician Jean Baptiste van Helmont undertook an experiment in which he added sweaty underwear to a bowl of wheat kernels. After a few weeks, he noticed that mice had appeared. These were completely normal adult mice which were capable at once of proliferating. His experiment was taken as confirmation of spontaneous generation.

Shortly afterwards, William Harvey deduced that all living organisms originate from an egg. However, he felt this view was completely consistent with the notion of spontaneous generation and was unaware of the apparent paradox in his reasoning. Other great thinkers, such as Copernicus, Galileo, Bacon, and Descartes, also were firmly convinced of spontaneous generation.

In the mid-seventeenth century the Tuscan physician Francesco Redi performed a different experiment. He placed freshly killed snake meat into a large vessel. After several days worms appeared all over the meat. When he placed muslin over his container, he noted that the worms developed into flies. A similar sort of experiment was performed again using fresh snake meat. This time Redi placed the muslin on top of the vessel shortly after placing the meat inside it. He noticed that the flies swarmed over the muslin and dropped eggs on top of the muslin. However, no worms appeared in the meat. From this, Redi concluded that flies had laid their eggs on the meat in the first experiment, and that the meat had served as food for the worms. But if no eggs were present, no worms appeared, and no flies developed. This experiment was extended to a variety of meats from different animals, always with the same results. It supported the view that life can only come from pre-existing life.

Early in the eighteenth century, John T. Needham attempted a new set of experiments. He prepared mutton broth in glass vessels, boiled the broth, and then sealed the vessels with corks. After several days, small organisms appeared. Shortly after this, Lazzaro Spallanzani performed a similar experiment, but he sealed the vessel before boiling the nutrient solution. When the flasks were broken open several days later, no microorganisms of any sort were found. Spallanzani concluded that Needham's findings were due to the passage of airborne living forms through the cork. Needham was not convinced of Spallanzani's apparently clear-cut results. Instead, he replied that Spallanzani's vigorous boiling had been quite brutal and had probably destroyed the vital force necessary for the formation of life. Apparent support of this view was provided by Gay-Lussac, who demonstrated that the air remaining in the flask after Spallanzani's extensive boiling contained no oxygen. Needham argued that it could not, therefore, support life.

Refutation of the theory came finally from the laboratory of the German physician Theodor Schwann. Schwann prepared a flask which had a piece of glass tubing extending from it. The tubing was shaped into a coil and passed over an alcohol lamp. When the infusion was boiled, steam was forced out of the tubing, taking contaminants with it. After the flask was allowed to cool, however, entering air had to pass over the coils which were heated by the alcohol lamp. This experiment was allowed to proceed for periods of up to six weeks. And under no conditions were living forms observed. Nonetheless, this experiment still did not convince some of the advocates of spontaneous generation.

It was not until the mid-nineteenth century, when the famous French microbiologist Louis Pasteur began his experiments, that the issue was finally closed.

First Pasteur demonstrated the presence of microorganisms in the air. He did this by sucking air through a tube which was plugged with cotton. After a period of 24 hours, the suction was removed, and the cotton plug was washed with either and alcohol. The particles which settled out were then examined under a microscope, and thousands of organisms were seen.

Pasteur performed his decisive experiments by using a flask in which the neck was bent out in the shape of an S. Sugared yeast water was boiled in the flask and then allowed to cool. The bends in the glass prevented the airborne microorganisms from getting to the solution. Instead their path caused them to collect along the sides of the S-shaped tubing. Thus, no living organisms appeared in the yeast water. Pasteur's experiment served to demonstrate that the source of microorganisms was the air. The organisms did not arise from the nutrients. He was also able to show that the heat had not destroyed any "vital force," for when he broke off the neck and exposed the nutrient to the air, microorganisms began to thrive in the solution. Nor had he blocked the formation of life by restricting air inflow.

It should be pointed out that this type of experiment refuted clearly the experiments which purported to demonstrate spontaneous generation. It does not of itself, however, repudiate the notion of spontaneous generation. In spite of his enormous contribution, it is interesting to note that Louis Pasteur's experiments and his standing in French science effectively stymied research into the origin of life and a number of other areas in microbiology for half a century.

In the late nineteenth and early twentieth centuries, a beginning was made in elucidating the chemical structure of living organisms. This work led the Russian biologist A. I. Oparin to propose a theory on the origin of life on earth. It is worth our while to examine this theory, first proposed in 1924, since it is the impetus for much of the research done in this field [see Table 1].

Table 1. *Oparin's Theory of the Origin of Life*

1. Hot earth formation
2. Scant, reducing atmosphere
3. $> 1,000°C$, formation of unsaturated hydrocarbons
4. $< 1,000°C$, addition of H_2O and NH_3
5. $< 100°C$, rains, formation of "hot dilute soup"
6. Formation of coacervates
7. Chemical evolution (abiogenesis, biopoesis)
8. First organism

Oparin believed that the earth had been very hot during its early history. He conjectured that its scant primitive atmosphere was quite different from that present today. Most of the gases which were present were easily lost into space. Any oxygen which might have been present would probably have reacted rapidly with hot metals on the surface of the earth, forming metal oxides. Thus, oxygen and carbon dioxide were rare. The early atmosphere was not an oxidizing atmosphere as ours is today, but a reducing one—that is, it contained basically only hydrogen.

After the cooling of the earth to temperatures slightly in excess of $1,000°$ Celsius, it became possible for stable bonds to form between carbon atoms on the earth's surface and hydrogen atoms. In the initial stages, with hydrogen quite limiting, one might envision that hydrogen and carbon formed a single bond, satisfying but one of the four valences of carbon:

$$H— \quad + \quad —\overset{\displaystyle |}{\underset{\displaystyle |}{C}}— \quad \longrightarrow \quad H—\overset{\displaystyle |}{\underset{\displaystyle |}{C}}—\text{radical}$$

$$2H—\overset{\displaystyle |}{\underset{\displaystyle |}{C}}— \quad \longrightarrow \quad H—C\equiv C—H$$
$$\text{(acetylene)}$$

This carbon-hydrogen radical is very reactive and in all probability underwent reaction with itself to form acetylene. Acetylene contains two hydrogen atoms, and three covalent bonds are shared between two carbon atoms. As the hydrogen concentration increased, two hydrogens could become attached to a carbon atom. The reaction of two such units would form ethylene and so forth until there was a whole host of short hydrocarbon compounds containing multiple carbon-carbon bonds, and perhaps some methane. As the temperature of the earth approached $1,000°$ Celsius, the nitrogen-hydrogen and oxygen-hydrogen bonds became stable, and water and ammonia formed in the atmosphere. These compounds react rapidly with hydrocarbons containing multiple bonds.

$$H - C \equiv C - H + H_2O \text{ or } NH_3$$
$$\text{(acetylene)} \qquad \text{(water)} \quad \text{(ammonia)}$$
$$\rightarrow H, C, N, O \text{ compounds}$$

Thus, there rapidly developed a whole host of compounds which contained not only hydrogen and carbon, but also nitrogen and oxygen.

Once the earth cooled to approximately $100°C$, water vapor condensed and rains occurred. These rains washed the organic compounds into pools and ultimately into oceans. The British scientist J. B. S. Haldane, a contemporary of Oparin, labeled this "the hot dilute soup."

The next stage of Oparin's theory envisions the formation of complex units which he calls *Coacervates* from this hot dilute soup. Hydrophobic molecules (nonpolar molecules which do not dissolve well in water) undergo a phase separation in water to first form collids. These colloids may associate to form droplets with even less bound water. These droplets, which contain clusters of colloids surrounded by a shell of water, are called coacervates (see Fig. 2). They have been shown by Oparin to possess selective permeability and reaction-rate enhancements for incorporated enzymes. Oparin has devoted his life to studying the function of coacervates prepared from contemporary biological materials.

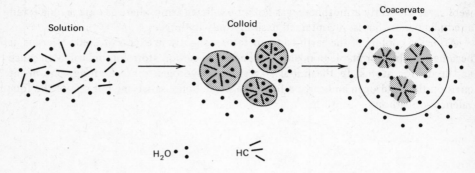

Fig. 2. Formation of coacervates, aggregates of colloidal droplets.

Oparin's theory next envisaged the long, lengthy development and selection of chemical reactions and molecules in these *protocell* coacervates. This period he labeled *chemical evolution*. It has also been referred to as abiogenesis or biopoesis. Oparin believed this period to have taken millions of years. The culmination of chemical evolution would presumably be a protocell which possessed the right kinds of reactions and the right kinds of compounds to be considered a living cell.

In light of some of the experimental findings that I shall discuss shortly, we will be able to come back and revise the Oparin theory in two specific areas. One is his notion of the formation of macromolecules in an aqueous environment; the second is the notion of a lengthy period of chemical evolution. But given Oparin's reasonable hypothesis for the origin of life on earth, it is now possible for us to begin a discussion of the experiments which bear on this subject.

In 1861, Butlcrov performed experiments which are meaningful in terms of the origin of life, although they were not performed in this context. He showed that the action of milk of lime on formaldehyde produced a series of simple sugars, although the chemistry of this reaction was not fully elucidated until 30 years later by Emil Fischer. In 1886, Low performed a similar sort of experiment using electric discharge. In 1904, Curtius used ultraviolet light from the sun to act on the ethyl ester of gylcine. This produced a fairly complex polypeptide product that demonstrated a biuret reaction.

In 1938, Groth and Suess reported that when a carbon dioxide—water atmosphere was irradiated with ultraviolet light, formaldehyde and glyoxal were major products. In 1951, Calvin and his associate used alpha particles generated by a cyclotron to bombard an atmosphere of carbon dioxide and hydrogen in the presence of ferrous ion. Formaldehyde, formic acid, and succinic acid were the products of this reaction.

Experiments producing nitrogen-containing compounds date back to 1898. Berthelot used electric discharge with simple alcohols and ethers in a nitrogen environment. In 1913, Loeb reported the formation of glycine from a mildly reducing atmosphere consisting of water vapor, ammonia, and carbon monoxide or carbon dioxide under the influence of electric discharge.

Several decades ago Harold C. Urey, while at the University of Chicago, argued for the kind of reducing atmosphere which had been discussed by Oparin, i.e., a hydrogen atmosphere with no carbon dioxide or carbon monoxide. In 1953, Urey's student, Stanley Miller, performed an experiment using the gaseous mixture of methane, ammonia, water, and hydrogen, under the influence of electric discharge. After a week of electric discharge, the reaction vessel was opened and found to contain a very large quantity of cyanide, evident both by its color and its odor, and aldehydes. Upon examination of this solution, four naturally occurring amino acids were found. Miller also noted the presence of fatty acids and some of their derivatives. He was unable to detect purines, pyrimidines, or other aromatic compounds among the reaction products. One point must be made: In the vicinity of electric discharge quite considerable temperatures may be reached. One, therefore, cannot completely rule out the notion of a thermal effect in Miller's experimental results.

Since then, many experiments of this sort have been performed under a variety of conditions. Abelson has extensively examined this system using an atmosphere of carbon dioxide, nitrogen, hydrogen, and water. Better yields of amino acids and unsaturated amino acids are obtained if nitrogen is replaced by ammonia. Unsaturated amino acids can be obtained only under conditions in which reducing atmospheres are prevalent. Grossenbacher and Knight in 1965, and more recently Ponnamperuma and Flores in 1966, exposed a mixture of methane, ammonia, and water to spark discharge and identified approximately ten amino acids from their experiments. In these experiments, 90 percent of the methane was converted into other compounds. Purines have also been detected in such experiments.

Two decades after the original Groth and Suess experiments of 1938, Groth and von Weyssenhoff reported the effects of ultraviolet light on a mixture of methane, ammonia, and water vapor. Traces of glycine and alanine were detected by paper chromatography. Terenin in 1959 reported the formation of the amino acids valine, norleucine, and a number of amines in an experiment similar to that performed by Groth and Suess.

In 1964, Harada and Fox examined the effects of short exposures to high temperatures of gaseous mixtures of methane, ammonia, and water. These gases were passed through a hot silicon tube at approximately 900° Celsius, collected, hydrolyzed, and analyzed. Fourteen of the 20 naturally occurring amino acids were detected in the solution. It is quite reasonable to suppose that the reaction pathway proceeded through a cyanide ion step. In 1968, Dose and Risi published the results of experiments in an ammonia, methane, carbon dioxide, and water atmosphere, following X-ray irradiation. A number of aliphatic acids and several amino acids were detected. Ponnamperuma has exposed methane, ammonia, and water vapor to 4.5 mev electrons from a linear accelerator and obtained amino acids, adenine, and some five- and six-carbon sugars.

Thus, it is possible to conclude that using a variety of irradiation sources, primitive atmospheres can be made to produce a variety of simple organic compounds. It should also be pointed out that forms of irradiation, such as X-rays, electric discharge, ultraviolet light, and alpha particles are considered to be ionizing radiations. They generally have a destructive effect rather than a constructive effect. Heat is the only form of radiation discussed so far which has a constructive effect. The speed with which compounds can be isolated from a primitive atmosphere, i.e., seconds, demonstrates the great difference between heat and other forms of irradiation. It should also be noted that heat cannot be ruled out in most of the experiments performed with other forms of irradiation and may actually be responsible for the formation of molecules. The low yields noted may be the result of extensive decomposition from the ionizing radiations.

One of the characteristics separating the nonliving from the living world is the presence of large molecules in biological systems. These macromolecules (proteins, nucleic acids, carbohydrates, and lipids) are formed from simple organic molecules, such as amino acids, nucleic acid bases, sugars, and fatty acids. Macromolecular formation requires a step in which the elements of water are lost in the process of forming the linking bond. Thus, it is difficult to imagine how these reactions may have occurred in the aqueous milieu. It is clear that some form of hypohydrous condition must obtain.

In 1958, Harada and Fox reported the panpolymerization of 18 common amino acids under hypohydrous conditions. They used two different approaches. The first was the thermal approach in which dry amino acids were heated above the boiling point of water to approximately 170° Celsius for periods of several hours to a day or two. The second set of conditions involved the use of the dehydrating agent, polyphosphoric acid. This is obtained by heating phosphoric acid for extended periods of time. With polyphosphoric acid, it is possible to use lower temperatures for the polymerization of amino acids, i.e., 50 to 60°C. The polymeric substance obtained from these experiments was labeled "proteinoid" to distinguish it from native protein, since a small number of nonnaturally occurring linkages were found in the preliminary work with the material. Molecular weights in the range of 4,000 to 10,000 were obtained. Table 2 summarizes some of the proteinoid's properties. It has been known for several decades that in basic media, cyanide ion will polymerize to form a high molecular weight material. It is generally believed, however, that this is a polynitrile and not truly a polypeptide. Upon acid hydrolysis, it is possible to detect amino acids; however, it is quite likely that the acid hydrolysis step introduces the elements of water to form amino acids through the decomposition of the polymer. Thus, this is not considered a relevant model.

Recently Krampitz and Fox have reported a new proteinoid synthesis. Amino acids are coupled to adenylic acid (AMP) in organic solvent with dehydrating agents. When these aminoacyl adenylates are placed in water, they undergo rapid polymerization to form a polymer which is primarily polypeptide. This material is called adenylate proteinoid to distinguish it from thermal proteinoid. This system bears a close resemblance to biological protein synthesis by the aminoacyl adenylate intermediate.

Similar experiments were reported by Schramm in 1956 for polynucleotide formation. Starting with nucleosides and ethylmetaphosphoric acid, Schramm

Table 2. *Common Properties of Thermal Proteinoid and Protein*

Qualitative composition
Range of quantitative composition
Limited Homogeneity
Molecular weights (4,000 to 10,000)
Color tests
Range of solubilities
Salting in and out
Precipitation by protein reagents
Hypochromicity
Infrared absorption
Amino acid recovery upon acid hydrolysis
Susceptibility to proteolytic enzymes
"Enzymelike" properties
Inactivatibility by heating in buffer
Nutritive quality
Hormonal activity
Self-assembly
Selective binding to polynucleotides

reported the formation of short units of nucleotides strung together. However, a number of laboratories have been unable to confirm these findings, and ethyl metaphosphoric acid is of dubious prebiological relevance. Schwartz, working in Fox's laboratory, was able to obtain polymers of cytidine by the thermal method.

Sanchez and Farris, working in Orgeil's laboratory, have reported on a number of polymerizations for polynucleotides. Many of the agents which they used have questionable relevance in the prebiological context. Therefore, we will not spend any time discussing the extensive work which has come from this laboratory. Oro has reported the production of polynucleotides by the action of ultraviolet light on nucleotides. Thus, once again, we see that the action of several forms of energy may bring about the reaction of simple molecules to form macromolecules. Fig. 3 summarizes these chemical formation reactions.

What then of higher levels of organization? How do these materials undergo self-assembly? Some answers to these questions have come from the laboratory. As outlined above, Harada and Fox previously demonstrated that a simple primitive atmosphere could produce simple macromolecules by the action of heat. They subsequently showed that amino acids can be coupled together to form a proteinoid material, again in the presence of heat. They have taken this one step further.

Adding proteinoid to water produces a new structure. This process can be aided by dissolving the proteinoid in warm solution and then allowing the solution to cool. Small units called *microspheres* form. These possess the properties of coacervates, but differ in that they are much more stable and more versatile.

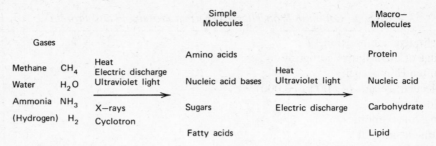

Fig. 3. Complex organic molecules may be formed using several energy sources.

Microspheres have been stored in the laboratory for up to three years. They possess what appear under electron microscopes to be double membranes, and these are selectively permeable. Simple laboratory experiments for biology students have been set up demonstrating selective retention of starch as opposed to glucose. These units possess a variety of weak catalytic activities, which is quite in accord with the notion that they are not the highly selected products of evolution. A variation of the acidic or basic residue content of the proteinoid used to make these microspheres produces a variation in their ability to stain gram positive or gram negative. Under the electron microscope, using standard staining techniques, they appear to have some degree of internal structure. One of their most surprising characteristics is their ability (under changes of salinity or pH) to undergo either budding or fission. In a saturated solution they may undergo heterotrophic poliferation. If amino acyladenylates are added to a solution of microspheres, the microspheres catalyze the production of more proteinoid which can be used for division of these microspheres.

The addition of proteinoid with varying amino acid composition to polynucleotides produces a variety of results depending on the nature of the material used. If DNA-like material is mixed with basic proteinoid, long threads reminiscent of chromatin are formed. When ribosomal RNA is mixed with basic proteinoids, microspheres not too dissimilar or shaped too dissimilar from ribosomes are formed.

Since these microspheres serve as such wonderful protocell models, and since they are composed only of the protein-like material, there has evolved a *protein-first* hypothesis. Virtually all of the properties of a living cell, save the nucleic acid-directed self-replication, are mimicked in these spheres. Table 3 lists a summary of microsphere properties.

It may, therefore, be considered that proteins were sufficient for the formation of protocells. Proteins in turn could eventually produce nucleic acid templates which would allow the control necessary for biological evolution. This theory runs contrary to what might be described as the *nucleic acid first* hypothesis, which was advocated by Hermann J. Muller and, more recently, by Alexander Rich. These gentlemen believed that for life to exist at all, nucleic acids must have been present first.

It is possible to state in conclusion that laboratory experiments have verified the feasibility of spontaneous generation of living forms on the primitive earth. A great

Table 3. *Properties of Thermal Proteinoid Microspheres*

Stability
Microscopic size
Variability in shape
Uniformity of size
Stainability
Osmotic sensitivity
Ultrastructure (electron microscope)
Double membrane
Selective permeability
Catalytic activities
Variation in association
Propagation (budding and fission)
Growth by accretion
Motility

variety of energy sources produce simple molecules, which in turn produce macromolecules. We have seen that these macromolecules may self-assemble to form protocell type units. Since this whole process may occur in a matter of seconds, the notion of spontaneous generation may be resurrected. One also sees that a great number of naturally occurring molecules are formed during these experiments; very few nonnaturally occurring molecules appear. In this context it must also be remembered that of all the molecules of organic chemistry, naturally occurring aromatic compounds are among the most stable known. Thus, these were probably selected in the very beginning.

The foregoing account allows us then to make two fundamental revisions in Oparin's theory. One: Hypohydrous conditions must have been necessary for the polymerization of simple molecules into macromolecules. The feasibility of these conditions occurring on the contemporary earth is quite well documented. Within an inch of the surface of virtually any active volcanic cinder cone in the world, one may find temperatures in excess of 200° Celsius.

Two: Millions of years were likely unnecessary for chemical evolution. Laboratory experiments again demonstrate that the reactions which occur in biological systems and the compounds which are formed are those which are most stable. Living systems have simply learned to control and regulate these to a more sophisticated degree than simple chemical systems would permit. One may envision the spontaneous generation of a complete but simple, unicellular living system on the primitive earth in a matter of seconds. Excessive biological contamination today makes it impossible to ever verify that such processes are recurring contemporaneously.

However, the search for extraterrestial life is one of the guiding principles of our space voyages.

In this regard, it is interesting to note the recent discoveries of amino acids in meteorites and the reports of alanine and glycine in moon dust samples. Besides strengthening the idea of a pattern for chemical evolution leading to the origin of life, the discoveries reinforce the possibility of chemical evolution leading to life elsewhere in the universe.

In this regard, it is interesting to note the recent discoveries of amino acids in moon dust samples and in meteorites. Analysis of Apollo 11 and 12 lunar fines has demonstrated the presence of glycine and alanine, two of nature's most prevalent amino acids. Hydrolysis of organic material on lunar fines yields also glutamic acid, aspartic acid, serine, and threonine, four more of the eighteen common amino acids. Analysis of the Murchison and Murray meteorites, carbonaceous chondrites, has demonstrated the presence of six amino acids: glycine, alanine, aspartic acid, glutamic acid, proline, and valine. An interesting observation relevant to these findings has been made in the laboratory. Heating formaldehyde and ammonia, which have recently been detected as interstellar gases, produces: glycine, alanine, aspartic acid, glutamic acid, serine, proline, and valine. Thus evidence is rapidly accumulating that the constituents necessary for life are found elsewhere in the universe and may be readily obtained from the simple gases found in the universe.

BIBLIOGRAPHY

1. "Apollo 11 Lunar Science Conference." *Science 167*; January 30, 1970.
2. Fox, S. W., *et al.* "Chemical Origins of Cells." *Chemical and Engineering News* 48:80-94; June 22, 1970.
3. Kenyon, D. H., and G. Steinman. *Biochemical Predestination.* McGraw-Hill Book Company, New York. 1969.
4. Keosian, J. *The Origin of Life.* Second Edition. Reinhold Book Corporation, New York, 1968.
5. Lipmann, F. In *The Origins of Prebiological Systems.* S. W. Fox, Editor. Academic Press, New York. 1965.
6. Oparin, A. I. *Origin of Life.* Second Edition. Dover Publications, New York. 1953.

21. *Antarctic Fossils and the Reconstruction of Gondwanaland*

EDWIN H. COLBERT

Gondwanaland refers to a hypothetical supercontinent that was once composed of the present continents of Africa, South America, Australia, India, Antarctica, and some major island. According to an old hypothesis, now rejuvinated with new evidence, pieces of Gondwanaland broke off and moved to their present position. This concept, called continental drift, has fascinated modern geologists and biologists alike, with adherents and opponents of the idea in both fields. This article briefly discusses the continental drift concept, and then presents fossil evidence that appears to support it. For example, fossils of the extinct amphibian *Lystrosaurius* have been found in southern Africa, India, and Antarctica. This unusual distribution is best explained if these continents were once part of a single landmass. Dr. Colbert recently retired as curator of vertebrate paleontology at the American Museum of Natural History. Largely through his efforts, a team from the American Museum discovered a new and important deposit of vertebrate fossils in Antarctica.

At the present time geology is experiencing a revolution as profound as the one that shook biology a century ago, when Charles Darwin and Alfred Russel Wallace propounded the theory of evolution. This geologic revolution has to do with the theory of continental drift, which postulates that the continents have been mobile throughout the immensity of geologic time, rather than the stable elements they were so long thought to be. This is a revolutionary idea indeed, as the theory of organic evolution was a revolutionary idea. And as the theory of the evolution of life through natural selection gave man a new view of nature and of his place therein, so the theory of drifting continents has given man a new view of the earth on which he lives.

The idea of the evolution of life had been "in the air" for some decades before Darwin's *Origin of Species* was published in 1859. Likewise, the idea of drifting continents has been in the air for several decades—since the early years of this century and, in some respects, even before that. Darwin and Wallace independently gave initial form to the theory of evolution, but it was largely through the detailed and massive work of Darwin that the theory became established. Frank Taylor, an American, and Alfred Wegener, a German, independently gave initial form to the theory of continental drift, in 1910 and 1912, but it was largely through the efforts of Wegener and his brilliant follower Alex Du Toit of South Africa that the theory was developed in considerable detail.

Source: Reprinted from *Natural History* Magazine, January 1972. Copyright © The American Museum of Natural History, 1972.

For years, however, many, perhaps the majority of geologists throughout the world strongly opposed the theory of continental drift. Wegener and Du Toit were ahead of their time; they had the concept, but they lacked the hard facts to give it a convincing basis. Now, within the past decade or so, facts have come to light in varied disciplines that have made continental drift not only a viable, convincing theory, but an exciting one as well. Continental drift is gaining ever wider acceptance among geologists the world around, and the modern geologic revolution is succeeding in a dramatic way.

To be valid, a theory must explain more or less satisfactorily all aspects of the phenomena with which it is concerned. For many years, numerous paleontologists— the students of ancient life on the earth—were not impressed by the theory of continental drift because they did not need it to explain the distributions of fossils on the continents. This was particularly true for the fossils of land-living vertebrates, backboned animals that moved from one place to another by dry-land routes. Paleontologists could explain the distributions of such animals through geologic time by postulating intercontinental movements across existing land bridges or across those that existed in the relatively recent geologic past: namely, the Panamanian Isthmus between the two Americas; the Bering bridge (presently interrupted by the relatively narrow and shallow Bering Strait) between the Eastern and Western Hemispheres; and of course the connections between Africa and the lands to the north. Australia, an island continent, was supposed to have had former connections with Asia. New Zealand and Madagascar, large islands near continents, were supposedly colonized by land-living vertebrates that adventitiously drifted to these isolated regions on masses of floating vegetation or logs. Such routes and means explained the distributions of ancient and present-day amphibians, reptiles, and mammals on the land masses of the earth.

Students of land-living vertebrates, however, largely ignored one continent—the island continent of Antarctica. It is true that today the edges of Antarctica are populated by such vertebrate animals as seals and penguins, as well as a few other birds, but the presence of these denizens of ocean and shore is readily explained. Aside from such marginal inhabitants, the absence of any true land-living vertebrates, recent or extinct, on the antarctic continent placed this great land mass, half again as large as the continental United States, generally outside the calculations of most students concerned with the distributions of ancient and recent tetrapods—the four-footed amphibians, reptiles, and mammals.

Then, in December, 1967, Peter J. Barrett, a New Zealand geologist working in the Transantarctic Mountains about 400 miles from the South Pole, discovered a small fragment of a fossil lower jaw on the slopes of Graphite Peak in rocks of early Triassic age.

The specimen was too incomplete for close identification, but there could be no doubt as to its general nature: it was a portion of the lower jaw of a labyrinthodont amphibian, one of the tetrapods that lived during late Paleozoic and early Mesozoic times, from about 350 million to 200 million years ago. Here was a fossil of great significance, and it immediately drew attention from paleontologists, geologists, and biologists, as well as from the general public. Here was some slight indication that in the distant past Antarctica had been inhabited by land-living vertebrates.

Immediately, questions were raised. Was it not possible that the owner of this piece of fossil jaw had reached Antarctica by swimming across the surrounding ocean? Modern amphibians cannot tolerate salt water; if we apply the same physiological standards to the extinct amphibians, they could not have swum to Antarctica. But perhaps the oceans were less salty 200 million yearr ago. Moreover, some early Triassic amphibians have been found in marine sediments in Spitzbergen, although whether these fossils represent animals that habitually lived in the sea is open to question. At any rate, the evidence of one small jaw fragment, although most significant, was somewhat equivocal. More evidence was needed.

So it was that in October, 1969, a group of us (William J. Breed of the Museum of Northern Arizona, James A. Jensen of Brigham Young University, Jon S. Powell of the University of Arizona, and myself) found ourselves at McMurdo Station in Antarctica, preparing to search for fossil vertebrates. We were part of a larger group of about twenty geologists and paleontologists, working under David H. Elliot, a geologist of note and a veteran antarctic explorer.

Our expedition was a gamble, and a costly one at that. We had no assurance that we would find fossils, and our chances for success seemed to diminsh every day at McMurdo, as we waited through the weeks for storms to abate. It was the stormiest antarctic spring in years. Each day, as the winds howled past our huts, driving clouds of snow across the great ice shelf and Ross Island, on which the base is located, our long-laid plans for a concerted fossil hunt became increasingly tenuous and dislocated.

At last, however, on November 22, we flew into our camp near Coalsack Bluff, a nunatak (the exposed top of an isolated mountain largely buried in ice) on the edge of an ice field some 30 miles west of the mighty Beardmore Glacier, and some 400 miles from the South Pole. Elliot had chosen this locality primarily because it was a good spot for supply planes to land. We were to have helicopter support, and we proposed, first of all, to fly across the Beardmore Glacier (itself some 30 miles in width) to Graphite Peak, to begin our search where Barrett had made his discovery. We wanted to begin at a place where we knew a fossil had been found.

At this point serendipity took over. Coalsack Bluff was about five miles away across the ice, and the helicopters had not as yet arrived, so on the first day in camp some of our group went over to Coalsack Bluff because it was there. Almost immediately we found fossil bones in some low cliffs, exposed on the far side of the nunatak. Before the day was out, nearly thirty fossils had been located along a half mile or more of cliff exposures. From that day until the end of our stay, we spent most of our time excavating fossils from the sandstone cliffs of Coalsack Bluff.

Something should be said about the locale at which we were excavating the fossil bones. A frequent question asked of us on our return was: "How did you find the fossils? Did you dig down through the ice for them?"

Antarctica is commonly pictured as a great, ice-covered continent, and so it is over much of its extent. But in the Transantarctic Mountains there are extensive cliff exposures where the high mountains rise above the level of the glaciers and ice fields. The mountains in large aspect form a continuous range across the continent, but at many places there are outlying nunataks, and Coalsack Bluff is one of these. The north side of Coalsack Bluff is a long slope, largely free of snow and ice. Its lower portion is composed of dark shales with layers of coal belonging to the Permian Buckley Formation, containing, in places, abundant fossil leaves of the characteristic Gondwana plant *Glossopteris*. Above these shales and coals is the Fremouw Formation of early Triassic age, an alternation of sandstones and shales. The sandstones, generally brown or gray in color, stand up as low cliffs, and the shales form the slopes between them. There are three such sandstone cliffs, one above the other, on the slopes of Coalsack Bluff. Finally, capping the nunatak and appearing on its slopes as intrusions, are thick volcanic rocks. These dense rocks, broken and weathered into highly polished slabs, cover much of the slope of Coalsack Bluff.

The weathering processes in Antarctica are unlike those in other parts of the world. Central Antarctica is a desert, with an amazingly scanty annual increment of moisture. Temperatures are low, so there is little thawing. Much of the erosion in the Transantarctic Mountains is effected by wind, wind that sweeps off the polar plateau in fierce gales, driving the dry snow in horizontal clouds. These are the ground blizzards of Antarctica. At extremely low temperatures the snow is so hard and dry that it acts very much like wind-driven sand. The force of these blizzards polishes the hard volcanic rocks and cuts them into weird shapes. On cliffs and exposed slopes, such as the long slope of Coalsack Bluff, the winds clear the snow away, leaving the rocks exposed—a fortunate circumstance for the fossil hunter. (In another sense the antarctic winds were anything but fortunate for us; they were our worst enemy in the field, frequently making our work difficult, and at times, impossible.)

The sandstone cliffs from which we collected the bones were the solidified remains of ancient stream channels. We were dealing with sediments laid down in streams, sediments containing the bones of amphibians and reptiles that had lived in and along the edges of the streams [see Fig. 1].

It soon became evident that we were finding the bones of labyrinthodont amphibians and mammallike reptiles. On December 4, a portion of a skull was discovered that proved to belong to the reptilian genus *Lystrosaurus*. The discovery of *Lystrosaurus* with other mammallike reptiles and with labyrinthodont amphibians indicated that we had found in the Transantarctic Mountains an association of amphibians and reptiles similar to that occurring in the Lower Triassic beds of South Africa, designated the *Lystrosaurus* fauna. The *Lystrosaurus* fauna has also been found in the Lower Triassic sediments of peninsular India, and in Sinkiang and Shansi, China.

We never did go to Graphite Peak, in part because we were completely busy at Coalsack Bluff, and in part because of helicopter troubles. Nor did we go, as originally planned, to McGregor Glacier, some 150 miles southwest of Beardmore Glacier, partly because of problems of logistical support, and partly because of the delays resulting from the bad weather that had plagued us. McGregor Glacier was reserved for the following season.

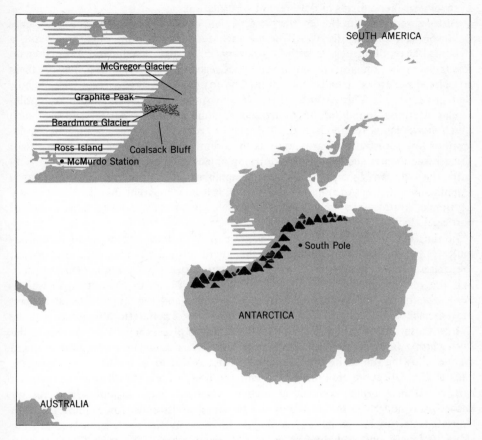

Fig. 1. Some of the places in Antarctica where fossils have been found are located on this map.

The next season came, the austral summer of 1970-71, and with it the campaign at McGregor Glacier. This time the fossil hunters were led by James W. Kitching of the Bernard Price Institute of Paleontology, Witwatersrand University, Johannesburg, assisted by John Ruben of the University of California and, for a short time, by Thomas Rich of Columbia University. Again David Elliot led the entire party working at McGregor Glacier. Kitching was the best possible man to continue the search for fossils in the Transantarctic Mountains. He has spent a lifetime working in the Permian and Triassic sediments of South Africa, and it is fair to say that no other paleontologist alive can equal his experience in the search for the Permo-Triassic amphibians and reptiles that occur so abundantly in the African Karroo sequence. We knew by then that the fossil tetrapods of Antarctica are of close African relationships;

Kitching was the logical man to look for additional and perhaps more complete *Lystrosaurus* fauna fossils in the Fremouw Formation.

History repeated itself. On the first day in camp at McGregor Glacier, James Collinson, a geologist, discovered in the rock a skeletal imprint of *Thrinaxodon,* a mammallike reptile associated with *Lystrosaurus* in the African sediments. From then on fossils were continually found in the McGregor Glacier region, many of them articulated skeletons or partial skeletons. The fossils found at McGregor Glacier show that in addition to *Thrinaxodon,* there were in Antarctica other mammallike reptiles similar to those found in the *Lystrosaurus* fauna of South Africa, and also such *Lystrosaurus* fauna tetrapods as the little reptile *Procolophon*; small eosuchian reptiles more or less ancestral to lizards; various thecodont reptiles especially characteristic of Triassic sediments; and labyrinthodont amphibians that may be compared not only with the African *Lystrosaurus* fauna amphibians but also with Lower Triassic amphibians found in Australia. Consequently, it is now apparent that there was a fully developed *Lystrosaurus* fauna living in Antarctica in early Triassic time, a fact of particular significance.

In the first place, the presence of a diversified *Lystrosaurus* fauna in Antarctica indicates beyond any reasonable doubt that there was a dry-land connection between the present south polar continent and southern Africa. *Lystrosaurus, Thrinaxodon,* and the other mammallike reptiles that have been found in Antarctica, as well as *Procolophon,* the thecodont and eosuchian reptiles, and the amphibians could have moved back and forth between what are now the Transarctic Mountains and the Karroo Basin only across a land route. In the second place, the broad spectrum of the *Lystrosaurus* fauna in Antarctica is almost certainly an indication of a wide dry-land avenue, allowing the entire fauna to spread from Africa to Antarctica (or vice versa). Such a complete representation of the fauna in both areas is strong evidence against a narrow isthmian bridge connecting ancinet Antarctica with ancient Africa, for we know from modern examples (from the Panamanian Isthmus, for example) that an elongated, narrow bridge acts as a zoological filter, permitting some animals to migrate along its length but excluding other animals from using it. No such filter effect is apparent in comparing the *Lystrosaurus* fauna fossils from the Transantarctic Mountains with those from South Africa. Indeed, the close resemblances between fossils in the two regions, extending down to a similarity of species among various genera, is evidence that in early Triassic time Antarctica and southern Africa were probably integral parts of a single continental land mass. The presence of the *Lystrosaurus* fauna in these two regions is probably a manifestation of a single fauna within the limits of its natural range. Again, the *Lystrosaurus* fauna, composed of various reptiles, some of them of considerable size, and of large amphibians as well, is obviously an assemblage of tropical or subtropical animals. This means that Triassic southern Africa and Antarctica probably were in latitudes lower than those they now occupy.

This brings us to the subject of Gondwanaland [Fig. 2] . The name was coined in the latter part of the nineteenth century by the Austrian geologist Eduard Suess to

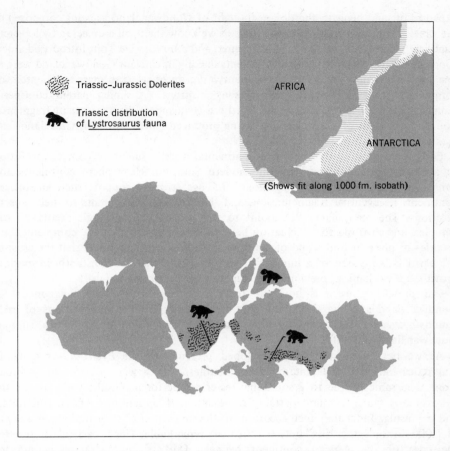

Fig. 2. The Gondwanaland of 200 million years ago included the now separate continents of Antarctica, Africa, and South America, along with India, Australia, and major islands. Whole groups of land animals could have easily moved from one region to another.

Within the figure:
Triassic–Jurassic Dolerites

Triassic distribution of Lystrosaurus fauna

AFRICA

ANTARCTICA

(Shows fit along 1000 fm. isobath)

designate a hypothetical gigantic continent, embracing the modern continents of the Southern Hemisphere, and extending across the Equator to include the peninsular portion of India as well. This great ancient continent was considered by many geologists as useful, and perhaps necessary, to explain many similarities among the rocks and fossils of the Southern Hemisphere continents and of peninsular India. Some of the early believers in Gondwanaland pictured it as an immense east to west land mass, including India and the Southern Hemisphere continents as they are now placed. Gondwanaland, they thought, subsequently disappeared as an entity by the foundering of large portions of land into the oceans, leaving the present continents as isolated remnants. Other students, who found it difficult to visualize the sinking of such great expanses of land beneath the ocean, thought of Gondwanaland as being composed of the present southern continents and India as we know them, all connected by long and relatively narrow land bridges. Then Wegener, and after him Du Toit, introduced a new concept, namely that the several continents making up ancient Gondwanaland were at one time contiguous, and that subsequently the ancestral land mass fragmented, its component parts drifting to their present positions. (A similar parent continent, Laurasia, has been proposed for the Northern Hemisphere, its subsequent fragmentation and the drift of the fragments having produced North America, Greenland, and most of Eurasia.)

Early opposition to the theory of continental drift included, of course, opposition to a Gondwanaland formed by the modern Southern Hemisphere continents and peninsular India. But modern geologic findings strongly support such an ancient continent, its eventual fragmentation, and the drift of its fragments to their present positions. The many facts that point to this sequence of geologic events are too complex and involved for elucidation here. Suffice it to say that the complementary theories of plate tectonics and of sea-floor spreading, which stipulate that the crust of the earth is composed of a number of gigantic plates that are constantly in motion, provide the mechanism, previously lacking, to explain continental drift.

Our present concern is how the fossil evidence accords with the concept of Gondwanaland and the theory of continental drift. Do the distributions of early land-living vertebrates, especially those that have been found in Antarctica, support Gondwanaland and drift?

As we have seen, the fully developed presence of the *Lystrosaurus* fauna in Antarctica indicates that Antarctica and southern Africa were joined along a broad front. The same is true to a somewhat lesser degree for peninsular India, where the *Lystrosaurus* fauna is found partially represented. If present-day Africa, Antarctica, and peninsular India are joined according to the similarities of their outlines at a depth of 1,000 fathoms, the "fits" between them are remarkable. This is particularly true for the edge of the African continent between Durban and Mozambique and for Antarctica along the Weddell Sea and the Princess Martha coast. Such a fit affords a broad connection between the two land masses, making of them essentially a single land.

And such a fit, together with fit of peninsular India between Antarctica and eastern Africa, brings the localities of the *Lystrosaurus* fauna in these now widely separated continents all within about 2,000 miles, or less, of each other. This distribution suggests a very reasonable range for a terrestrial vertebrate fauna, as judged by modern standards. A single species of *Lystrosaurus* is present in Antarctica, Africa, and India [see Figs. 3 and 4]: it seems quite probable that this species on the modern continents represents the disruption of what was once a relatively compact range of distribution. (The presence of elements of the *Lystrosaurus* fauna in China has as yet to be explained, but at the moment it would appear that facts are accululating that will account very satisfactorily for the Chinese occurrences of these early Triassic tetrapods.)

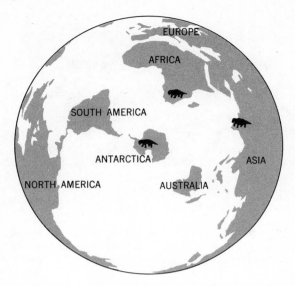

Fig. 3. The present locations of the southern continents are shown in this view from above the Southern Hemisphere. The reptile symbols indicate where *Lystrosaurus* fossils have been found.

So it is that the discovery of the Lower Triassic *Lystrosaurus* fauna in the Transantarctic Mountains is a paleontological development of prime importance. It helps prove the close connection of Antarctica and southern Africa in Triassic times. From this demonstration of faunal and continental relationships one proceeds to the conclusions that there was such an entity as Gondwanaland, that Gondwanaland was broken asunder, that its fragments drifted apart, and that Antarctica, once the habitat of tropical amphibians and reptiles (and abundant plants as well), came to occupy a position in a climate quite inimical to the life that had once flourished in benign temperatures. Other geologic and paleontological facts support the conclusions drawn from continental outlines and the occurrences of the *Lystrosaurus* fauna, such as the general expression of Permian and Triassic geology in southern Africa and Antarctica, the presence of extensive volcanic rocks in the two continents, and the development of fossil plants in these areas. But the occurrences of the *Lystrosaurus* fauna are also important; they give solid evidence for land connections. The evidence of geophysics involves certain assumptions, as does that of geology. The evidence of the fossil plants is strong, but there is always the possibility (although according to the paleobotanists, a very slim one) that these plants may have been distributed in part by windborne transportation of seeds. The evidence of the land-living tetrapods, present in the two regions as fully developed faunas, cannot be denied. These animal assemblages most surely had to move from the one region to the other on dry land.

The recogniton of a *Lystrosaurus* fauna in the Transantarctic Mountains is of significance not only because it adds a large dimension to our knowledge of ancient

Fig. 4. This is the mammallike reptile *Lystrosaurus* as it might have looked in Antarctica 200 million years ago. When they were fully grown, most species reached about the size of a large present-day dog.

life on what is now the South Polar continent but also, as we have seen, because of the strong confirmation it lends to continental drift and to the former existence of Gondwanaland. Important as the discoveries of the past two years are, however, they have merely scratched the surface of antarctic paleontological riches. For riches there are, in the form of numerous untouched exposures of the Fremouw Formation containing abundant fossils.

Much progress has been made in the elucidation of ancient life on Antarctica since those tragic days sixty years ago, when Scott and his companions unsuccessfully struggled back from the South Pole, dragging their sledge loaded with survival gear and with some 25 pounds of precious fossil plants. Even more progress lies ahead. In Antarctica surely are paleontological answers to many questions regarding the evolution and distribution of life on the ancient continent of Gondwanaland.

22. *Human Activities and Megafaunal Extinctions*

GROVER S. KRANTZ

What caused the extinction of large groups of seemingly very successful animals in the evolutionary past? The answers will probably never be known for very ancient groups such as dinosaurs. For more recent evolutionary times, however, as in late Pleistocene, considerable evidence is accumulating that primitive man played a major role in the extinction of, at least, some of the large mammals of his time. Discussion of his possible modes form the body of this paper. Although the author seems to say that our early ancestors abused the environment even in their times, the evidence probably emphasizes more significantly that man is and always has been an integral part of nature. Grover Krantz, an anthropologist interested in the origin and evolution of man, is an associate professor at Washington State University, Pullman, Washington.

One of the most vexing and most discussed problems that researchers in both paleontology and archeology are concerned with is that of the extinction of large animals in the Late Pleistocene. For all the efforts to clarify the cause or causes, there remain two sharply opposed schools of thought. Some authorities hold that the hunting activity of prehistoric man directly eliminated many species; others believe their disappearance is primarily a result of abrupt climatic changes. Both views are well represented in the recent volume edited by Martin and Wright on *Pleistocene Extinctions* (1967). The present paper offers a new interpretation of the data, which might have had a place in that book had the study been undertaken a few years earlier.

Man may well have caused many of these extinctions, I believe, but not by the direct means of physical elimination through excessive predation. I shall describe below three possible indirect means: (1) competition for a particular food supply, (2) causing one herbivore to exterminate other herbivores, and (3) agricultural practices altering ecological relationships. The first and third methods have been touched upon by others, while the second, to the best of my knowledge, is set forth here for the first time.

First let us examine the evidence that prehistoric man have have hunted out these animals. The most convincing argument is the coincidence in time of the arrival of man with these extinctions, or, as in Africa and South Asia, the supposed first development of big game hunting. Martin (1967: p. 111) gives the approximate dates for these events in each area as follows:

Africa, South Asia	40,000 to 50,000 years B.P.
Australia	13,000 years B.P.

Source: Reprinted by permission of the publisher and author from *American Scientist,* 58:164-170, 1970, the journal of The Society of Sigma Xi.

Northern Eurasia	11,000 to 13,000 B.P.
North America	11,000 B.P.
South America	10,000 B.P.
West Indies	"Mid-Postglacial"
New Zealand	900 B.P.
Madagascar	800 B.P.

Another argument advanced against any nonhuman or natural cause is the lack of replacement of the extinct forms by ecological equivalents. Martin (1958) and Jelinek (1967) see this as indicating human agency, which alone could eliminate a species leaving its ecological niche open. However, Guilday (1967) shows how the same phenomenon can occur under the stress of temporary climate change, as in the altithermal.

The last major argument is not so much support for man's being the exterminating agency as denial of the most obvious alternative explanation—climatic change. The nearly perfect coincidence of so many mammal extinctions in Northern Eurasia, North and South America, and Australia with the final withdrawl of the Wisconsin-Wurm ice sheet has led many authorities to conclude that abrupt changes in climate were the major causal factor (Guilday 1967; Slaughter 1967; Hester 1967; and Kowalski 1967). This conclusion has been challenged by those who point out that no comparable extinctions occurred at the ends of any of the previous glacial advances (Edwards 1967: p. 144; Jelinek 1967: p. 193; and Martin 1967: p. 81).

Other, less often stressed, arguments include the comparative absence of extinctions among smaller mammals in the Late Pleistocene as well as no notable loss of plant species (Martin 1967: p. 78, and Leopold 1967). Had climatic change been the predominant factor, it is argued, plants and small mammals ought to have suffered to some degree as well.

The above arguments could be extended and documented at length, but the same question always arises: some new factor must have been present at each time and place of massive extinctions—and the only new factor that appears to be consistently present was man. The conclusion seems inevitable that prehistoric man must have hunted and killed these animals on such a scale that they were totally eliminated.

However reasonable this conclusion may appear in the light of one class of evidence, it fails utterly when questioned from another angle: just how was it possible for prehistoric hunters to accomplish this feat? Even Martin acknowledges the problem, stating (1967: p. 115), "We must beg the question of just how and why prehistoric man obliterated his prey," and further, "The thought that prehistoric hunters ten to fifteen thousand years ago exterminated far more large animals than has modern man with modern weapons and advanced technology is certainly provocative and perhaps even deeply disturbing." It is more than disturbing, it is impossible.

In order to clarify this problem somewhat one must consider the presumed causes of extinctions, in particular the predator-prey relationship.

When a hunting species specializes in taking a particular species as its food source, the population of the hunter is directly dependent upon that of its prey. Only a certain number of individuals of a prey species can be "harvested" without altering its population level. This "harvest," in turn, will support only a certain population of the hunting species. If the prey declines in numbers, so will the hunters, who can exist only in proportion to their food supply. Normally a stable balance is maintained easily and automatically. Some arctic species undergo drastic fluctuations in population, with cycles of several years' duration. Here too, the numbers of the hunting species rise and fall correspondingly, always in response to the availability of the prey.

This relationship still holds true even if the hunter is dependent on a particular prey species only during one brief season of the year. However great a variety of food they may consume in all other seasons, the hunter's population will be limited to those who can be fed in the season of scarcest game.

Clearly no extinctions will occur under these circumstances. The scarcer the prey, the scarcer the hunter, and predation pressure eases off automatically.

If the hunter species is less particular and takes its food from a number of game sources in all seasons, then a slightly different relationship follows. Should one of the prey species be hunted until its number are significantly reduced, individuals of this species will be less often encountered. If the hunter is to maintain his population independent of this reduction in one kind of prey, he must concentrate correspondingly more on other species. As the hunter takes what are available of his several game species, those fewest in number will be the least often met and least often taken, thus tending automatically to be preserved from extinction.

Vulnerability to predation is thus not an inevitable or even a likely cause of a species' demise. Extinction threatens only if the numbers are so reduced that mating encounters utilize but a small portion of their reproductive potential. It is only with the advent of "civilized man" that this becomes a reasonable possibility.

These predator-prey relationships hold when primitive man is the predator, and in some respects even more surely. Man's hunting practices are entirely learned by each generation from the preceding one. Any break in this cultural continuity would leave man without any instinctive behavior to depend upon. Also, the special activities and knowledge required for hunting each species of game will differ somewhat from those needed for every other. Should one game animal become scarce and thus more rarely encountered, human hunters, in concentrating more on other species (or upon vegetable gathering), will quickly lose the learned skills pertinent to taking the scarce game. Unlike inherent behavior programming, learned skills can disappear in a generation, and even seriously deteriorate in just a few years if not reinforced by practice.

In addition to the above points, nonliterate hunters are also known for deliberate game conservation beyond the automatic system which applies to other carnivores (see Heizer 1955 for numerous examples). All primitive hunters are quite aware of the importance of females and young for future game supplies and will often treat them accordingly—something no natural carnivore will do.

The seeming wastefulness of cliff drives may not be as significant as may appear. In the first place, such drives are not known to have been used earlier than 9,000 years ago (Hester 1967: p. 181), somewhat too late to be involved in most of the major extinctions. Also, such stampedes are not age-selective in their destruction; they take young and old alike, a point which will be seen later to have some significance.

Direct evidence also exists that, in North America at least, early hunting man did not significantly prey upon those animals he supposedly exterminated. Many archeological kill sites which are now well documented include the remains of animals killed by Paleo-Indians in the Late Pleistocene. These data, summarized by Hester (1967: p. 180), show that only two genera were hunted in significant numbers, mammoth and bison, one extinct and the other not. In 25 of the 34 documented sites bison was the exclusive or predominant game found, mammoth ran a poor second as the sole or major game in 7, and at only 2 were various other animals emphasized. The virtual absence of most extinct animals from kill sites has been noted as a serious objection to the idea of man's role in their demise (Jelinek 1967: p. 198; and Guilday 1967: p. 137).

At this point we seem to be faced with two "facts" that are reasonably well proven. First, the advent of big-game-hunting man alone correlates with the times of massive extinctions. Second, hunting man could not and did not exterminate these animals. These two "facts" appear to be contradictory. Other researchers have proceeded on the assumption that if one of these "facts" is true the other must somehow be explained away or ignored.

I shall attempt to show here that there is not necessarily any contradiction between the two "facts." At the root of the matter is the assumption made by most authorities that man could have exterminated a species *only* by directly destroying virtually all of its individuals. There are other methods.

Modern studies of ecology indicate that the major, if not the sole, cause of extinction is the removal of some essential aspect of the environment, and not predation. This sort of thing has been suggested but little explained by some who refer simply to the "indirect" influence of man (Simpson 1965: p. 228) or to the upset of a nicely balanced state of nature (Romer 1933; and Colbert 1942). The problem at hand is thus changed to answering the question: What actions of man could have so altered the "balance of nature" as to deprive numerous species of some ecological necessity?

Direct Competition

One distinct category of extinctions was, in a sense, caused by man: the extermination of certain large carnivorous mammals. Several writers have pointed out how difficult it would have been for early man to seek out and destroy such dangerous animals as extinct lions, cave hyenas, and saber-toothed cats (Eiseley 1943a; and Butzer 1964: p. 400). It is all the more unlikely that man could have directly eliminated any such carnivores in view of the predator-prey relationships noted above, considering the carnivores as prey and man as the predator.

At least two writers saw why these carnivores disappeared, though without suggesting the hand of man in the process. Guilday (1967: p. 122) and Vereshchagin (1967: p. 392) noted that with the loss of their large herbivorous prey the carnivores similarly vanished. Edwards (1967: p. 146) quite simply explains that as man can be a more efficient predator, the others are eliminated by direct competition (see also Martin and Guilday 1967: p. 33-34 on *Smilodon*). This becomes a case of ecological replacement: man need not kill a single animal to eliminate the species; he needs merely to deprive them of their normal food supply by getting it first. This would be especially effective at times of seasonal crisis, like northern winters, when food resources are reduced to a minimum and competition is most intense.

Probably most large carnivore extinctions since the Middle Pleistocene were caused by the direct competition of man's increasing hunting abilities. The early disappearance of saber-tooths from most of the Old World was not from actual extinction of their prey species, some of which still exist, but rather from a scarcity of these individual animals (young, lame, ill, and aged) which they normally preyed upon. These were being increasingly taken by early human hunters. The late occurrence of *Machairodus* in the latest glacial stage in Britain (Flint 1957: p. 456) indicates their survival only in the coldest northern climates, which man could not inhabit up to that time. Hunting man and sabertooths competed so closely that when the former was present these great cats were not. Similarly, *Smilodon* vanished as soon as big game hunting began in the New World, and shortly before its prey species died out.

Indirect Causes

So far I have discussed the role human agency may have played in extinctions by somehow removing the food of a species. The next method to be described is even less direct: man may have caused certain herbivores to exterminate other species with similar ecological requirements. In this section I will concentrate mainly on the well documented extinctions which occurred in North America at the end of the Pleistocene. These include many grazing mammals which did not significantly compete with early man for food resources.

The number of genera (33) named by Martin (1967) as recently becoming extinct in North America by "overkill" is perhaps exaggerated. Mason (1962: p. 243) lists only 15 genera with a total of 20 species, and Newell (1963: p. 48) gives 12 genera and a total of only 16 species. Removing carnivores from these lists will reduce Martin's list by 4 genera, Mason's by 2, and Newell's by 1.

The accuracy of these figures is perhaps worthy of a short digression. They are based essentially on a comparison of living fauna with the fossil fauna of the Upper Pleistocene. Earlier extinctions are based on comparisons of one fossil fauna with another fossil fauna. Since living and fossil animals have generally been classified according to different procedures and by different people, there is good reason to question any comparison of one with the other. Kurten has noted (1968: p. 40) that the tendency now in paleontology is to "recognize broad, inclusive species with a span of morphological variation similar to that found in related species today," but that this has not been the practice in the past. There were probably far fewer extinctions at the end of the Pleistocene than is usually supposed because there were actually not so many species which could become extinct.

It is likely that the American bison was a single species at any one time, and that changes observed are a combination of evolutionary change and movements of subspecies in response to climate shifts (Eiseley 1943b; and Cornwall 1968: p. 184). In my opinion *Canis dirus* is probably not specifically distinct from the living timber wolf except possibly as an ancestral species. The number of species of elephants reported in the literature is probably also too many.

In spite of all such reservations it is still clear that a number of distinct types of large herbivores did become extinct within a short period of time. These include mammoth, mastodon, horse, camel, several ground sloths, at least one pronghorn, and possibly some cattle.

The archeological record in North America of early man's associations with these mammals, as noted above, is most informative. Only the mammoth and, especially, the bison occur in great numbers as the victims of hunting activities of Paleo-Indians. The absence, in most cases, of the remains of horse, camel, mastodon, or pronghorn from kill sites and their occurrence in natural deposits hardly support the contention that man killed them off. In order for man, or any other predator, to destroy a species he must dispatch the vast majority of the individuals himself, and almost all of these at an early age prior to their reproduction. Clearly this did not happen, as the species in question are well represented by adult animals in deposits unrelated to any human activity.

If one accepts the archeological and paleontological evidence at face value, an interesting conclusion seems to be indicated: that heavy hunting by early man *preserved* at least one type while unhunted species became extinct. This is just the opposite of over kill as it has been presented. It appears that while man preyed heavily on the bison it *increased* in numbers and range. To suggest that hunting by man actually *caused* such an expansion of the bison may seem odd, but this can be demonstrated to be a likely possibility. The demonstration involves principally a comparison of human predation practices with those of other carnivores. Demographic data on age distributions within a prey population often show sharp differences depending on the type of predator that is chiefly concerned.

No satisfactory detailed figures seem to be available for bison, but the fate of other animals in similar circumstances illustrates the point quite well. Large natural carnivores prey mainly on two age categories of their game, with clear seasonal emphases. In the warmer half of the year, beginning with foaling or calving, the young herbivores are the major part of their diet. In the colder months it is mainly the oldest individuals, especially diseased and injured ones, that are taken. At all times it is the prey that is easiest to catch that is eaten.

This kind of seasonal variation in predation habits can be shown on a graph of age distributions of the prey population [Fig. 1]. A "survivorship curve" would show how many individuals remain at the end of each year from an original 100 percent born in a given spring. Data for a typical large herbivore survivorship curve under natural predation come from a recent study of the wolves and moose of Isle Royale in Lake Superior (Mech 1966). Here, well over half (62%) of each year's crop of young moose

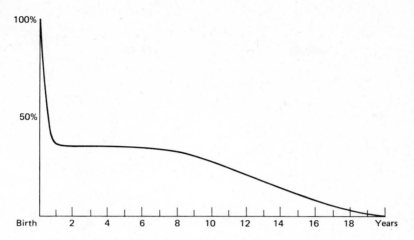

Fig. 1. Typical survivorship curve of large herbivore under natural predation.

is taken by the wolves in that year before winter sets in. After that, the young adults can better defend themselves and rarely fall victim for the next several years. Finally with increasing age the moose are again unable to escape the wolves.

In his work on the wolves of Mt. McKinley, Murie (1944) found the bighorn sheep had a similar survivorship distribution, as based on ages of collected skulls. Murie's figures did not show as large a proportion of first-year kills; but it is clear, as he notes, that skulls of very young sheep would be more apt to disintegrate or be eaten than older skulls. With some correction for this factor, the age distribution of these sheep agrees with that for Mech's moose.

Other data which are less easily graphed all indicate a massive carnivore kill of each year's young, followed by a plateau of little predation, and finally the gradual elimination of the aging animals. In terms of the yearly cycle, it is during the winter that hunting man, if present, is pressing hardest for largely the same victims in lieu of significant vegetable foods. Competition with man in this season alone could be sufficient cause for local disappearance of those carnivores whose prey happened to be the same as man's.

In the absence of the natural carnivores, the herbivorous species would have a survivorship curve [Fig. 2], which follows mainly from human predation. This becomes a rather different picture, especially after the development of projectile weapons with which a man can dispatch adult animals almost as easily as young ones. Human hunters may also, in many cases, emphasize the young and aged game, but generally not to the same degree as most carnivores. Data on human game kills tend to indicate a more nearly equal take at all ages of the prey species. If such predation is severe enough, the surviving numbers of reproducing adults may be no more than when natural carnivores are involved, but in many instances this is not what happens.

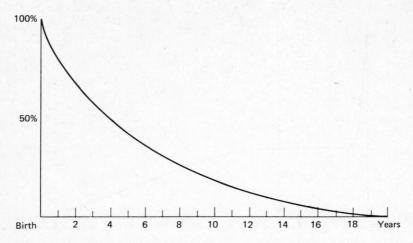

Fig. 2. Typical survivorship curve of large herbivore under human predation.

Age distribution of prey populations under human predation is clearly shown in one case Boudiere (1956: p. 293) records the extensive take of chamois by one hunter in the Alps and Pyrenees. Here there is no killing of the youngest animals, but the rest of the survivorship curve is distinctly different from that resulting from carnivore kills. There is no leveling off or plateau at the young adult stage.

This chamois curve representing the take of one hunter also shows the age distribution of the population from which they were taken according to availability. The age distribution was caused, in turn, by this and previous hunting which also took animals at all ages with no emphasis on the young and aged and no avoidance of young adults. The curve might be completed by adding an estimate, probably high, of 20 percent more as having died in their first year.

Other recent data on age distributions of human kills include Friley on otter (1949a) and on beaver (1949b). These records are graded not by age in years but rather by degrees of maturity; still they indicate the same curve of survivorship, with no plateau at any stage. More data, especially on kills by primitive hunters, would be useful to illustrate this point.

Some archeological data are also available on the proportions of young individuals recovered from occupation and butchering sites of ancient man. Butzer (1964: p. 382) records 19 percent as being "juveniles" among the reindeer taken at the Lake Mousterian site of Salzgitter-Lebenstedt. Soergel (1922, as cited by Butzer 1964: p. 388) mentions proportions of immature individuals of numerous species of game ranging from 25 to 35 percent in various European Paleolithic sites. In America, Hester (1967: p. 181) notes that Paleo-Indians took game of all ages with no particular emphasis. While some of these figures indicate a rather high mortality of young game animals, they are all still far below the over 50 percent which is indicated for natural carnivores.

A comparison can now be made of the two survivorship curves with natural carnivore predation (based on Mech and Murie) and with human predation (based on Bourliere, Friley, and others) [Fig. 3]. The wide variation possible in each of these curves could, in some instances, negate the observations that follow. In general, however, the difference between human and other carnivore predation practices tends to cause a marked contrast in the age distributions of the prey population.

Most conspicuous is the difference in numbers of surviving young adults. Under human predation these individuals, the major breeding stock, are far more numerous because their numbers were not so drastically depleted when they were still immature. The numbers of surviving older adults might be greater under natural predation, but this would not generally be enough to outweigh the breeding potential of the younger ones. It can thus be seen that when natural predators are replaced by hunting man the most likely expectation is an *increase* in the population of the prey. If all else remains equal, this prey species ought then to exert considerable population pressure against any competing species in the area with similar ecological requirements.

A general picture of a series of biological events can now be suggested:

1. Human hunters enter an area and specialize in taking a particular species of game animal.

2. The local carnivore (or carnivores) specializing in the same prey are starved out during winter competition and become extinct.

3. With man taking proportionally fewer of the young, the prey species expands in numbers and presumably in territory.

4. Other herbivores are exterminated by pressure from the "favored prey" through competition for such things as food under stress conditions.

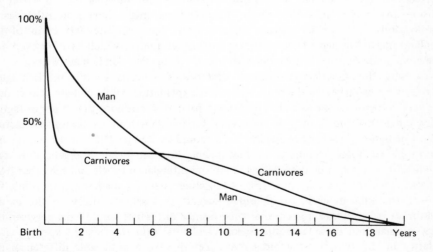

Fig. 3. Superimposed survivorship curves showing differences between natural and human predation.

5. Still other carnivores may next become extinct as their prey disappears, if they cannot depend on the expanding type.

Of course, some rather close predator-prey relationships must already have been in existence for this sequence of events to occur. If the predator which man competes with has another major food supply during seasons of stress, it will not become extinct. If other competing carnivores are taking significant numbers of the young of this game species, its population will not expand. Even with only moderate predator-prey specializations the chain of events outlined should occur to some degree.

In the case of the prominent faunal extinctions in North America, a reconstruction of the particular events, according to this scheme, might have been the following:

1. The first hunting by man with projectile weapons in North America began about 12,000 years ago (Haynes 1967), with emphasis on killing mammoths and especially bison.

2. The saber-toothed cats which depended upon the same two species, especially on the aged animals in winter, were competed out of existence.

3. Bison greatly increased in numbers because their young were no longer being killed by sabertooths. (That this did not apply to the mammoth will be discussed below.)

4. Bison exterminated many other species which were in close ecological competition. Thus the demise of such other plains herbivores as horse, extinct pronghorn, various cattle, camel, and perhaps also the mammoth.

5. The jaguar, presumably a predator on horse, camel, or cattle (or all of them), then disappeared from the area.

The above is probably a minimal description of events. Other species may have been involved through ecological relationships which are not obvious to this writer. In environments other than the plains, Paleo-Indians might have had other game specializations with similar effects on still other species. The extinction of the woodland mastodon suggests competition from other animals which have survived, but there are no data on this comparable with that on plains Paleo-Indian activities.

Just why the American mammoth disappeared remains a problem. Ecological pressure from bison seems the most reasonable explanation, while climate change may have been a major factor as well, and human predation no more than a minor factor.

Some light may be thrown on the reason for the American mammoth's extinction by a set of figures given by Soergel in 1922 (quoted by Zeuner 1963: p. 17) on the age distribution of fossil elephants of Pleistocene Europe. From Soergel's data two survivorship curves can be drawn—one from first interglacial fossils and the other from fossils of third interglacial times [Fig. 4]. Zeuner assumes man was responsible for most of the kills in each group, but I suggest that sabertooths were the earlier predator. It is an educated guess that only a small difference would exist between the ability of man and of sabertooths to kill elephants; thus one might expect a close similarity in the two survivorship curves. Yet there is a noticeable difference, the earlier predation pattern being more like the natural one, with a greater "take" of young animals and a slight tendency to level off in most of the adult years.

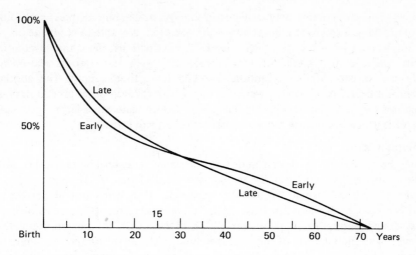

Fig. 4. Suvivorship curves of Pleistocene European elephants under human predation (late), and presumed natural predation (early).

Unlike most herbivores, elephants begin breeding at a relatively late age—about sixteen years. While there is some separation between the two curves, representing more elephants alive in early adulthood under human predation, this is more than canceled by the late onset of reproduction. These rather scanty data suggest that elephants, unlike the bison, would *not* increase in population after human predation replaced that of saber-tooth cats.

In Europe a chain of events similar to that conjectured for North America seems to have occurred, though with less drastic consequences in terms of numbers of kinds of animals exterminated. Throughout the Upper Paleolithic there was an increase in emphasis on reindeer hunting in Western Europe and a decrease in hunting mammoth and woolly rhinoceros. By the last stage, the Magdalenian, there was an almost total concentration on reindeer as the game animal. One might conclude that man gradually shifted his hunting emphasis solely according to availability of game species. It seems more likely that the increasing abundance of reindeer was largely a result rather than a cause of man's concentration of this species.

In discussing the Meiendorf site of 13,500 B.P., Butzer (1964: p. 410) stated this proposition indirectly: "There is no ready explanation why bison and woolly mammoth should be completely absent in the Hamburg area other than by reason of deliberate specialization on reindeer by prehistoric man." Perhaps this was not intended to mean that man's specialization on one species was the *cause* of the absence of others in the area (not just the site), but that is what I am suggesting.

In some European regions a corresponding emphasis on horse hunting seems to have occurred. If we could identify the natural predators involved we might find a similar situation. The same events may have occurred in Australia, but adequate archeological data are lacking. Gill (1955), in reviewing the situation, was able to conclude only that man could not have been the direct cause of the extinctions. Here, annual variation in rainfall, rather than winter cold, is the probable source of the stress season during which one species eliminates another by taking its food supply. The aborigine's concentration on kangaroos and wallabies might have relieved other predation pressure and permitted these medium-sized marsupials to exterminate other larger species. More evidence is needed to pursue the Australian situation any further.

Agricultural Man

The third method by which man has been able to exterminate species without direct killing is by agriculturally related practices. Three of Martin's associations between the advent of man and the disappearance of big game animals are the more recent cases of the West Indies, New Zealand, and Madagascar (see also Hooijer 1967; Battistini and Verin 1967; and Walker 1967). In each case, the human population at the time of extinctions was of the Neolithic type, with the subsistence economy based on agriculture. There is little doubt that these poeple were responsible for the disappearance of numerous species, but not necessarily in the sense of hunters destroying their prey.

There are several ways in which man, since the Neolithic revolution, has upset natural environments. Farming may remove local vegetation which was a food supply for some animals. Fencing will restrict some animals' movements as well as deny them access to the planted crops. Forest clearing will remove habitats as well as certain food sources, substituting others. Draining swamps and irrigating dry lands likewise alters vegetation and environment. Stock raising introduces major competitors for certain food resources. The physical presence of man with his activities and constructions can interfere with mating, nesting, or maternal behavior, or simply induce animals to leave an area to avoid such disturbances.

The possibility that some of these agricultural activities have caused exterminations has been pointed out by Guilday (1967: pp 122, 126, 158) for much of the megafauna in North Africa and the Middle East, and by Hester (1967: p. 178) for North America since European colonization.

In addition to his great ability to affect native fauna indirectly, the agriculturalist differs from the hunter in his lack of dependence upon that fauna. The extinction of numerous large native herbivores will have little or no ill effect on that farmer who depends on his crops and/or herds; he may even welcome their disappearance. The hunter, on the other hand, would perish without them.

While many writers have missed the point by emphasizing the Paleo-Indian's dependence on big game hunting to account for his supposed destructiveness, Edwards (1967: pp. 147-48) realized that only by reacquiring plant-gathering habits would primitive man have been able to afford the luxury of exterminating his game. This, of course, does not explain how such extermination was possible on a hunting and gathering level, but it does illustrate the point that man can destroy only what he does not depend upon, without destroying himself in the process.

The recent destruction of megafauna in New Zealand, Madagascar, and the West Indies followed from the activities of agricultural man rather than of hunters. This points up a meaningful dichotomy of human societies in terms of their effect on the environment. The contrast between "prehistoric" and "modern" man in this connection is not of much value. The "modern" Plains Indians and African Bushmen have had a very different effect on their environments than did the simple grain farmers of the Eastern Mediterranean some eight or ten thousand years ago. Agriculturally based societies differ from one another only in the degree to which they disturb the natural environment, while they all differ in kind from hunting and gathering societies.

There are some additional peculiarities of agricultural societies, in terms of size of population and commerce, which open up the possibility, at least, that they could have exterminated certain species by direct overpredation. Given a farming subsistence base, almost any area will support and feed a population ten to a hundred times larger than it could maintain by hunting and gathering alone. With such numbers available, especially for seasonal hunting, serious decimation of nonessential animal populations is certainly possible.

Even more important, perhaps, are such things as fur trapping, ivory hunting, sport and trophy collecting, and other ventures mainly for the commercial market. Items in this broad category will have an exchange value based, in part, upon their scarcity. Hunters may thus concentrate on certain animals not for their own use but for sale or trade with agriculturally based peoples. In these cases a decline in the supply of a particular animal may raise its price and serve to increase predation pressure. Just as the African elephant now has the added threat of ivory hunters, it may be possible that the last Siberian mammoth was killed for the Chinese ivory market.

Let me add that the above arguments are not proposed as the entire explanation of megafaunal extinctions. Climate change cannot be ignored, because the combination of an unusually abrupt end to the last glaciation with the subsequent altithermal warming may have been unique in the Pleistocene. Human agency, by the means suggested here, is proposed to account for extinctions only where climate change is an inadequate explanation.

BIBLIOGRAPHY

Battistini, R., and P. Verin. 1967. Ecological changes in Protohistoric Madagascar, in Martin and Wright, pp. 407-24.
Bourlière, François. 1956. *The Natural History of Mammals.* New York: Knopf.
Butzer, Karl W. 1964. *Environment and Archeology.* Chicago: Aldine.
Colbert, Edwin H. 1942. The association of man with extinct mammals in the Western hemisphere, in *Proceedings, Eighth American Scientific Congress,* 2: 27.
Cornwall, I. W. 1968. *Prehistoric Animals and Their Hunters.* London: Faber & Faber.
Edwards, William Ellis. 1967. The Late-Pleistocene extinction and diminution in size of many mammalian species, in Martin and Wright, pp. 141-54.

Eiseley, Loren C. 1943a. Archaeological observations on the problem of postglacial extinction, in *American Antiquity, 8*: 209-17.

──── 1943b. Did the Folsom bison survive in Canada? In *Scientific Monthly, 56*: 468-72.

Flint, Richard F. 1957. *Glacial and Pleistocene Geology.* New York: Wiley.

Friley, C. E., Jr. 1949a. Age determination, by use of the baculum in the river otter, *Lutra c. canadensis,* in *Journal of Mammalogy, 30*: 102-10.

──── 1949b. Use of the baculum in age determination of Michigan beaver, in *Journal of Mammalogy, 30*: 261-67.

Gill, Edmund D. 1955. The problem of extinctions, with special reference to Australia's marsupials, in *Evolution, 9*: 87-92.

Guilday, John E. 1967. Differential extinction during Late-Pleistocene and Recent times, in Martin and Wright, pp. 121-40.

Haynes, C. Vance, Jr. 1967. Carbon-14 dates and early man in the New World, in Martin and Wright, pp. 267-86.

Heizer, Robert F. 1955. Primitive man as an ecologic factor, in *Kroeber Anthropological Society Papers, 13*: 1-31.

Hester, James A. 1967. The agency of man in animal extinctions, in Martin and Wright, pp. 169-92.

Hooijer, D. A. 1967. Pleistocene vertebrates of the Netherlands Antilles, in Martin and Wright, pp. 399-406.

Jelinek, Arthur J. 1967. Man's role in the extinction of Pleistocene faunas, in Martin and Wright, pp. 193-200.

Kowalski, Kazimierz. 1967. The Pleistocene extinction of mammals in Europe, in Martin and Wright, pp. 349-64.

Kurtén, Bjorn. 1968. *Pleistocene Mammals of Europe.* Chicago: Aldine.

Leopold, Estella B. 1967. Late-Cenozoic patterns of plant extinction, in Martin and Wright, pp. 203-46.

Martin, Paul S. 1958. Pleistocene ecology and biogeography of North America, in *Zoogeography,* C. L. Hubbs, ed., American Association for the Advancement of Science, Publication 51, pp. 375-420.

──── 1967. Prehistoric overkill, in Martin and Wright, pp. 75-120.

Martin, Paul S., and John E. Guilday. 1967. A Bestiary for Pleistocene biologists, in Martin and Wright, pp. 1-62.

Martin, Paul S., and Herbert E. Wright, Jr., eds. 1967. *Pleistocene Extinctions,* vol. 6, Proceedings of the Eighth Congress of the International Association for Quaternary Research. New Haven: Yale University Press.

Mason, R. J. 1962. The Paleo-Indian tradition in Eastern North America, in *Current Anthropology, 3*: 227-78.

Mech, L. David. 1966. *The Wolves of Isle Royale,* Fauna of the National Parks of the United States, Fauna Series 7. U.S. Government Printing Office.

Murie, A. 1944. *The Wolves of Mount McKinley,* Fauna of the National Parks of the U.S., Fauna Series 5. U.S. Govt. Printing Office.

Newell, Norman D. 1963. Crises in the history of life, in *Human Variations and Origins,* W. S. Laughlin and R. H. Osborne, eds. San Francisco: Freeman. (*Scientific American,* Feb. 1963.)

Romer, Alfred S. 1933. Pleistocene vertebrates and their bearing on the problem of human antiquity in North America, in *The American Aborigines,* ed. by D. Jenness, pp. 76-77.

Simpson, George G. 1965. *The Geography of Evolution.* Philadelphia. Chilton Books.

Slaughter, Bob H. 1967. Animal ranges as a clue to Late-Pleistocene extinction, in Martin and Wright, pp. 155-68.

Vereshchagin, N. K. 1967. Primitive hunters and Pleistocene extinction in the Soviet Union, in Martin and Wright, pp. 365-98.

Walker, Alan. 1967. Patterns of extinction among the subfossil Madagascan Lemuroids, in Martin and Wright, pp. 425-32.

Zeuner, F. E. 1963. *A History of Domesticated Animals.* London: Hutchinson.

23. *The Nature of the Darwinian Revolution*

ERNST MAYR

In this section a world-renowned evolutionist and taxonomist reviews the development of Darwin's evolution concepts, the intellectual atmosphere during Darwin's times, and the social impact of his *Origin of Species*. As the author so well points out, Darwinism was a unique type of revolution in the sense that it replaced a number of other theories, took place over an extended period of time and, in fact, is still having an enormous impact on the lives of man. Dr. Ernst Mayr is Alexander Agassiz Professor of Zoology at the Museum of Comparative Zoology, Harvard University. This paper was delivered at the annual meeting of the American Association for the Advancement of Science in Philadelphia in 1971.

The road on which science advances is not a smoothly rising ramp; there are periods of stagnation, and periods of accelerated progress. Some historians of science have recently emphasized that there are occasional breakthroughs, scientific revolutions (*1*), consisting of rather drastic revisions or previously maintained assumptions and concepts. The actual nature of these revolutions, however, has remained highly controversial (*2*). When we look at those of the so-called scientific revolutions that are most frequently mentioned, we find that they are identified with the names Copernicus, Newton, Lavoisier, Darwin, Planck, Einstein, and Heisenberg; in other words, with one exception, all of them are revolutions in the physical sciences.

Does this focus on the physical sciences affect the interpretation of the concept "scientific revolution"? I am taking a new look at the Darwinian revolution of 1859, perhaps the most fundamental of all intellectual revolutions in the history of mankind. It not only eliminated man's anthropocentrism, but affected every metaphysical and ethical concept, if consistently applied. The earlier prevailing concept of a created, and subsequently static, world was miles apart from Darwin's picture of a steadily evolving world. Kuhn (*1*) maintains that scientific revolutions are characterized by the replacement of an outworn paradigm by a new one. But a paradigm is, so to speak, a bundle of separate concepts, and not all of these are changed at the same time. In this analysis of the Darwinian revolution, I am attempting to dissect the total change of thinking involved in the Darwinian revolution into the major changing concepts, to determine the relative chronology of these changes, and to test the resistance to these changes among Darwin's contemporaries.

The idea of evolution had been widespread for more than 100 years before 1859. Evolutionary interpretations were advanced increasingly often in the second half of the

18th and the first half of the 19th centuries, only to be ignored, ridiculed, or maligned. What were the reasons for this determined resistance?

The history of evolutionism has long been a favorite subject among historians of science (3-5). Their main emphasis, however, has been on Darwin's forerunners, and on any and every trace of evolutionary thinking prior to 1859, or on the emergence of evolutionary concepts in Darwin's own thinking. These are legitimate approaches, but it seems to me that nothing brings out better the revolutionary nature of some of Darwin's concepts (6) than does an analysis of the arguments of contemporary antirevolutionists.

Cuvier, Lyell, and Louis Agassiz, the leading opponents of organic evolution, were fully aware of many facts favoring an evolutionary interpretation, and likewise of the Lamarckian and other theories of transmutation. They devoted a great deal of energy to refute evolutionism (7-10) and supported instead what, to a modern student, would seem a less defensible position. What induced them to do so?

It is sometimes stated that they had no other legitimate choice, because—it is claimed—not enough evidence in favor of evolution was available before 1859. The facts refute this assertion. Lovejoy (11), in a superb analysis of this question, asks: "At what date can the evidence in favor of the theory of organic evolution ... be said to have been fairly complete?" Here, one can perhaps distinguish two periods. During an earlier one, lasting from about 1745 to 1830, much became known that suggested evolution or, at least, a temporalized scale of perfection (12). Names like Maupertuis (1745), de Maillet (1749), Buffon (1749), Diderot (1769), Erasmus Darwin (1794), Lamarck (1809), and E. Geoffrey St. Hilaire (1818) characterize this period. Enough evidence from the fields of biogeography, systematics, paleontology, comparative anatomy, and animal and plant breeding, was already available by about 1812 (date of Cuvier's *Ossemens Fossiles*) to have made it possible to develop some of the arguments later made by Darwin in the *Origin of Species* (6). Soon afterward, however, much new evidence was produced by paleontology and stratigraphy, as well as by biogeography and comparative anatomy, with which only the evolutionary hypothesis was consistent; these new facts "reduced the rival hypothesis to a grotesque absurdity" (11). Yet, only a handful of authors [including Meckel (1821), Chambers (1844), Unger (1852), Schaaffhausen (1853), Wallace (1855)] adopted the concept of evolution while such leading authorities as Lyell, R. Owen, and Louis Agassiz vehemently opposed it.

Time does not permit me to marshal the abundant evidence in favor of evolution which existed by 1830. A comprehensive listing has been provided by Lovejoy (11), although the findings of systematics and biogeography must be added to his tabulation. The patterns of animal distribution were particularly decisive evidence, and it is no coincidence that Darwin devoted to it two entire chapters in the *Origin*. In spite of this massive evidence, creationism remained "the hypothesis tenaciously held by most men of science for at least twenty years before 1859" (11). It was not a lack of supporting facts, then, that prevented the acceptance of the theory of evolution, but rather the power of the opposing ideas.

The Nature of the Darwinian Revolution   281

Curiously, a number of nonscientists, particularly Robert Chambers (*13*) and Herbert Spencer saw the light well before the professionals. Chambers, the author of the *Vestiges of the Natural History of Creation*, developed quite a consistent and logical argument for evolutionism, and was instrumental in converting A. R. Wallace, R. W. Emerson, and A. Schopenhauer to evolutionism. As was the case with Diderot and Erasmus Darwin, these well-informed and broadly educated lay people looked at the problem in a "holistic" way, and thus perceived the truth more readily than did the professionals who were committed to certain well-established dogmas. A view from the distance is sometimes more revealing, for the understanding of broad issues, than the myopic scrutiny of the specialist.

Power of Retarding Concepts

Why were the professional geologists and biologists so blind when the manifestations of evolution were staring them in the face from all directions? Darwin's friend Hewett Watson put it this way in 1860 (*14*, p. 226): "How could Sir Lyell . . . for thirty years read, write, and think on the subject of species *and their succession*, and yet constantly look down the wrong road? Indeed, how could he? And the same question can be asked for Louis Agassiz, Richard Owen, almost all of Lyell's geological colleagues, and all of Darwin's botanist friends from Joseph Hooker on down. They all displayed a nearly complete resistance to drawing what to us would seem to be the inevitable conclusion from the vast amount of evidence in favor of evolution.

Historians of science are familiar with this phenomenon; it happens almost invariably when new facts cast doubt on a generally accepted theory. The prevailing concepts, although more difficult to defend, have such a powerful hold over the thinking of all investigators, that they find it difficult, if not impossible, to free themselves of these ideas. To illustrate this by merely one example, I would like to quote a statement by Lyell: "It is idle . . . to dispute about the abstract possibility of the conversion of one species into another, when there are known causes, so much more active in their nature, which must always intervene and prevent the actual accomplishment of such conversions" (*9*, p. 162). Actually one searches in vain for a demonstration of such "known causes" and any proof that they "must" always intervene. The cogency of the argument relied entirely on the validity of silent assumptions.

In the particular case of the Darwinian revolution, what were the dominant ideas that formed roadblocks against the advance of evolutionary thinking? To name these concepts is by no means easy because they are silent assumptions, never fully articulated. When these assumptions rest on religious beliefs or on the acceptance of certain philosophies, they are particularly difficult to reconstruct. This is the major reason why there is so much difference of opinion in the interpretation of this period. Was theology responsible for the lag, or was it the authority of Cuvier or Lyell, or the acceptance of catastrophism (with progressionism), or the absence of a reasonable explanatory scheme? All of these interpretations and several others have been advanced, and all presumably played some role. Others, particularly the role of essentialism, have so far been rather neglected by the historians.

Natural Theology and Creationism

The period from 1800 to the middle of that century witnessed the greatest flowering of natural theology in Great Britain (5, 15). It was the age of Paley and the Bridgewater Treatises, and virtually all British scientists accepted the traditional Christian conception of a Creator God. The industrial revolution was in full swing, the poor workingman was exploited unmercifully, and the goodness and wisdom of the Creator was emphasized constantly to sooth guilty consciences. It became a moral obligation for the scientist to find additional proofs for the wisdom and constant attention of the Creator. When Chambers in his *Vestiges* (13) dared to replace direct intervention of the Creator by the action of secondary causes (natural laws), he was roundly condemned. Although the attacks were ostensibly directed against errors of fact, virtually all reviewers were horrified that Chambers had "annulled all distinction between physical and moral," and that he had degraded man by ranking him as a descendant of the apes and by interpreting the universe as "the progression and development of a rank, unbending, and degrading materialism" (5, p. 150; 16). It is not surprising that in this intellectual climate Chambers had taken the precaution of publishing anonymously. Yet the modern reader finds little that is objectionable in Chambers' endeavor to replace supernatural explanations by scientific ones.

To a greater or lesser extent, all the scientists of that period resorted, in their explanatory schemes, to frequent interventions by the Creator (in the running of His world). Indeed, proofs of such interventions were considered the foremost evidence for His existence. Agassiz quite frankly describes the obligations of the naturalist in these words: "Our task is . . . complete as soon as we have proved His existence" (10, p. 132). To him the *Essay on Classification* was nothing but another Bridgewater Treatise in which the relationship of animals supplied a particularly elaborate and, for Agassiz, irrefutable demonstration of His existence.

Natural theology equally pervades Lyell's *Principles of Geology*. After discussing various remarkable instincts, such as pointing and retrieving, which are found in races of the dog, Lyell states: "When such remarkable habits appear in races of this species, we may reasonably conjecture that they were given with no other view than for the use of man and the preservation of the dog which thus obtains protection" (9, p. 455). Even though cultivated plants and domestic animals may have been created long before man, "some of the qualities of particular animals and plants may have been given solely with a view to the connection which, it was foreseen, would exist between them and man" (9, p. 456). Like Agassiz, Lyell believed that everything in nature is planned, designed, and has a predetermined end. "The St. Helena plants and insects [which are now dying out] may have lasted for their allotted term" (9, p. 9). The harmony of living nature and all the marvelous adaptations of animals and plants to each other and to their environment seemed to him thus fully and satisfactorily explained.

Creationism and the Advances of Geological Science

At the beginning of the 18th century, the concept of a created world seemed internally consistent as long as this world was considered only recently created (in

4004 B.C.), static, and unchanging. The "ladder of perfection" (part of God's plan) accounted for the "higher" and "lower" organization of animals and man, and Noah's flood for the existence of fossils. All this could be readily accommodated within the framework of a literal Biblical interpretation.

The discovery of the great age of the earth (5, 17) and of an ever-increasing number of distinct fossil faunas in different geological strata necessitated abandoning the idea of a single creation. Repeated creations had to be postulated, and the necessary number of such interventions had to be constantly revised upward. Agassiz was willing to accept 50 or 80 total extinctions of life and an equal number of new creations. Paradoxically, the advance of scientific knowledge necessitated an increasing recourse to the supernatural for explanation. Even such a sober and cautious person as Charles Lyell frequently explained natural phenomena as due to "creation" and, of course, a carefully thought-out creation. The fact that the brain of the human embryo successively passes through stages resembling the brains of fish, reptile, and lower mammal discloses "in a highly interesting manner, the unity of plan that runs through the organization of the whole series of vertebrated animals; but they lend no support whatever to the notion of a gradual transmutation of one species into another; least of all of the passage, in the course of many generations, from an animal of a more simple to one of a more complex structure" (9, p. 20). When a species becomes extinct it is replaced "by new creations" (9, p. 45). Nothing is impossible in creation. "Creation seems to require omnipotence, therefore we cannot estimate it" (18, p. 4). "Each species may have had its origin in a single pair, or individual where an individual was sufficient, and species may have been created in succession at such times and in such places as to enable them to multiply and endure for an *appointed* period, and occupy an *appointed* space of the globe?" (italics mine) (9, pp. 99-100). Everything is done according to plan. Since species are fixed and unchangeable, everything about them, such as the area of distribution, the ecological context, adaptations to cope with competitors and enemies, and even the date of extinction, was previously "appointed," that is, predetermined.

This constant appeal to the supernatural amounted to a denial of all sound scientific methods, and to the adoption of explanations that could neither be proven nor refuted. Chambers saw this quite clearly (13). When there is a choice between two theories, either special creation or the operation of general laws instituted by the Creator, he exclaimed, "I would say that the latter [theory] is greatly preferable, as it implies a far grander view of the Divine power and dignity than the other" (13, p. 117). Indeed, the increasing knowledge of geological sequences, and of the facts of comparative anatomy and geographic distribution, made the picture of special creation more ludicrous every day (11, p. 413).

Essentialism and a Static World

Thus, theological considerations clearly played a large role in the resistance to the adoption of evolutionary views in England (and also in France). Equally influential, or perhaps even more so, was a philosophical concept. Philosophy and natural history during the first half of the 19th century, particularly in continental Europe, were strongly dominated by typological thinking [designated "essentialism" by Potter (19, 20)]. This presumes that the changeable world of appearances is based on underlying immutable essences, and that all members of a class represent the same essence. This idea was first clearly enunciated in Plato's concept of the *eidos*. Later it became a dominant element in the teachings of Thomism (21), and of all idealistic philosophy. The enormous role of essentialism in retarding the acceptance of evolutionism was long overlooked (22, 23). The observed vast variability of the world has no more reality, according to this philosophy, than the shadows of an object on a cave wall, as Plato expressed it in his allegory. The only things that are permanent, real, and sharply discontinuous from each other are the fixed, unchangeable "ideas" underlying the observed variability. Discontinuity and fixity are, according to the essentialist, as much the properties of the living as of the inanimate world.

As Reiser (24) has said, a belief in discontinuous, immutable essences is incompatible with a belief in evolution. Agassiz was an extreme representative of this philosophy (23). To a lesser extent the same can be demonstrated for all of the other opponents of evolutionism, including Lyell. When rejecting Lamarck's claim that species and genera intergrade with each other, Lyell proposes that the following laws "prevail in the economy of the animate creation. . . . Thirdly, that there are fixed limits beyond which the descendants from common parents can never deviate from a certain type; fourthly, that each species springs from one original stock, and can never be permanently confounded by intermixing with the progeny of any other stock; fifthly, that each species shall endure for a considerable period of time" (9, p. 433). All nature consists, according to Lyell, of fixed types created at a definite time. To him these types were morphological entities, and he was rather shocked by Lamarck's idea that changes in behavior could have any effect on morphology.

As an essentialist, Lyell showed no understanding of the nature of genetic variation. Strictly in the scholastic tradition, he believed implicitly that essential characters could not change; this could occur only with nonessential characters. If an animal is brought into a new environment, "a short period of time is generally sufficient to effect nearly the whole change which an alteration of external circumstances can bring about in the habits of a species, . . . such capacity of accommodation to new circumstances is enjoyed in very different degrees by different species" (9, p. 464). For instance, if we look at the races of dogs, they show many superficial differences "but, if we look for some of those essential changes which would be required to lend even the semblance of a foundation for the theory of Lamarck, respecting the growth of new organs and the gradual obliteration of others, we find nothing of the kind" (9, p. 438). This forces Lyell to question even Lamarck's conjecture "that the wolf may have been the original of the dog." The fact that in the (geologically speaking) incredibly short time since the dog was domesticated, such drastically different races as the Eskimo dog, the hairless Chihuahua, the greyhound, and other extremes evolved is glossed over.

Lyell's Species Concept

Holding a species concept that allowed for no essential variation, Lyell credited species with little plasticity and adaptability. This led him to an interpretation of the fossil record that is very different from that of Lamarck. Anyone studying the continuous changes in the earth's surface, states Lyell, "will immediately perceive that, amidst the vicissitudes of the earth's surface, species cannot be immortal, but must perish, one after the other, like the individuals which compose them. There is no possibility of escaping from this conclusion, without resorting to some hypothesis as violent as that of Lamarck who imagined ... that species are each of them endowed with indefinite powers of modifying their organization, in conformity to the endless changes of circumstances to which they are exposed" (9, pp. 155-156).

The concept of a steady extermination of species and their replacement by newly created ones, as proposed by Lyell, comes close to being a kind of microcatastrophism, as far as organic nature is concerned. Lyell differed from Cuvier merely in pulverizing the catastrophes into events relating to single species, rather than to entire faunas. In the truly decisive point, the rejection of any possible continuity between species in progressive time sequences, Lyell entirely agreed with Cuvier. When he traced the history of a species backward, Lyell inexorably arrived at an original ancestral pair, at the original center of creation. There is a total absence in his arguments of any thinking in terms of populations.

The enormous power of essentialism is in part explainable by the fact that it fitted the tenets of creationism so well; the two dogmas strongly reinforced each other. Nothing in Lyell's geological experience seriously contradicted his essentialism. It was not shaken until nearly 25 years later when Lyell visited the Canary Islands (from December 1853 to March 1854) and became acquainted with the same kind of phenomena that, in the Galapagos, had made Darwin an evolutionist and which, in the East Indian Archipelago, gave concrete form to the incipient evolutionism of A. R. Wallace. Wilson (18) has portrayed the growth of doubt which led Lyell to publicly confess his conversion to evolutionism in 1862. The adoption of population thinking by him was a slow process, and even years after his memorable discussion with Darwin (16 April 1856), Lyell spoke in his notebooks of "variation or selection" as the important factor in evolution in spite of the fact that Darwin's entire argument was founded on the need for both factors as the basis of a satisfactory theory.

Lyell and Uniformitarianism

It is a long-standing tradition in biological historiography that Lyell's revival of Hutton's theory of uniformitarianism was a major factor in the eventual adoption of evolutionary thinking. This thesis seems to be a great oversimplification; it is worthwhile to look at the argument a little more critically (25). When the discovery of a series of different fossil faunas, separated by unconformities, made the story of a single flood totally inadequate, Cuvier and others drew the completely correct conclusion that these faunas, particularly the alternation of marine and terrestrial faunas, demonstrated a frequent alternation of rises of the sea above the land and the subsequent reemergence of land above the sea. The discovery of mammoths frozen into the ice of Siberia favored the additional thesis that such changes could happen very rapidly. Cuvier was exceedingly cautious in his formulation of the nature of these "revolutions" and "catastrophes," but he did admit, "The breaking to pieces and overturning of the strata, which happened in former catastrophes, show plainly enough that they were sudden and violent like the last [which killed the mammoths and embedded them in ice]" (26, p. 16). He implied that most of these events were local rather than universal phenomena, and he did not maintain that a new creation had been required to produce the species existing today. He said merely "that they [modern species] did not anciently occupy their present locations and that they must have come there from elsewhere" (26, pp. 125-126).

Cuvier's successors did not maintain his caution. The school of the so-called progressionists (27) postulated that each fauna was totally exterminated by a catastrophe at the end of each geologic period, followed by the special creation of an entirely new organic world. Progressionism, therefore, was intellectually a backward step from the widespread 18th-century belief that the running of the universe required only occasional, but definitely not incessant, active intervention by the Creator: He maintained stability largely through the laws that He had decreed at the beginning, and which allowed for certain planetary and other perturbations. This same reasoning could have easily been applied to the organic world, and this indeed is what was done by Chambers in 1844, and by many other devout Christians after 1859.

Catastrophism was not as great an obstacle to evolutionism as often claimed. It admitted, indeed it emphasized, the advance which each new creation showed over the preceding one. By also conceding that there had been 30, 50, or even more than 100 extinctions and new creations, it made the concept of these destructions increasingly absurd, and what was finally left, after the absurd destructions had been abandoned, was the story of the constant progression of faunas (28). As soon as one rejected reliance on supernatural forces, this progression automatically became evidence in favor of evolution. The only other assumption one had to make was that many of the catastrophes and extinctions had been localized events. This was, perhaps, not too far from Cuvier's original viewpoint.

The reason why catastrophism was adopted by virtually all of the truly productive leading geologists in the first half of the 19th century is that the facts seemed to support it. Breaks in fossil strata, the occurrence of vast lava flows, a replacement of terrestrial deposits by marine ones and the reverse, and many other phenomena of a similar, reasonably violent nature (including the turning upside down of whole fossil sequences) all rather decisively refuted a rigid uniformitarian interpretation. This is why Cuvier, Sedgwick, Buckland, Murchison, Conybeare, Agassiz, and de Beaumont, to mention a few prominent geologists, adopted more or less catastrophist interpretations.

Charles Lyell was the implacable foe of the "catastrophists," as his opponents were designated by Whewell (29). In his *Principles of Geology* (9), Lyell promoted a "steady state" concept of the world, best characterized by Hutton's motto, "no vestige of a beginning—no prospect of an end." Whewell coined the term "uniformitarianism" (30) for this school of thought, a term which unfortunately had many different meanings. The most important meaning was that it postulated that no forces had been active in the past history of the earth that are not also working today. Yet, even this would permit two rather different interpretations. Even if one includes supernatural agencies among forces and causes, one can still be a consistent uniformitarian, provided one postulates that the Creator continues to reshape the world actively even at the present. Rather candidly, Lyell refers to this interpretaion, accepted by him, as "the perpetual intervention hypothesis" (*18*, p. 89).

Almost diametrically opposed to this were the conclusions of those who excluded all recourse to supernatural interventions. Uniformitarianism to them meant simply the consistent application of natural laws not only to inanimate nature (as was done by Lyell) but also to the living world (as proposed by Chambers). The important component in their argument was the rejection of supernatural intervention rather than a lip service to the word uniformity.

It is important to remember that Lyell applied his uniformitarianism in a consistent manner only to inanimate nature, but left the door open for special creation in the living world. Indeed as Lovejoy (*11*) states justly, when it came to the origin of new species, Lyell, the great champion of uniformitarianism, embraced "the one doctrine with which uniformitarianism was wholly incompatible—the theory of numerous and discontinuous miraculous special creations." Lyell himself did not see it that way. As he wrote to Herschel (*31*), he considered his notion "of a succession of extinction of species, and creation of new ones, going on perpetually now . . . the grandest which I had ever conceived, so far as regards the attributes of the Presiding Mind." There is evidence, however, that Lyell considered these creations not always as miracles, but sometimes as occurring "through the intervention of intermediate causes" thus being "a natural, in contradistinction to a miraculous process." By July 1856, after having read Wallace's 1855 paper, and after having discussed evolution with Darwin (16 April 1856), Lyell had become completely converted to believing that the introduction of new species was "governed by laws in the same sense as the Universe is governed by laws" (*18*, p. 123).

Only the steady-state concept of uniformitarianism was novel in Lyell's interpretation. The insistence that nature operates according to eternal laws, with the same forces acting at all times was, from Aristotle on, the standard explanation among most of those who did not postulate a totally static world, for instance, among the French naturalists preceding Cuvier. Consequently, acceptance of uniformitarianism did not, as Lyell himself clearly demonstrated, require the acceptance of evolutionism. If one believed in a steady-state world, as did Lyell, uniformitarianism was incompatible with evolution. Only if it was combined with the concept of a steadily changing world, as it was in Lamarck's thinking, did it encourage a belief in evolution. It is obvious, then, that the statement "uniformitarianism is the pacemaker of evolutionism," is an exaggeration, if not a myth.

But what effect did Lyell have on Darwin? Everyone agrees that it was profound; there was no other person whom Darwin admired as greatly as Lyell. *Principles of Geology,* by Lyell, was Darwin's favorite reading on the *Beagle* and gave his geological interests new direction. After the return of the *Beagle* to England, Darwin received more stimulation and encouragement from Lyell than from any other of his friends. Indeed, Lyell became a father figure for him and stayed so for the rest of his life. Darwin's whole way of writing, particularly in the *Origin of Species,* was modeled after the *Principles.* There is no dispute over these facts.

But, what was Lyell's impact on Darwin's evolutionary ideas? There is much to indicate that the influence was largely negative. Knowing how firmly Lyell was opposed to the possibility of a transmutation of species, as documented by his devastating critique of Lamarck, Darwin was very careful in what he revealed to Lyell. He admitted that he doubted the fixity of species, but after that the two friends apparently avoided a further discussion of the subject. Darwin was far more outspoken with Hooker to whom he confessed as early as January 1844, "I am almost convinced . . . that species are not (it is like confessing murder) immutable" (*14,* p. 23). It was not until 1856 that Darwin fully outlined his theory of evolution to Lyell (*18,* p. xlix). This reticence of Darwin was not due to any intolerance on Lyell's part (or else Lyell would not have, after 1856, encouraged Darwin so actively to publish his heretical views), but rather to an unconscious fear on Darwin's part that his case was not sufficiently persuasive to convert such a formidable opponent as Lyell. There has been much speculation as to why Darwin had been so tardy about publishing his evolutionary views. Several factors were involved (one being the reception of the *Vestiges*), but I am rather convinced that his awe of Lyell's opposition to the transmutation of species was a much more weighty reason than has been hitherto admitted. It is no coincidence that Darwin finally began to write his great work within 3 months after Lyell took the initiative to consult him and to encourage him. Lovejoy summarizes the effect of Lyell's opposition to evolution in these words: "It was . . . his example and influence, more than the logical force of his arguments, that so long helped to sustain the prevalent belief that transformism was not a scientifically respectable theory" (*11*). I entirely agree with this evaluation.

Unsuccessful Refutations Owing to Wrong Choice of Alternatives

Creationism, essentialism, and Lyell's authority were not, however, the only reasons for the delay in the acceptance of evolution; others were important weaknesses in the scientific methodology of the period. There was still a demand for conclusive proofs. "Show me the breed of dogs with an entirely new organ," Lyell seems to say, "and I will believe in evolution." That much of science consists merely in showing that one interpretation is more probable than another one, or consistent with more facts than another one, was far less realized at that period than it is now (*32*).

That victory over one's opponent consists in the refutation of his arguments, however, was taken for granted. Cuvier's, Lyell's, Agassiz's, and Darwin's detailed argumentations were all attempts to "falsify," as Popper (*33*) has called it, the statements of their opponents. This method, however, has a number of weaknesses. For instance, it is often quite uncertain what kind of evidence or argument truly represents a falsification. More fatal is the frequently made assumption that there are only two alternatives in a dispute. Indeed, the whole concept of "alternative" is rather ambiguous, as I shall try to illustrate with some examples from pre-Darwinian controversies.

We can find numerous illustrations in the antievolutionary writings of Charles Lyell and Louis Agassiz of the limitation to only two alternatives when actually there was at least a third possible choice. Louis Agassiz, for instance, never seriously considered the possibility of true evolution, that is, of descent with modification. For him the world was either planned by the Creator, or was the accidental product of blind physical causes (in which case evolution would be the concatenation of such accidents). He reiterates this singularly simple-minded choice throughout the *Essay on Classification* (*10*): "physical laws" versus "plan of creation" (p. 10), "spontaneous generation" versus "divine plan" (p. 36), "physical agents" versus "plan ordained from the beginning" (p. 37), "physical causes" versus "supreme intellect" (p. 64), and "physical causes" versus "reflective mind" (p. 127). By this choice he not only excluded the possibility of evolution as envisioned by Darwin, but even as postulated by Lamarck. Nowhere does Agassiz attempt to refute Lamarckian evolution. His physical causes, in turn, are an exceedingly narrow definition of natural causes, since it is fully apparent that Agassiz had a very simple-minded Cartesian conception of physical causes as motions and mechanical forces. "I am at a loss to conceive how the origin of parasites can be ascribed to physical causes" (*10*, p. 126). "How can physical causes be responsible for the form of animals when so many totally different animal types live in the same area subjected to identical physical causes?" (*10*, pp. 13-14). The abundant regularities in nature demonstrate "the plan of a Divine Intelligence" since they cannot be the result of blind physical forces. (This indeed was a standard argument among adherents of natural theology.) It never occurred to Agassiz that none of his arguments excluded a third possibility, the gradual evolution of these regularities by processes that can be daily observed in nature. This is why the publication of Darwin's *Origin* was such a shock to him. The entire evidence against evolution, which Agassiz had marshaled so assiduously in his *Essay on Classification,* had become irrelevant. He had failed completely to provide arguments against a third possibility, the one advanced by Darwin.

The concept of evolution, at that period, still evoked in most naturalists the image of the *scala naturae*, the ladder of perfection. No one was more opposed to this concept than Lyell, the champion of a steady-state world. Any finding that contradicted a steady progression from the simple toward the more perfect refuted the validity of evolution, he thought. Indeed, the fact that mammals appeared in the fossil record before birds, and that primates appeared in the Eocene considerably earlier than some of the orders of "lower" mammals were, to him, as decisive a refutation of the evolutionary theory as was to Agassiz the fact that the four great types of animals appeared simultaneously in the earliest fossil-bearing strata.

The assumption that refuting the *scala naturae* would refute once and for all any evolutionary theory is another illustration of insufficient alternatives. Lyell was quite convinced that the concept of a steady-state world would be validated (including regular special creations), if it could be shown that those mechanisms were improbable or impossible which Lamarck had proposed to account for evolutionary change.

But there were also other violations of sound scientific method; for instance, the failure to see that both of two alternatives might be valid. In these cases, the pre-Darwinians arrived at erroneous conclusions because they were convinced that they had to make a choice between two processes which, in reality, occur simultaneously. For example, neither Lamarck nor Lyell understood speciation (the multiplication of species), but this failure led them to opposite conclusions. When looking at fossil faunas, Lamarck, a great believer in the adaptability of natural species, concluded that all the contained species must have evolved into very different descendants. Lyell, as an essentialist, rejected the possibility of a change in species and therefore he believed, like Cuvier, that all of the species had become extinct, with replacements provided by special creation. Neither Lamarck nor Lyell imagined that both processes, speciation and extinction, could occur simultaneously. That the turnover of faunas could be a balance of both processes never entered their minds.

Failure To Separate Distinct Phenomena

A third type of violation of scientific logic was particularly harmful to the acceptance of evolutionary thinking. This was the erroneous assumption that certain characteristics are inseparably combined. For instance, both Linnaeus and Darwin assumed, as I pointed out at an earlier occasion (*34*), that if one admitted the *reality* of species in nature, one would also have to postulate their immutable *fixity*. Lyell, as a good essentialist, unhesitatingly endorsed the same thesis: "From the above considerations, it appears that species have a real existence in nautre; and that each was endowed, at the time of its creation, with the attributes and organization by which it is now distinguished" (*9*, p. 21). He is even more specific about this in his notebooks (*18*, p. 92). That species could have full "reality" in the nondimensional situation (*34*) and yet evolve continuously was unthinkable to him. Reality and constancy of species were to him inseparable attributes.

Impact of the Origin of Species

The situation changed drastically and permanently with the publication of the *Origin of Species* in 1859. Darwin marshaled the evidence in favor of a transmutation of species so skillfully that from that point on the eventual acceptance of evolutionism was no longer in question. But he did more than that. In natural selection he proposed a mechanism that was far less vulnerable than any other previously proposed. The result was an entirely different concept of evolution. Instead of endorsing the 18th-century concept of a drive toward perfection, Darwin merely postulated change. He saw quite clearly that each species is forever being buffeted around by the capriciousness of the constantly changing environment. "Never use the word(s) higher and lower" (*35*) Darwin reminded himself. By chance this process of adaptation sometimes results in changes that can be interpreted as progress, but there is no intrinsic mechanism generating inevitable advance.

Virtually all the arguments of Cuvier, Lyell, and the progressionists became irrelevant overnight. Essentialism had been the major stumbling block, and the development of a new concept of species was the way to overcome this obstacle. Lyell himself eventually (after 1856) understood that the species problem was the crux of the whole problem of evolution, and that its solution had potentially the most far-reaching consequences: "The ordinary naturalist is not sufficiently aware that, when dogmatizing on what species are, he is grappling with the whole question of the organic world and its connection with a time past and with man" (*18*, p. 1). And, since he came to this conclusion after studying speciation in the Canary Islands, he added: "A group of islands, therefore, is the fittest place for Nature's trial of such permanent variety-making and where the problem of species-making may best be solved" (*18*, p. 93). This is what Darwin had discovered 20 years earlier.

Special Aspects of the Darwinian Revolution

No matter how one defines a scientific revolution, the Darwinian revolution of 1859 will have to be included. Who would want to question that, by destroying the anthropocentric concept of the universe, it caused a greater upheaval in man's thinking than any other scientific advance since the rebirth of science in the Renaissance? And yet, in other ways, it does not fit at all the picture of a revolution. Or else, how could H. J. Muller have exclaimed as late as 1959: "One hundred years without Darwinism are enough!" (*36*)? And how could books such as Barzun's *Darwin, Marx, Wagner* (1941) and Himmelfarb's *Darwin and the Darwinian Revolution* (1959), both displaying an abyss of ignorance and misunderstanding, have been published relatively recently? Why has this revolution in some ways made such extraordinarily slow headway?

A scientific revolution is supposedly characterized by the replacement of an old explanatory model by an incompatible new one (*1*). In the case of the theory of evolution, the concept of an instantaneously created world was replaced by that of a slowly evolving world, with man being part of the evolutionary stream. Why did the full acceptance of the new explanation take so long? The reason is that this short description is incomplete, and therefore misleading, as far as the Darwinian revolution is concerned.

Before analyzing this more fully, the question of the date of the Darwinian revolution must be raised. That the year 1859 was a crucial one in its history is not questioned. Yet, this still leaves a great deal of leeway to interpretation. On one hand, one might assert that the age of evolutionism started even before Buffon, and that the publication of *Origin* in 1859 was merely the last straw that broke the camel's back. On the other hand, one might go to the opposite extreme, and claim that not much had changed in the thinking of naturalists between the time of Ray and Tournefort and the year 1858, and that the publication of the *Origin* signified a drastic, almost violent revolution. The truth is somewhere near the middle; although there was a steady, and ever-increasing, groundswell of evolutionary ideas since the beginning of the 18th century, Darwin added so many new ideas (particularly an acceptable mechanism) that the year 1859 surely deserves the special attention it has received.

Two components of the Darwinian revolution must thus be distinguished: the slow accumulation of evolutionary facts and theories since early in the 18th century, and the decisive contribution which Darwin made in 1859. Together these two components constitute the Darwinian revolution.

The long time span is due to the fact that not simply the acceptance of one new theory was involved, as in some other scientific revolutions, but of an entirely new conceptual world, consisting of numerous separate concepts and beliefs (*37*). And not only were scientific theories involved, but also a whole set of metascientific credos. Let me prove my point by specifying the complex nature of the revolution: I distinguish six major elements in this revolution, but it is probable that additional ones should be recognized. (*32*).

The first three elements concern scientific replacements:

1. *Age of the earth.* The revolution began when it became obvious that the earth was very ancient rather than having been created only 6000 years ago (*17*). This finding was the snowball that started the whole avalanche.

2. *Refutation of both catastrophism (progressionism) and of a steady-state world.* The evolutionists, from Lamarck on, had claimed that the concept of a more or less steadily evolving world, was in better agreement with the facts than either the catastrophism of the progressionists or Lyell's particular version of a steady-state world. Darwin helped this contention of the evolutionists to its final victory.

3. *Refutation of the concept of an automatic upward evolution.* Every evolutionist before Darwin had taken it for granted that there was a steady progress of perfection in the living world. This belief was a straight line continuation of the (static) concept of a scale of perfection, which was maintained even by the progressionists for whom each new creation represented a further advance in the plan of the Creator.

Darwin's conclusion, to some extent anticipated by Lamarck, was that evolutionary change through adaptation and specialization by no means necessitated continuous betterment. This view proved very unpopular, and is even today largely ignored by nonbiologists. This neglect is well illustrated by the teachings of the school of evolutionary anthropology, or those of Bergson and Teilhard de Chardin.

The last three elements concern metascientific consequences. The main reason why evolutionism, particularly in its Darwinian form, was such slow progress is that it was the replacement of one entire *weltanschauung* by a different one. This involved religion, philosophy, and humanism.

4. *The rejection of creationism.* Every antievolutionist prior to 1859 allowed for the intermittent, if not constant, interference by the Creator. The natural causes postulated by the evolutionists completely separated God from his creation, for all practical purposes. The new explanatory model replaced planned teleology by the haphazard process of natural selection. This required a new concept of God and a new basis for religion.

5. *The replacement of essentialism and nominalism by population thinking.* None of Darwin's new ideas was quite so revolutionary as the replacement of essentialism by population thinking (*19-23, 38*). It was this concept that made the introduction of

natural selection possible. Because it is such a novel concept, its acceptance has been slow, particularly on the European continent and outside biology. Indeed, even today it has by no means universally replaced essentialism.

6. *The abolition of anthropocentrism.* Making man part of the evolutionary stream was particularly distasteful to the Victorians, and is still distasteful to many people.

Nature of the Darwinian Revolution

It is now clear why the Darwinian revolution is so different from all other scientific revolutions. It required not merely the replacement of once scientific theory by a new one, but, in fact, the rejection of at least six widely held basic beliefs [together with some methodological innovations (*32*)].

Furthermore, it had a far greater relevance outside of science than any of the revolutions in the physical sciences. Einstein's theory of relativity, or Heisenberg's of statistical prediction, could hardly have had any effect on anybody's personal beliefs. The Copernican revolution and Newton's world view required some revision of traditional beliefs. None of these physical theories, however, raised as many new questions concerning religion and ethics as did Darwin's theory of evolution through natural selection.

In a way, the publication of the *Origin* in 1859 was the midpoint of the so-called Darwinian revolution rather than its beginning. Stirrings of evolutionary thinking preceded the *Origin* by more than 100 years, reaching an earlier peak in Lamarck's *Philosophie Zoologique* in 1809. The final breakthrough in 1859 was the climax in a long process of erosion, which was not fully completed until 1883 when Weismann rejected the possibility of an inheritance of acquired characters.

As in any scientific revolution, some of the older opponents, such as Agassiz, never became converted. But the Darwinian revolution differed by the large number of workers who accepted only part of the package. Many zoologists, botanists, and paleontologists eventually accepted gradual evolution through natural causes, but not through natural selection. Indeed, on a worldwide basis, those who continued to reject natural selection as the prime cause of evolutionary change were probably well in the majority until the 1930s.

Two conclusions emerge from this analysis. First, the Darwinian and quite likely other scientific revolutions consist of the replacement of a considerable number of concepts. This requires a lengthy period of time, since the new concepts will not all be proposed simultaneously. Second, the mere summation of new concepts is not enough; it is their constellation that counts. Uniformitarianism, when combined with the belief in a static essentialistic world, leads to the steady-state concept of Lyell, while when combined with a concept of change, it leads to the evolutionism of Lamarck. The observation of evolutionary changes, combined with essentialist thinking, leads to various saltationist or progressionist theories, but, combined with population thinking, it leads to Darwin's theory of evolution by natural selection.

It is now evident that the Darwinian revolution does not conform to the simple model of a scientific revolution, as described, for instance, by T. S. Kuhn (*1*). It is actually a complex movement that started nearly 250 years ago; its many major components were proposed at different times, and became victorious independently of each other. Even though a revolutionary climax occurred unquestionably in 1859, the gradual acceptance of evolutionism, with all of its ramifications, covered a period of nearly 250 years (*37*).

REFERENCES AND NOTES

1. T. S. Kuhn, *The Structure of Scientific Revolutions* (Univ. of Chicago Press, Chicago, 1962).
2. S. Toulmin, *Boston Stud. Phil. Sci.* 3, 333 (1966); I. Lakatos and A. Musgrave, Eds., *Criticism and the Growth of Knowledge* (Cambridge Univ. Press, Cambridge, England, 1970), reviewed by D. Shapere, *Science 172*, 706 (1971).
3. To cite only a few: L. Eiseley, *Darwin's Century* (Doubleday, New York, 1958): B. Glass, O. Temkin, W. L. Straus, Jr., Eds., *Forerunners of Darwin, 1745-1859* (Johns Hopkins Press, Baltimore, 1959); J. C. Greene, *The Death of Adam* (Iowa State Univ. Press, Ames, 1959); W. Zimmermann, *Evolution, Geschichte ihrer Probleme und Erkenntnisse* (Alber, Freiburg, West Germany, 1953); J. C. Greene, "The Kuhnian paradigm and the Darwinian revolution in natural history," in *Perspectives in the History of Science and Technology*, D. H. D. Roller, Ed. (Univ. of Oklahoma Press, Norman, 1971); M. T. Ghiselin, *New Lit. Hist. 3*, 113 (1971).
4. G. de Beer, *Charles Darwin* Doubleday, Garden City, N.Y., 1964).
5. C. C. Gillispie, *Genesis and Geology* (Harvard Univ. Press, Cambridge, Mass., 1951; rev. ed., Harper & Row, New York, 1959).
6. C. Darwin, *On the Origin of Species by Means of Natural Selection* (1859).
7. G. Cuvier, *Essay on the Theory of the Earth* (Edinburgh, ed. 3, 1817); much of it is an implicit refutation of Lamarck's ideas.
8. W. Coleman, *Georges Cuvier, Zoologist* (Harvard Univ. Press, Cambridge, Mass., 1964).
9. C. Lyell, *Principles of Geology* (John Murray, London, 1835), vols. 2 and 3 (I have used ed. 4). Important for British biology because it was the first presentation of Lamarck's theories to the English-speaking world (Book III, chap. I-XI). Darwin had previously heard about Lamarck from R. E. Grant in Edinburgh in 1827 [see (*4*, p. 28)].
10. L. Agassiz, *Essay on Classification* (Little, Brown, Boston, 1857; reprint, Belknap, Cambridge, Mass., 1962).
11. A. O. Lovejoy, "The argument for organic evolution before the *Origin of Species, 1830-1858*," in *Forerunners of Darwin, 1745-1859*, B. Glass, O. Temkin, W. L. Straus, Jr., Eds. (Johns Hopkins Press, Baltimore, 1959), pp. 356-414.
12. A. O. Lovejoy, *The Great Chain of Being* (Harvard Univ. Press, Cambridge, Mass., 1963), Lecture 9.
13. The authorship of the anonymously published *Vestiges of the Natural History of*

Creation (1844) did not become known until after the death of Robert Chambers [M. Millhauser, *Just Before Darwin* (Wesleyan Univ. Press, Middletown, Conn., 1959); see also (*5* chap. 6)]. A sympathetic analysis of the *Vestiges,* that does not concentrate on Chambers' errors and his gullibility, is still wanting [F. N. Egerton, *Stud. Hist. Phil. Sci. 1,* 176 (1971)].

14. F. Darwin, Ed., *Life and Letters of Charles Darwin* (Sources of Science Ser. No. 102; reprint of 1888 ed., Johnson Reprints, New York, 1969), vol. 2.
15. H. Fruchtbaum, "Natural theology and the rise of science," thesis, Harvard University (1964).
16. A. Sedgwick, *Edinburgh Rev. 82,* 3 (1845).
17. F. C. Haber, in *Forerunners of Darwin, 1745-1859,* B. Glass, O. Temkin, W. L. Straus, Jr., Eds. (Johns Hopkins Press, Baltimore, 1959), pp. 222-261.
18. L. G. Wilson, Ed., *Sir Charles Lyell's Scientific Journals on the Species Question* (Yale Univ. Press, New Haven, Conn., 1970).
19. K. R. Popper, *The Open Society and its Enemies* (Routledge & Kegan Paul, London, 1945).
20. D. Hull, *Brit. J. Phil. Sci. 15,* 314 (1964); *ibid.* 16, 1 (1965).
21. Aristotle is traditionally included among the essentialists, but newer researches cast considerable doubt on this. There is a growing suspicion that much of the late medieval thought labeled as Aristotelianism, had little to do with Aristotle's actual thinking. See, for instance, M. Delbrück, in *Of Microbes and Life,* J. Monod and E. Borek, Eds. (Columbia Univ. Press, New York, 1971), pp. 50-55.
22. E. Mayr, in *Evolution and Anthropology: A Centennial Appraisal* (Anthropological Society of Washington, Washington, D.C., 1959).
23. _____ , *Harvard Libr. Bull. 13,* 165 (1959).
24. O. L. Reiser, in *A Book that Shook the World,* R. Buchsbaum, Ed. (Univ. of Pittsburgh Press, Pittsburgh, Pa., 1958).
25. It is impossible in the limited space available to give a full documentation for the refutation of this thesis [see W. Coleman, *Biology in the Nineteenth Century* (Wiley, New York, 1971), p. 63].
26. G. Cuvier, *Essay on the Theory of the Earth,* R. Jameson, Transl. (Edinburgh, ed. 3, 1817). It is frequently stated that Cuvier believed in large-scale creations, necessary to repopulate the globe after major catastrophes, and this may well be true. However, I have been unable to find an unequivocal statement to this effect in Cuvier's writings [see also (*8,* p. 136)].
27. Progressionism was the curious theory according to which evolution did not take place in the organisms but rather in the mind of the Creator, who—after each catastrophic extinction—created a new fauna in the more advanced state to which His plan of creation had progressed in the meantime. This thought was promoted in Britain particularly by Hugh Miller (*Footprints,* 1847), Sedgwick (*Discourse,* 1850), and Murchison (*Siluria,* 1854), and in America by L. Agassiz (*Essay,* 1857); see *3*).
28. The difference between catastrophism and uniformitarianism became smaller, as it was realized that many of the "catastrophes" had been rather minor events, and that contemporary geological phenomena (earthquakes, volcanic erupt-

ions, tidal waves, glaciation) could have rather catastrophic effects [S. Toulmin, in *Criticism and the Growth of Knowledge,* I. Lakatos and A. Musgrave, Eds. (Cambridge Univ. Press, Cambridge, England, 1970), p. 42].

29. [W. Whewell] *Brit. Critic 9,* 180 (1831); *Quart. Rev. 47,* 103 (1832). For a full discussion of catastrophism see (5). A new interpretation of the traditional geological theories which Lyell opposed is given by M. J. S. Rudwick, "Uniformity and progression," in *Perspectives in the History of Science and Technology,* D. H. Roller, Ed. (Univ. of Oklahoma Press, Norman, 1971), pp. 209-237.

30. The term uniformitarianism was applied to at least four different concepts, and this caused considerable confusion, to put it mildly. For recent reviews see R. Hooykaas, *Natural Law and Divine Miracle* (E. J. Brill, Leiden, the Netherlands, 1959); S. J. Gould, *Amer. J. Sci. 263,* 233 (1965); C. C. Albritton, Jr., Ed., "Uniformity and simplicity," *Geol. Soc. Amer. Spec. Pap. No. 89* (1967); M. S. J. Rudwick, *Proc. Amer. Phil. Soc. 111,* 272 (1967); G. G. Simpson, "Uniformitarianism," in *Essays in Evolution and Genetics* (Appleton-Century-Crofts, New York, 1970), pp. 43-96. The most important interpretations of uniformitarianism are: the same processes act now as in the past, the magnitude of geological events is as great now as in the past, and the earth a steady-state system.

31. Mrs. Lyell, Ed., *Life, Letters and Journals of Sir Charles Lyell* (John Murray, London, 1881), vol. 1, pp. 467-469 (letter of 1 June 1836 to J. W. Herschel).

32. Darwin's *Origin* was one of the first scientific treatises in which the hypothetico-deductive method was rather consistently employed [M. Ghiselin, *The Triumph of the Darwinian Method* (Univ. of California Press, Berkeley, 1969)]. Equally important, and even more novel, was Darwin's demonstration that deterministic prediction is not a necessary component of causality [M. Scriven, *Science 130,* 477 (1959)]. Perhaps this can be considered a corollary of population thinking, but it is further evidence for the extraordinary complexity of the Darwinian revolution.

33. K. R. Popper, *The Logic of Scientific Discovery* (Hutchison, London, 1959).

34. E. Mayr, Ed., *The Species Problem* (AAAS, Washington, D.C., 1957), p. 2.

35. F. Darwin and A. C. Seward, Eds., *More Letters of Charles Darwin* (reprint of 1903 ed., Johnson Reprints, New York, 1971), vol. 1, p. 114.

36. H. J. Muller, "One hundred years without Darwinism are enough," *School Sci. Math. 1959,* 304 (1959).

37. It remains to be determined to what extent a similar claim can also be made for some of the physical sciences, for instance, the Corpernican revolution.

38. In all recent discussions of natural selection, the assumption is made that the concept traces back to the tradition of Adam Smith, Malthus, and Ricardo, with the emphasis on competition and progress. This interpretation overlooks the point that the elimination of "degradations of the type" as the essentialists would call it, does not lead to progress. For the typologist, natural selection is merely the elimination of inferior types, an interpretation again revived by the mutationists (after 1900). Darwin was the first to see clearly that a second factor was necessary, the production of new variation. (This leads to population thinking.) Selection can be creative only when such

new individual variation is abundantly available.

39. I greatly benefited from stimulating discussions with S. J. Gould and F. Sulloway, who read a draft of this essay, and from a series of most valuable critical comments, received from Prof. L. G. Wilson, which helped me to correct several errors. My own interpretation, however, still differs in some crucial points from that of Prof. Wilson. Some of the analysis was prepared while I served as Visiting Fellow at the Institute for Advanced Study, Princeton, N.J., in 1970.

A view of the Pawnee National Grasslands.

24. *Biogeochemical Cycles*

GENE E. LIKENS AND F. HERBERT BORMANN

Biogeochemical cycling refers to the passage of minerals through an ecosystem by means of organisms (biological), by processes such as weathering and erosion (geological), and by various chemical reactions that occur during the cycling. How does one go about doing research on such a complex system? The authors of this article did it by analyzing water samples taken from streams and snow melts in a forest ecosystem in New Hampshire. The various minerals and their quantities in the water reflected the types of cycles taking place there. As described in the paper, measurements of minerals in the water runoff from a deforested part of the ecosystem were drastically different from those in the rest of the ecosystem. In other words, the cycling process was disrupted by the loss of vegetation. Practices such as clear-cutting of forests, road building, and mass use of herbicides should be reevaluated in view of the drastic effects they may produce in the mineral cycling in an ecosystem. Dr. Likens is an associate professor of biology at Cornell University, and Dr. Herbert Bormann is professor of forest ecology at Yale University.

Professor Eugene Odum has referred to ecological systems as the "functional units of nature." Although this definition is oversimplified and general, it has appeal in our attempts to piece together the complex components of the biosphere. [See Fig. 1] An ecosystem may be simply defined as any unit of nature in which there is a functional and dependent interchange of energy and nutrients between living and nonliving components. An ecosystem may be very small, such as a puddle; or it may be large, such as a forest or an ocean. It also may have precise or real boundaries, such as those of a lake or an island, but more frequently it doesn't. Ecosystem boundaries on land are often rationalized and established by the investigator to facilitate conceptualization and study. Ecosystems are the building blocks or functioning units that make up the biosphere.

The survival or ecological systems depends upon a continuous input of energy and nutrients. An ecological system is richly endowed with a budget of inputs and outputs, and one of the reasons why it is difficult to assess the impact of human activity on nature is the lack of precise information about these inputs and outputs and about the delicate adjustments that maintain a balance. We should also remember that since all ecosystems interact, changes in one may cause changes in others by a kind of chain reaction. Projects such as the logging of a forest or the building of an interstate highway may have consequences far beyond the individual project. Similarly,

Source: Reprinted with permission of the publisher and authors from *The Science Teacher*, *39*(4):15–20, April 1972.

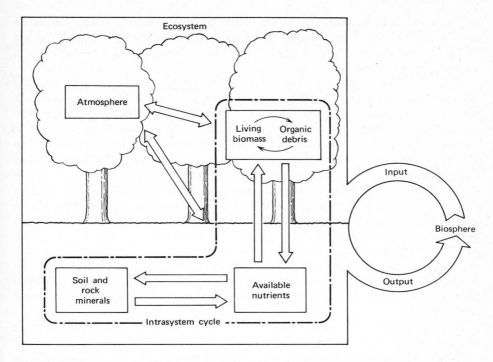

Fig. 1. Simplified nutrient relationships for a terrestrial ecosystem, showing sites of accumulation and major pathways. Inputs and outputs link individual ecosystems with the rest of the biosphere.

technological solutions frequently do not solve environmental problems; rather, the problems are often merely shifted from one level to another or from one ecosystem to another. There are a great many examples of this; two should make the point.

Air pollution in Great Britain and in the Ruhr Valley has been greatly alleviated by a very simple technological change. Taller smokestacks and chimneys push the smoke and gases higher into the atmosphere. However, this so-called "cure" has merely shifted the problem to somebody else. Wind carries the smog from these industrial areas primarily to Scandinavia. Here the increased addition of sulfur compounds, mainly sulfur dioxide and hydrogen sulfide, in the atmosphere has led to a striking increase in the acidity of rainfall. Since 1956 the acidity of rainwater has increased more than 200 times. This in turn has led to the acidification of lakes and streams in Scandinavia—to the extent that reduced growth and even kills of salmon and trout have been observed. There have also been effects on the soil (loss of fertility), on forestry (reduced growth), and corrosion of buildings and other structures. It is estimated that damage from acid corrosion in Sweden alone may amount to as much as 1 percent of its gross national product. This is analogous to chemical warfare, and the Swedes are not particularly happy about it, as you can imagine. What with fishing recently banned in more than 50 of their southern lakes because of mercury pollution and with DDT concentrations in Baltic fish now 10 times higher than in fish from the North Sea, the Swedes have rapidly become aware of the biospheric complexity of environmental problems.

It is frequently stated that through technology we may be able to dump our solid trash into the ocean with a minimum of ecological disturbance. One current plan calls for grinding all trash and garbage from the Baltimore and Washington metropolitan area, making a slurry of it, and then transporting it by pipeline to the edge of the Continental Shelf, some one hundred miles offshore. Here again, the problem is merely shifted from trash on land to trash in the ocean. And it is clear that there are far too few data to support the claim that this is really a solution to the problem.

Recently we heard an interesting suggestion for the trash problem. It was that we continue to collect solid trash and garbage from individuals and organizations. After it is collected, it could be compressed, sterilized, wrapped in a waterproof container, and then given back to the individual. He could do anything he wanted with it: put it in his attic, in his garage, basement, or warehouse—just as long as he didn't remove it from his property. Obviously this solution has immediate appeal in that it puts the problem where it belongs, with the individual or the industry who generates the waste, but also obviously it is unworkable. For one thing, like so many proposals, it penalizes urbanites.

These examples clearly point out our vital need to reuse and recycle material much in the same way that natural ecosystems do and to make our decisions on the basis of an honest accounting in which all costs are included. After all, this is a basic ecologic principle in the functioning of natural ecosystems. There is a distinct lack of good ecological information upon which to prejudge the effects of proposed technological developments and manipulations. Sound, quantitative information is needed to assess environmental problems from a holistic view and to develop honest balance sheets.

Let us look more closely at one of the basic functions of an ecosystem, the flow of nutrients and other materials. First we must distinguish between flux and cycling. Flux refers to the directional input and output that occurs across ecosystem boundaries. From such information we can determine budgets and observe balancing mechanisms. Cycling refers to the two-way exchange between living and non-living components within the ecosystem, but is tempered by external flux. Within the boundaries of an ecosystem, chemicals are continually withdrawn from the abiotic reservoir of the ecosystem (e.g., the air, the water, or the land), utilized by and circulated through the biotic portion, and returned in one form or another to the abiotic reservoir.

Because chemicals cycle between living and nonliving components, the terms "biogeochemical cycling" or "nutrient cycling" are frequently used to describe this process or basic ecosystem function. The term "mineral cycling" has been used widely, but strictly speaking, minerals do not cycle within ecosystems. Also it should be apparent that the movement of water and air is vital to the transport of chemicals such as N, C, H, O, P, S, etc., between and within ecosystems.

There are few quantitative data available for the flow and cycling of these materials in natural ecosystems. There is a large amount of data on various aspects, but it is exceedingly difficult to assemble this information in terms of an entire ecosystem and all of its interacting parts.

We began studies about eight years ago to determine the magnitude of the natural biogeochemical flux and cycling of nutrients for a forest ecosystem in New Hampshire. Like a city, the forest ecosystem survives because of a specific and continuous input and output of nutrients, and we wanted to quantify these fluxes as well as the rates of internal recycling. Our ecosystem unit is a watershed or drainage area. Therefore, its boundaries are defined by the drainage of water.

Our studies have been done in cooperation with the U.S. Forest Service within the Hubbard Brook Experimental Forest in West Thornton, New Hampshire. Several students and colleagues from various universities and federal laboratories also are involved in the overall project.

The forest is characterized by beech, yellow birch, and sugar maple trees. The bedrock is granitic and is watertight. This latter point is exceedingly important for these quantitative studies. If water were lost through deep rock strata rather than through the drainage stream, it would be almost impossible to know how much water was lost in that way and also what amount of nutrients that water carried out of the system. Since our ecosystem is watertight, all of the liquid water leaving the system is measured at gauging wiers. Water vapor is lost by evaporation and transpiration.

Precipitation collection stations scattered throughout the forest provide data on the input of both rain and snow throughout the year. By and large the monthly input of precipitation is constant throughout the year. However, loss of water by stream flow varies enormously. The largest runoff occurs in the spring when the heavy accumulation of snow melts. Approximately 57 percent of the total yearly runoff occurs in the months of March, April, and May. In contrast very small amounts of water are lost by stream drainage during the summer months when most of the water evaporates directly to the atmosphere from the forest vegetation. Yearly, about 123 cm of water come into the ecosystem in precipitation; 58 percent runs off in streams, and 42 percent is lost by evaporation and transpiration.

Precipitation chemistry is highly variable. First of all, we found that precipitation is surprisingly acid. The average pH of rain and snow is about 4, and hydrogen and sulfate ions are the dominant ions. Also, significant amounts of nitrate, ammonium, calcium, magnesium, phosphate, chloride, sodium, and potassium fall onto the ecosystem in rain and snow. Exceedingly clear, pure mountain streams drain these forested watershed-ecosystems. The major ions in stream water are calcium and sulfate, but in relatively low concentrations. (Table 1)

To determine the magnitude of nutrient flux across the ecosystem boundaries, we measured the concentration of chemicals in both the precipitation and the stream water. These data multiplied by accurate data for precipitation and stream runoff, provided by the Forest Service, allow us to calculate budgets of nutrient inputs and outputs. We have made measurements of this sort on six watershed-ecosystems of the Hubbard Brook Experimental Forest since 1963. (Table 2) Normally there are net

Table 1. *Weighted Average Concentrations of Various Dissolved Substances in Bulk Precipitation and Stream Water for Undisturbed Watersheds 1 through 6 of the Hubbard Brook Experimental Forest During 1963-1969.*

Chemical	Precipitation	Stream Water
	mg/l	*mg/l*
Calcium	0.21	1.58
Magnesium	0.06	0.39
Potassium	0.09	0.23
Sodium	0.12	0.92
Aluminum	a	0.24
Ammonium	0.22	0.05
Hydrogen	0.07	0.01
Sulfate	3.1	6.4
Nitrate	1.31	1.14
Chloride	0.42	0.64
Bicarbonate	a	1.9[b]
Dissolved silica	a	4.61

[a]Not determined, but very low.
[b]Watershed 4 only.

losses of calcium, magnesium, aluminum, sodium, and silica from the ecosystem via streams each year. However, losses of chloride and potassium in stream water about equal the inputs in precipitation. The net loss of sulfate, although usually consistent, is small in relation to the amount input in precipitation. Of particular interest is the finding that more nitrogen and phosphorus are brought into these undisturbed ecosystems, dissolved in rain and snow, than are lost in drainage waters. Thus, generally small amounts of nutrients are lost from these undisturbed forested ecosystems in drainage waters relative to the amounts input, stored, or cycled.

Man's activities can completely reverse these relationships. It was apparent from the beginning of our study that the opportunity to study experimentally some of man's effects on an entire ecosystem was exciting, scientifically sound, and a most powerful approach to unraveling the complexity of ecosystems.

Our first experiment of this type was deforestation of an entire watershed. In the autumn of 1965 every tree and shrub on the watershed was cut and reduced to ground level. No roads were built, and none of the logs were taken out of the system since this wasn't a typical logging operation. Later the system was treated with herbicides to inhibit all regrowth of vegetation. The ecosystem was maintained in this condition until 1969 when herbicide treatment was stopped, and the forest was allowed to regrow.

As a result of this experimental deforestation the runoff of water increased 40

Table 2. *Average Chemical Input and Output for Undisturbed, Forested Watershed Ecosystems 1 Through 6 of the Hubbard Brook Experimental Forest during 1963-1969. Output Data for W2 After Treatment Are Not Included.*

Chemical	Number of Watershed Years[a]	Input	Output	Mean Values (kg/ha-yr) Net Loss	Net Gain
SiO_2-Si	10	[b]	16.4	−16.4	
Ca	30-32	2.6	11.7	− 9.1	
SO_4-S	10-15	12.7	16.2	− 3.5	
Na	30-32	1.5	6.8	− 5.3	
Mg	30-32	0.7	2.8	− 2.1	
Al	10	[b]	1.8	− 1.8	
K	30-32	1.1	1.7	− 0.6	
NO_3-N	10-15	3.7	2.0		+1.7
NH_4-N	10-15	2.1	0.3		+1.8
Cl	8-12	5.2	4.9		+0.3
P	2	0.02[d]	0.01		+0.01
HCO_3-C	4[c]	[b]	2.9	− 2.9	

[a]No. of watersheds times years of data.
[b]Not measured, but very small.
[c]Watershed No. 4 only.
[d]Estimated concentration of 2 μg/liter.

percent the first year, 28 percent the second year, and 26 percent the third year. This means that the increased runoff loss would amount to a quantity of water about 35 cm deep over the entire watershed during the first year. Most of this water would have been lost by transpiration had the vegetation not been cut or killed. In summer, when the runoff is normally low, the increase in stream discharge was orders of magnitude greater per unit area than on adjacent uncut areas.

However, chemical concentrations in stream water increased unexpectedly. (Fig. 2) In particular, nitrate concentrations increased spectacularly, reaching a maximum concentration of almost 90 mg per liter. Normally concentrations are less than 2 ppm in the undisturbed watersheds. These increased concentrations plus the increased discharge of water had a very great effect on the nutrient budgets. (Table 3) Typically there is a relatively small net loss of calcium from the undisturbed forest (8 to 10 kg/ha-yr in 1966-69). In contrast 75 kg/ha were lost from the deforested watershed the first year and 90 kg/ha the second. For potassium, the same relationship was observed although the change was even larger.

The change in the nitrate-nitrogen budget was very large. Normally nitrogen compounds are conserved by these mature forest ecosystems. In other words, more is

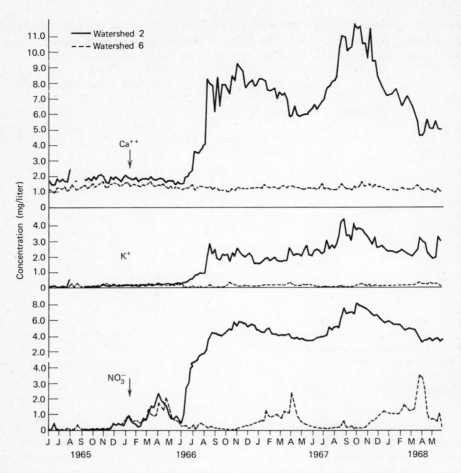

Fig. 2. Measured stream water concentrations for calcium, potassium, and nitrate ions in watersheds 2 (deforested) and 6. Note the change in scale for the nitrate concentration. The arrow indicates the completion of cutting in watershed 2.

added in precipitation with rain and snow than is lost from the system in drainage water; there is an annual net gain for the system. In contrast, the deforested system lost 97 kg/ha the first year and 142 kg/ha the second year, as is calculated from Table 3.[1] The average loss of nitrate nitrogen over a three-year period was about 110 kg/ha-yr. This is more than double the amount of nitrogen taken up by the undisturbed system each year. If all of the nitrogen were to come in only from precipitation and if there were no losses whatsoever, it would take about 43 years for the system to make up the amount of nitrogen lost in stream waters during these three years. If there were normal losses each year, as in the undisturbed situation, it would take about 100 years to replenish the nitrogen which was lost as a result of cutting

Table 3. *Comparative Net Losses or Gains of Dissolved Solids in Runoff from Undisturbed (W6) and Deforested (W2) Watersheds of the Hubbard Brook Experimental Forest. Values in Metric Tons/km²-yr (Metric Tons/km² x 10 = kg/ha)*

Chemical	1966-1967		1967-1968		1968-1969	
	W6	W2	W6	W2	W6	W2
Ca	−0.8	−7.5	−0.9	−9.0	−1.0	−6.8
K	−0.1	−2.3	−0.2	−3.6	−0.2	−3.3
Al	−0.1	−1.7	−0.3	−2.4	−0.3	−2.1
Mg	−0.3	−1.6	−0.3	−1.8	−0.3	−1.3
Na	−0.6	−1.7	−0.7	−1.7	−0.6	−1.2
NH_4	+0.2	+0.1	+0.3	+0.2	+0.3	+0.2
H	+0.1	+0.07	+0.09	+0.05	+0.08	+0.04
NO_3	+1.5	−43.0	+1.1	−62.8	+0.5	−45.5
SO_4	−0.8	−0.5	−1.0	0	−1.9	−2.0
HCO_3	−0.2	−0.1	−0.3	0	0	0
Cl	+0.2	−0.1	0	−0.4	−0.1	−0.1
SiO_2-aq.	−3.6	−6.6	−3.6	−6.9	−2.9	−5.9
Total	−4.6	−65.0	−5.9	−88.4	−6.2	−68.0

down the trees and preventing their regrowth. What are the reasons for this?

In the undisturbed forest, decay of organic debris such as leaves, twigs, bark, etc. produces ammonium compounds. This ammonium will either be taken up directly by the vegetation or it may be converted by microorganisms to nitrate and then taken up by the vegetation and held within the system. In the absence of vegetation the ammonium compounds are converted to nitrate and leached rapidly from the system. By removing this one component, the vegetation, a major change in the basic nitrogen cycle occurred. Alteration of the basic nitrogen cycle was the major change in ecosystem function, but there were many others. (Fig. 3) Some were expected, but many were completely unexpected.

The system maintains integrity by relatively simple biotic factors. These include such things as dead leaves which plaster over stream banks, or small rootlets which help to hold the stream bank in place and keep it from breaking apart and washing away. Without living vegetation to continuously resupply these dead leaves and to renew the small rootlets, the biotic regulation of erosion and transport was greatly reduced so that the output of particulate matter from the system was increased about 11 times after three years.

Along with this, the acidity of stream water greatly increased: Its pH decreased from about 5.1 to 4.3. Stream water might be characterized as a very dilute solution of sulfuric acid in the undisturbed situation, but it changed to a stronger solution of nitric acid as a result of deforestation.

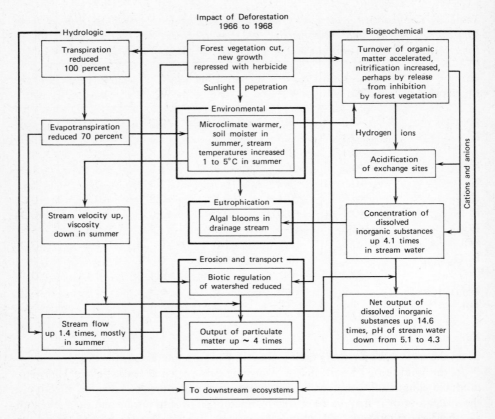

Fig. 3. A summary of some of the ecological effects of deforestation in the Hubbard Brook Experimental Forest. The forest was cut in the autumn of 1965.

The stream water from the deforested watershed appeared to be just as clear and potable as that from adjacent, undisturbed watersheds. However between August 1966 and January 1970 the nitrate concentration almost continuously exceeded, and at times doubled, the maximum concentration recommended as allowable for drinking water. Thus the deforestation treatment resulted in significant pollution of the drainage stream.

Increased sunlight striking the stream in the absence of a forest canopy, the high nutrient concentrations, and higher stream temperature, all resulted in significant eutrophication. Whereas the undisturbed streams contain few algae, a dense bloom of green algae occurred in the stream of the deforested watershed each summer. This is a good example of how a change in one component of an ecosystem alters the structure and function, often unexpectedly, in other parts of the same ecosystem or in an interrelated ecosystem. Unless these interrelationships are understood, naive management practices can produce unexpected and possibly widespread deleterious results.

These results pose some very important questions for forestry management. Clear-cutting of forest vegetation is practiced widely in the United States, but there are few data on its total effect on the ecosystem. To our knowledge there are no other quantitative data on chemical losses from cleared forests. Since the deforestation treatment was done as an experiment, logging roads were not constructed, timber was not removed, and herbicides were used to prevent vegetation regrowth. However preliminary results from studies now in progress on a number of commercially logged areas in northern New England clearly show the same pattern of nutrient loss and eutrophication in drainage waters as that which occurred in the experimentally deforested watershed.

If recharge of nitrogen is solely dependent upon input in rain and snow, decades would be required to replenish the nitrogen lost as a result of clear-cutting. How does this affect long-term productivity of the forest? Are there other mechanisms of nitrogen replenishment? We don't know. Neither do we have good information on regrowth in old roadbeds or even good information on the role of noncommercial forest species in the function of many forest ecosystems.

Perhaps the greatest damage done to lands is that done by road building. Breaking of forest cover by road building is a double threat. Roads open land to erosion, which is the primary enemy of the forest ecosystem, and space consumed by roads is taken out of production, or subsequent production may be very low on abandoned roadbeds. As much as 20 percent of a forest ecosystem may be consumed in road building. This could remove 20 percent of the area from future production.

The application of herbicides to the deforested watershed now has been stopped, and the vegetation is becoming re-established. Studies of revegetation in combination with the study of nutrient flux using the watershed-ecosystem concept are providing some very valuable ecological information. Ecosystem recovery, including the maintenance or re-establishment of regulating mechanisms, is of vital interest to managers and conservationists. We need to know how, and at what speed, ecosystems can recover, or regain ecological balance, after serious external manipulations.

In cooperation with the U.S. Forest Service, we have undertaken a number of other experiments to elucidate the behavior of an ecosystem under stress, in an attempt to develop sound management practices for landscapes. One watershed was commercially clear-cut and another was commercially strip-cut during 1970. Alternate sets of horizontal strips cut on the latter watershed will clear the watershed of harvestable timber in four years. A small buffer strip remains uncut along the stream. This is an attempt to develop a forest management scheme without the nutrient-dumping effects associated with clear-cutting. These experiments will evaluate effects on production of stream water, and regeneration of the forest.

The studies at Hubbard Brook illustrate ecological interactions of the structure and function of a northern deciduous forest ecosystem. They also reinforce the suggestion of many ecologists that the ecosystem concept is a powerful analytical tool for producing and testing management strategy. The ecosystem concept provides the best way of seeing nature whole and provides a realistic scheme for ecologic bookkeeping.

BIBLIOGRAPHY

1. Bormann, F.H., and G.E. Likens. "Nutrient Cycling." *Science 155*:424-429; January 27, 1967.
2. Hornbeck, J.W., R.S. Pierce, and C.A. Federer. "Streamflow Changes After Forest Clearing in New England." *Water Resources Research 6:*1124-1132; 1970.
3. Likens, G.E., *et al.* "Effects of Forest Cutting and Herbicide Treatment on Nutrient Budgets in the Hubbard Brook Watershed-Ecosystem." *Ecological Monographs 40:*23-47; 1970.
4. Odum, E.P.. *Fundamentals of Ecology,* Second Edition. W.B. Saunders Co., Philadelphia, Pennsylvania. 1959.

ACKNOWLEDGEMENTS:

This is contribution No. 47 of the Hubbard Brook Ecosystem Study. Financial support for the field study at Hubbard Brook was provided by the National Science Foundation and was done through the cooperation of the Northeastern Forest Experiment Station, Forest Service, U.S. Department of Agriculture, Upper Darby, Pennsylvania.

25. The Williamstown Study of Critical Environmental Problems

A SUMMARY PREPARED BY THE EDITORS OF THE *BULLETIN OF THE ATOMIC SCIENTISTS*

"During July 1970, about 100 scientists and professionals conducted a month-long Study of Critical Environmental Problems (SCEP) on the campus of Williams College, Williamstown, Mass. The project was sponsored by the Massachusetts Institute of Technology and supported by nine federal agencies and six private institutions and foundations.

"The focus of this effort was trained on environmental problems of worldwide significance. The Study was concerned mainly with the consequences of pollution in terms of changes in climate, ocean ecology and large, terrestrial, eco-systems.

"In general, local and regional environmental problems, such as the thermal pollution of lakes and waterways, and the direct health effects of polution on man were not considered. Nor did the Study examine in any detail the problems of radioactive waste disposal and the effects of nuclear power and Plowshare activities.

"The pollutants it considered were those arising chiefly from the chemical processes and waste disposal practices of civilization. These included carbon dioxide (CO_2), sulfur dioxide (SO_2), oxides of nitrogen (NO), chlorinated hydrocarbons, other hydrocarbons, heavy metals and oil. Within this range, the Study produced an authoritative and thoughtful report. It may well become a launching platform for national and international programs of pollution control.

"One of the striking aspects of SCEP is its humility. It does not attempt to ring the doomsday bell. Rather, it stands as a sober, reflective and careful statement of what man knows about the effects of these pollutants on the environment and what he has yet to learn. It thus becomes an assessment of the state of knowledge about these problems in the year 1970.

"But SCEP does not quite stop there. It goes on to suggest what man can do about the problems he does understand and how he can acquire essential information about those he doesn't.

"The Williamstown Study was made by as distinguished an assembly of scientists and professionals in law and social studies as ever addressed itself to the global pollution problem. There were experts in meteorology, atmospheric chemistry, oceanography, biology, ecology, geology, physics, engineering, economics and other social sciences. They were drawn from 17 universities, 13 federal departments and agencies, three national laboratories and 11 nonprofit and industrial corporations. About 40 worked the entire month of July on the study and 30 participated part-time. In addition, another 30 experts served as consultants."

Bulletin of the Atomic Scientists

Source: Reprinted by permission of Science and Public Affairs, the *Bulletin of the Atomic Scientists,* October 1970. Copyright © by the Educational Foundation for Nuclear Science.

The Problems Studied

The global environmental problems studied by SCEP were:

Climatic effects of increasing carbon dioxide content of the atmosphere.

Climatic effects of the particle load of the atmosphere.

Climatic effects of contamination of the troposphere and stratosphere by sub-sonic and super-sonic transport aircraft.

Ecological effects of DDT and other toxic persistent pesticides.

Ecological effects of mercury and other toxic heavy metals.

Ecological effects of petroleum oil in the oceans.

Ecological effects of nutrients in estuaries, lakes and rivers.

For these topics, the following general questions were addressed:

• What can we now authoritatively say on the subject?

• What are the gaps in knowledge which limit our confidence in the assessments we can now make?

• What must be done to improve the data and our understanding of its significance so that better assessments may be made in the future?

• What programs of focused research, monitoring and/or action are needed?

• What are the characteristics of the national and/or international action needed to implement the recommendations of the Study?

Recommendations

Carbon Dioxide in the Atmosphere

Findings. All combustion of fossil fuels produces CO_2. It has been steadily increasing in the atmosphere at 0.2 per cent per year. Half of the amount man puts into the atmosphere stays and produces this rise in concentration. The other half goes into the biosphere and the oceans, but we don't know the partition in uptake, as between these two reservoirs.

The amount of CO_2 from fossil fuels is a small part of the natural CO_2 which is constantly being exchanged between the atmosphere/oceans and the atmosphere/forests. We have very little knowledge of such amounts.

The projected 18 per cent increase resulting from fossil fuel combustion to the year 2000 might increase the surface temperature of the earth .5°C; a doubling of the CO_2 might increase mean annual surface temperatures 2°C. Surface temperature changes of 2°C could lead to long-term warming of the planet. These estimates are based on a relatively primitive computer model with no consideration of important motions in the atmosphere, and hence are very uncertain but they are the best we have.

If we had to stop producing CO_2, no coal, oil or gas could be burned, and all modern societies would come to a halt. The only possible alternative is nuclear energy, whose by-products may cause serious environmental effects. Also, we don't have electric motor vehicles to be propelled by electricity from nuclear energy.

SCEP believes that the likelihood of direct climate change in this century resulting from CO_2 is small, but its long term potential consequences are so large that much more must be learned about future trends of climate change if society is to have time to adjust to changes which may be necessary.

Recommendations

1. Improvement of our estimates of future combustion of fossil fuels and the resulting emissions.

2. Study of changes in the mass of living matter and decaying products.

3. Continuous measurement and study of the carbon dioxide content of the atmosphere in a few areas remote from known sources—specifically four stations and some aircraft flights. We particularly recommend that the existing record at Mauna Loa Observatory be continued indefinitely.

4. Systematic study of the partition of carbon dioxide between the atmosphere and the oceans and biomass.

5. Development of comprehensive global computer models which include atmospheric motions and ocean-atmosphere interaction to study: circulation, clouds, precipitation and temperature patterns for expected CO_2 levels; and effects of stratospheric cooling.

Fine Particles in the Atmosphere

Findings. Fine particles change the heat balance of the earth because they both reflect and absorb radiation from the sun and the earth. Large amounts of such particles enter the troposphere (the zone up to 40,000 feet) from natural sources such as sea spray, wind blown dust, volcanoes and from the conversion of naturally occurring gases—SO_2, NO_x and hydrocarbons—into particles.

Man puts large quantities of sulfates, nitrates and hydrocarbons into the atmosphere which become fine particles and include special species, such as urban smog.

Particle levels have been increasing as observed at stations in Europe, North America and the North Atlantic, but not over the Central Pacific.

We do not know enough about the optical properties (reflection versus absorption) of particles to know whether they produce warming or cooling of the earth surface.

Recommendations

1. Studies to determine optical properties of fine particles, their sources, transport, and amounts in both troposphere and stratosphere, and their effects on cloud reflectivity.

2. Extending and improving solar radiation measurements.

3. Study of feasibility of satellite measurements of particle concentration and distribution.

4. Monitoring from ground and aircraft—10 fixed long-term stations and 100 stations for short-lived particles.

5. Develop atmospheric computer models which include particles.

Thermal Pollution

Although by the year 2000 we expect global thermal power output to be six times the present level, we do not expect it to affect global climate. Over cities it does already create "heat islands" and as these grow larger, they may have regional climatic effects and they should be studied.

Atmospheric Oxygen: Non-Problem

Atmospheric oxygen is practically constant. It varies neither over time (since 1910) nor regionally and is always very close to 20.946 per cent. Calculations show that depletion of oxygen by burning all the recoverable fossil fuels in the world would reduce it only to 20.800 per cent. It should probably be measured every 10 years to make sure that it is remaining constant.

Effects of Present Jet Aircraft

Observers all over the world have watched a jet contrail spread out to form a cirrus cloud. Observations at Denver and Salt Lake City show a systematic increase in such clouds since the advent of jets. Although they seem to be only regional, there is a possibility that they may have broader effects and should be studied.

SST's in the Stratosphere

Findings. The stratosphere where supersonic jet transports will fly at 65,000 feet is a very rarified region with little vertical mixing. Gases and particles produced by jet exhaust may remain for one to three years before disappearing.

Using FAA estimates of 500 SST's operating in 1985-90 mostly in the Northern Hemisphere, flying seven hours a day, at 65,000 feet, propelled by 1,700 engines like the General Electric-4 being developed for the Boeing 2707-300, we have estimated the steady state amounts of combustion products using GE calculations of the amount of such products because no test measurements exist. We have compared such amounts of a steady state basis with the natural levels of water vapor, sulfates, nitrates, hydrocarbon and soot. All are believed to form fine particles. We have also compared these levels with the amounts of particles put into the atmosphere by the volcano eruption of Mt. Agung in Bali in 1963.

In our calculations we used jet fuel of 0.05 per cent sulfur. We are told that a specification of 0.01 per cent sulfur could be met in the future at higher cost.

We do not believe that CO_2 resulting from such operations is likely to affect the climate. We are genuinely concerned about the possibility of increased stratospheric cloudiness, and about the fine particles, even using the calculated amounts given us by GE.

Clouds are known to form in the winter polar stratosphere. Two factors will increase the future likelihood of greater cloudiness in the stratosphere due to moisture added by the SST. First is the increased stratospheric cooling due to the increasing CO_2 content of the atmosphere. Second is the closer approach to saturation indicated by the observed increase of stratospheric moisture.

The largest engine whose combustion products have been actually measured in static ground tests was the Pratt and Whitney-JT9D used on the Boeing 747. It's fuel consumption rate is one-third that of the GE-4. Combustion products from such test of the JT9D, leading to particles, were much greater than the calculated values for the GE-4.

It is claimed that the particle formation is very small at 65,000 feet. Very, very little is known about reactions under such conditions. One guess is now as good as another.

Depending upon the actual particle formation, the effects of 500 SST's could range from a small, widespread continuous "Agung" effect to one as big as "Agung."

The temperature of the equatorial stratosphere (a belt around the globe) increased 6 to 7°C and remained at 2 to 3°C above its pre-Agung level for several years. No apparent temperature change was found in the lower troposphere.

Clearly such consequences are on a global scale even though the most pronounced effects would be felt where the highest density of traffic existed, i.e., the North Atlantic Ocean.

Conclusions. SCEP concludes with respect to contamination of the stratosphere by products of SST's that:

1. CO_2 creates no problem.
2. Global water vapor may increase 10 per cent; increases in regions of dense traffic may go up 60 per cent.
3. Particles from SO_2, hydrocarbons and soot may double pre-Agung global averages and peak at ten times those levels where there is dense traffic.
4. Effects of climate could be increased clouds from water vapor, and increased temperatures in the stratosphere with possible increase in surface temperatures.
5. A feeling of genuine concern has emerged from the above set of conclusions. The projected SST's can have a clearly measurable effect in a large region of the world and quite possibly on a global scale. We must emphasize that we cannot be certain about the magnitude of the various consequences.

Recommendations

1. That uncertainties about SST contamination and its effects be resolved before large scale operation of SST's begins.
2. That the following program of action be commenced as soon as possible:

    a. Begin to monitor the lower stratosphere for water vapor and particles and develop means to measure SO_2, NO_x and hydrocarbons.

    b. Determine whether additional cloudiness will occur in the stratosphere and the effects of such changes.

    c. Obtain better estimates of emission of combustion products under simulated flight conditions and under real flight conditions at the earliest opportunity.

    d. Using data resulting from a, b and c, estimate effects on weather and climate.

DDT and Related Persistent Toxic Pesticides

The ecological effects which have been identified with DDT are both general and specific. In general, the use of pesticides on crops generally requires continued and increased use of different and stronger pesticides. This is the result of a complex ecological system in which the reduction of one pest and innocuous (to man) predators allows new pests to become dominant. As one example, the egg shells of many birds are becoming thinner reducing hatching success. In several species, these effects now seriously threaten reproductive capabilities. Damage to these predators in the ecological system tends to create a situation in which pest outbreaks are likely to occur.

The concentrations and effects of DDT in the open oceans are not known. There are no reliable estimates and no direct measurements have been made. It is known that large amounts leave the area of application through the atmosphere and are transmitted through the world and some portion of this falls into the oceans.

DDT collects in marine organisms. Detrimental effects have not been observed in the open ocean, but DDT residues in mackeral caught off of California have already exceeded permissible tolerance levels for human consumption. It is known that reproduction of fresh-water game fish are being threatened. The effect of DDT on the ability of ocean phytoplankton to convert carbon dioxide into oxygen is not considered significant. The concentration necessary to induce significant inhibition exceeds expected concentrations in the open ocean by ten times its solubility (1 ppb) in water.

As DDT is being phased out, control measures must be accomplished through a variety of existing techniques such as biological controls and environmental management in addition to the use of degradable insecticides to offset food and health problems in the developing countries.

Recommendations

1. We recommend a drastic reduction in the use of DDT as soon as possible *and* that subsidies be furnished to developing countries to enable them to afford to use nonpersistent, but more expensive, pesticides and other pest control techniques.

2. In order to obtain information about the concentrations and effects of DDT in the marine environment, a baseline program of measurement should be initiated. This might involve taking about 1,000 samples at selected locations and analyzing them over the course of a year. A full-scale monitoring program should await the results of such a program.

3. We recommend greatly increased effort and support for the research and development of integrated pest control, combining a minimal use of pesticides with maximal use of biological control.

Mercury and Other Toxic Heavy Metals

Findings. Many heavy metals are highly toxic to specific life stages of a variety of organisms, especially shellfish. Most are concentrated in terrestrial and marine organisms by factors ranging from a few hundred to several hundred thousand times the concentrations in the surrounding environment.

The major sources of mercury are industrial processes and biocides. Although the use of mercury in biocides is relatively small, it is a direct input into the environment. There are many other possible routes but little data exists about the rates of release to the environment.

Recommendations

1. Pesticidal and biocidal use of mercury should continue to be drastically curtailed, particularly where safer less persistent substitutes can be used.

2. Industrial wastes and emissions of mercury should be controlled and recovered to the greatest extent possible, using available control and recovery methods.

3. World production, uses and waste products should be carefully monitored.

Oil in the Ocean

Findings. It is likely that up to 1.5 million tons of oil are introduced into the oceans every year through ocean shipping, offshore drilling and accidents. In addition, as much as two to three times this amount could eventually be introduced into waterways and eventually the oceans as a result of emission and wasteful practices on land.

Very little is known about the effects of oil in the oceans on marine life. Present results are conflicting. The effects of one oil spill which have been carefully observed indicate severe damage to marine organisms. Observations of other spills have not shown such a marked degree of damage. Different kinds of damage have been observed for different spills.

Potential effects include direct kill of organisms through coating, asphyxiation, or contact poisoning; direct kill through exposure to the water soluble toxic components of oil; destruction of the food sources of organisms; and incorporation of sublethal amounts of oil and oil products into organisms, resulting in reduced resistance to infection and other stresses or in reproductive successes.

Recommendations

1. Much more extensive research is required to determine the effect of oil in the ocean. Future oil spills should be systematically studied beginning immediately after they occur so that a comprehensive analysis of the effects can be developed over time. Sites of previous spills should be reexamined to study the effects in sediments.

2. Political and legal possibilities should be explored which would accomplish the conversion to Load-On-Top techniques by that 20 per cent of the world's tankers which do not use this method.

3. The possibility of recycling used oil should be explored.

Nutrients

Findings. Eutrophication of waters through over-fertilization (principally with nitrogen and phosphorus) produces an excess of organic matter which decomposes, removes oxygen and kills the fish. Estuaries are increasingly being eutrophied. Pollution of in-shore regions eliminates the nursery grounds of fish including many commercial species which inhabit the oceans.

Most (probably between 60 per cent and 70 per cent) of the phosphorus causing over-enrichment of water bodies comes from municipal wastes. In the United States about 75 per cent of the total phosphorus in these waters comes from detergents. Urban and rural land runoff contributes the remainder (approximately 30 per cent to 40 per cent). A major contributor is runoff from feedlots, manured lands and eroding soil.

Trends in both nutrient use and loss are rising. Fertilizer consumption is expected to increase greatly in both developed and developing countries in the next decade increasing the nutrient runoff from agricultural lands. Concentration of animal production will continue with the result that losses of nutrients from feedlot runoff will quadruple by 2000. Urban waste production is expected to quadruple by 2000, which means greater potential loss of nutrients directly into coastal waters.

Recommendations

1. Develop technology and encourage its application to reclaim and recycle nutrients in areas of high concentrations, such as sewage treatment plants and feedlots.

2. Avoid use of nutrients in products which are discharged in large quantities into air or water. For example, reformulate detergents to eliminate or reduce waste phosphates, but be certain the substitutes degrade and do not poison the ecosystem.

3. Effect control of nutrient discharges in natural regions, such as river basins, estuaries and coastal oceans, through appropriate institutions.

Wastes from Nuclear Energy

It has taken our full efforts to probe in some depth a few questions. We decided deliberately to omit consideration of others of great importance. One of these is the problem of perpetual management of the large quantities of radioactive wastes which are by-products of nuclear power.

No other environmental pollutant has been so carefully monitored and contained. Yet, as we look back on our intense examination of the effects of the products of fossil fuel combustion, we have become aware of our neglect of a different class of pollutants which will grow greatly in quantity in the next 30 years.

We call to the attention of one of the sponsors of SCEP, the AEC, our decision to omit this item from our agenda and our concern about the subject.

Recommendations

That an independent, intensive, multi-disciplinary study be made of the trade-offs in national energy policy between fossil fuel and nuclear sources, with a special focus on problems of safe management of the radioactive by-products of nuclear energy leading to recommendations concerning the content and scale and urgency of needed programs.

General Conclusions

In studying the specific problems outlined above, SCEP reaffirmed the conclusions and principles which underlie the ecological, social and political implications of most critical environmental problems. Efforts to examine or ameliorate the effects of these problems should include explicit recognition of the considerations.

Ecological Considerations

An estimate is needed of the ecological demand, a summation of all of man's demands upon the environment, such as the extraction of resources and the return of waste. Such demand-producing activities as agriculture, mining and industry have global annual rates of increase of 3 per cent, 5 per cent and 7 per cent, respectively.

An integrated rate of increase is estimated to lie between 5 per cent and 6 per cent per year, in comparison with a population rate of annual increase of only 2 per cent.

Natural ecosystems still provide us many free services. At least 99 per cent of the potential pests of man are held to very low densities by natural control. Insects pollinate most of the vegetables, fruits, berries and flowers, whether they be wild or cultivated. Commercial fish are produced almost entirely in natural ecosystems. Vegetation reduces floods, prevents erosion and air conditions the landscape. Fungi and minute soil animals work jointly on plant debris and weathered rocks to produce soil. Natural ecosystems cycle matter through green plants, animals and decomposers, eliminating wastes. Organisms regulate the amount of nitrates, ammonia and methane in the environment. On a geological time scale, life regulates the amount of carbon dioxide and oxygen in the atmosphere.

The functions of ecological systems connect the impact of man upon the environment with the services supplied by nature. Ecological impairment eventually leads to a loss of such services. The health and vigor of ecological systems are easily reduced if (1) general and widespread damage occurs to the predators, (2) substantial numbers of species are lost, or (3) general biological activity is depressed. Most pollutants that affect life have some effect on all three processes.

To prevent further deterioration of the biosphere, and to repair some of the present damage, action is urgently needed. In addition to a variety of specific recommendations such as those accompanying the specific problem areas, SCEP recommends that the following activities be developed in national and international programs:

a. *Technology Assessment.* An information center that centralizes data on products of industry and agriculture, especially new products and new increases in production. Such a center will also identify potentially hazardous materials, and promote research on their toxicity and persistence in nature.

b. *Environmental Assessment.* An information center that centralizes data on the distributions of pollutants, and on the health and pollution loads of organisms.

c. *Problem Evaluation.* A think-center to evaluate problems on the basis of the above information, to determine the urgency for action and to identify options.

d. *Public Education.* A service center to present the results of the above in simple form, and to distribute such materials to educational institutions and the news media.

Social and Political Changes

SCEP has concentrated on a few global problems. The main thrust of our recommendations is to gather more information about pollution of the planet. This information would improve our understanding of the impact of man's activities on the earth's resources of air, water and those on land: that is, the ecological demand of man's activities. Relevant data on critical global problems is very poor and this seriously limits our understanding of their meaning.

We have tried to estimate scales of world activities to the year 2000. In very few areas is there reliable data for projections. Indeed, much data about world activities today in areas of importance to this Study have been found fragmentary and contradictory. Far better estimates well into the 21st Century are needed in order to assess the expected impact of man on the world ecological system to give us time to take action to avoid crisis or catastrophe.

We have looked beyond the gathering of data and its interpretation to the question of how remedial action may be taken. Unless information leads to action to abate or control pollution it is largely useless.

Earlier in our history, the prevailing value system assigned an overriding priority to the first order effects of applied science and technology: the goods and services produced. We took the side effects—pollution—in stride. A shift in values appears to be under way that assigns a much higher priority than before to the control of the side effects. This does not necessarily impart a reduced interest in production and consumption. When the crunch comes—when the implications of remedial action and the choices that must be made become clear—will we have second thoughts? Or will we bog down in confusion and frustration? Will we hold to our course, insisting that our society make a more thorough and imaginative use of its resources of science and technology, its organizational skills and its financial resources in an effort to achieve an optimal balance between the production we need and the side effects which we must bring under control? We hope the answer to these question is in the affirmative.

The problem of action is compounded because contributions to global pollution come from activities in countries all over the world. Action to control depends upon agreed data on amounts of pollution and their harmful effects. Actual control will depend upon national action by governments. It is not enough that the United States exercise control. If others pollute our common resources of the air and oceans, the perils remain. This challenge will be before the U. N. Conference on Man and the Environment in Stockholm in 1972. We hope that the SCEP reports will form useful inputs to that Conference, and that the SCEP study model may be applied to other critical problems of the environment.

NASA.

26. The New Biology: What Price Relieving Man's Estate?

LEON R. KASS

The expression *new biology* refers primarily to new techniques capable or potentially capable of making humans almost disease-free and thus that can lead to an extended, useful life. If, however, these techniques (genetic engineering, for example) become a reality, what then will man do with this gift? The author of this article believes that we should ponder this question carefully or else we may become dehumanized, degraded, and perhaps slaves of technology. He suggests several ways in which we can use our new biomedical powers with humility and restraint. Doctor Kass, formerly executive secretary of the Committee on the Life Sciences and Social Policy, National Research Council of the National Academy of Sciences, is now a tutor, St. John's College, Annapolis, Maryland.

Recent advances in biology and medicine suggest that we may be rapidly acquiring the power to modify and control the capacities and activities of men by direct intervention and manipulation of their bodies and minds. Certain means are already in use or at hand, others await the solution of relatively minor technical problems, while yet others, those offering perhaps the most precise kind of control, depend upon further basic research. Biologists who have considered these matters disagree on the question of how much how soon, but all agree that the power for "human engineering," to borrow from the jargon, is coming and that it will probably have profound social consequences.

These developments have been viewed both with enthusiasm and with alarm; they are only just beginning to receive serious attention. Several biologists have undertaken to inform the public about the technical possibilities, present and future. Practitioners of social science "futurology" are attempting to predict and describe the likely social consequences of and public responses to the new technologies. Lawyers and legislators are exploring institutional innovations for assessing new technologies. All of these activities are based upon the hope that we can harness the new technology of man for the betterment of mankind.

Yet this commendable aspiration points to another set of questions, which are, in my view, sorely neglected—questions that inquire into the meaning of phrases such as the "betterment of mankind." A *full* understanding of the new technology of man requires an exploration of ends, values, standards. What ends will or should the new techniques serve? What values should guide society's adjustments? By what standards

Source: Copyright 1971 American Association for the Advancement of Science. Reprinted with permission of the publisher and author from *Science, 174*:779-788, 19 November 1971.

should the assessment agencies assess? Behind these questions lie others: what is a good man, what is a good life for man, what is a good community? This article is an attempt to provoke discussion of these neglected and important questions.

While these questions about ends and ultimate ends are never unimportant or irrelevant, they have rarely been more important or more relevant. That this is so can be seen once we recognize that we are dealing here with a group of technologies that are in a decisive respect unique: the object upon which they operate is man himself. The technologies of energy or food production, of communication, of manufacture, and of motion greatly alter the implements available to man and the conditions in which he uses them. In contrast, the biomedical technology works to change the user himself. To be sure, the printing press, the automobile, the television, and the jet airplane have greatly altered the conditions under which and the way in which men live; but men as biological beings have remained largely unchanged. They have been, and remain, able to accept or reject, to use and abuse these technologies; they choose, whether wisely or foolishly, the ends to which these technologies are means. Biomedical technology may make it possible to change the inherent capacity for choice itself. Indeed, both those who welcome and those who fear the advent of "human engineering" ground their hopes and fears in the same prospect: *that man can for the first time recreate himself.*

Engineering the engineer seems to differ in kind from engineering his engine. Some have argued, however, that biomedical engineering does not differ qualitatively from toilet training, education, and moral teachings—all of which are forms of so-called "social engineering," which has man as its object, and is used by one generation to mold the next. In reply, it must at least be said that the techniques which have hitherto been employed are feeble and inefficient when compared to those on the horizon. This quantitative difference rests in part on a qualitative difference in the means of intervention. The traditional influences operate by speech or by symbolic deeds. They pay tribute to man as the animal who lives by speech and who understands the meanings of actions. Also, their effects are, in general, reversible, or at least subject to attempts at reversal. Each person has greater or lesser power to accept or reject or abandon them. In contrast, biomedical engineering circumvents the human context of speech and meaning, bypasses choice, and goes directly to work to modify the human material itself. Moreover, the changes wrought may be irreversible.

In addition, there is an important practical reason for considering the biomedical technology apart from other technologies. The advances we shall examine are fruits of a large, humane project dedicated to the conquest of disease and the relief of human suffering. The biologist and physician, regardless of their private motives, are seen, with justification, to be the well-wishers and benefactors of mankind. Thus, in a time in which technological advance is more carefully scrutinized and increasingly criticized, biomedical developments are still viewed by most people as benefits largely without qualification. The price we pay for these developments is thus more likely to go unrecognized. For this reason, I shall consider only the dangers and costs of biomedical advance. As the benefits are well known, there is no need to dwell upon them here. My discussion is deliberately partial.

I begin with a survey of the pertinent technologies. Next, I will consider some of the basic ethical and social problems in the use of these technologies. Then, I will briefly raise some fundamental questions to which these problems point. Finally, I shall offer some very general reflections on what is to be done.

The Biomedical Technologies

The biomedical technologies can be usefully organized into three groups, according to their major purpose: (i) control of death and life, (ii) control of human potentialities, and (iii) control of human achievement. The corresponding technologies are (i) medicine, especially the arts of prolonging life and of controlling reproduction, (ii) genetic engineering, and (iii) neurological and psychological manipulation. I shall briefly summarize each group of techniques.

1. *Control of Death and Life.* Previous medical triumphs have greatly increased average life expectancy. Yet other developments, such as organ transplantation or replacement and research into aging, hold forth the promise of increasing not just the average, but also the maximum life expectancy. Indeed, medicine seems to be sharpening its tools to do battle with death itself, as if death were just one more disease.

More immediately and concretely, available techniques of prolonging life—respirators, cardiac pacemakers, artificial kidneys—are already in the lists against death. Ironically, the success of these devices in forestalling death has introduced confusion in determining that death has, in fact, occurred. The traditional signs of life—heartbeat and respiration—can now be maintained entirely by machines. Some physicians are now busily trying to devise so-called "new definitions of death," while other maintain that the technical advances show that death is not a concrete event at all, but rather a gradual process, like twilight, incapable of precise temporal localization.

The real challenge to death will come from research into aging and senescence, a field just entering puberty. Recent studies suggest that aging is a genetically controlled process, distinct from disease, but one that can be manipulated and altered by diet or drugs. Extrapolating from animal studies, some scientists have suggested that a decrease in the rate of aging might also be achieved simply by effecting a very small decrease in human body temperature. According to some estimates, by the year 2000 it may be technically possible to add from 20 to 40 useful years to the period of middle life.

Medicine's success in extending life is already a major cause of excessive population growth: death control points to birth control. Although we are already technically competent, new techniques for lowering fertility and chemical agents for inducing abortion will greatly enhance our powers over conception and gestation. Problems of definition have been raised here as well. The need to determine when individuals acquire enforceable legal rights gives society an interest in the definition of human life and of the time when it begins. These matters are too familiar to need elaboration.

Technologies to conquer infertility proceed alongside those to promote it. The first successful laboratory fertilization of human egg by human sperm was reported in 1969 *(1)*. In 1970, British scientists learned how to grow human embryos in the laboratory up to at least the blastocyst stage [that is, to the age of 1 week *(2)*]. We may soon hear about the next stage, the successful reimplantation of such an embryo into a woman previously infertile because of oviduct disease. The development of an artificial placenta, now under investigation, will make possible full laboratory control of fertilization and gestation. In addition, sophisticated biochemical and cytological techniques of monitoring the "quality" of the fetus have been and are being developed and used. These developments not only give us more power over the generation of human life, but make it possible to manipulate and to modify the quality of the human material.

2. *Control of Human Potentialities.* Genetic engineering, when fully developed, will wield two powers not shared by ordinary medical practice. Medicine treats existing individuals and seeks to correct deviations from a norm of health. Genetic engineering, in contrast, will be able to make changes that can be transmitted to succeeding generations and will be able to create new capacities, and hence to establish new norms of health and fitness.

Nevertheless, one of the major interests in genetic manipulation is strictly medical: to develop treatments for individuals with inherited diseases. Genetic disease is prevalent and increasing, thanks partly to medical advances that enable those affected to survive and perpetuate their mutant genes. The hope is that normal copies of the appropriate gene, obtained biologically or synthesized chemically, can be introduced into defective individuals to correct their deficiencies. This *therapeutic* use of genetic technology appears to be far in the future. Moreover, there is some doubt that it will ever be practical, since the same end could be more easily achieved by transplanting cells or organs that could compensate for the missing or defective gene product.

Far less remote are technologies that could serve *eugenic* ends. Their development has been endorsed by those concerned about a general deterioration of the human gene pool and by others who believe that even an undeteriorated human gene pool needs upgrading. Artificial insemination with selected donors, the eugenic proposal of Herman Muller *(3)*, has been possible for several years because of the perfection of methods for long-term storage of human spermatozoa. The successful maturation of human oocytes in the laboratory and their subsequent fertilization now make it possible to select donors of ova as well. But a far more suitable technique for eugenic purposes will soon be upon us—namely, nuclear transplantation, or cloning. Bypassing the lottery of sexual recombination, nuclear transplantation permits the asexual reproduction or copying of an already developed individual. The nucleus of a mature but unfertilized egg is replaced by a nucleus obtained from a specialized cell or an adult organism or embryo (for example, a cell from the intestines or the skin). The egg with its transplanted nucleus develops as if it had been fertilized and, barring complications, will give rise to a normal adult organism. Since almost all the hereditary material (DNA) of a cell is contained within its nucleus, the renucleated egg and the individual into which it develops are genetically identical to the adult organism that was the source of the donor nucleus. Cloning could be used to produce sets of unlimited numbers of genetically identical individuals, each set derived from a single parent. Cloning has been successful in amphibians and is now being tried in mice; its extension to man merely requires the solution of certain technical problems.

Production of man-animal chimeras by the introduction of selected nonhuman material into developing human embryos is also expected. Fusion of human and nonhuman cells in tissue culture has already been achieved.

Other, less direct means for influencing the gene pool are already available, thanks to our increasing ability to identify and diagnose genetic diseases. Genetic counselors can now detect biochemically and cytologically a variety of severe genetic defects (for example, Mongolism, Tay-Sachs disease) while the fetus is still in utero. Since treatments are at present largely unavailable, diagnosis is often followed by abortion of the affected fetus. In the future, more sensitive tests will also permit the detection of heterozygote carriers, the unaffected individuals who carry but a single dose of a given deleterious gene. The eradication of a given genetic disease might then be attempted by aborting all such carriers. In fact, it was recently suggested that the fairly common disease cystic fibrosis could be completely eliminated over the next 40 years by screening all pregnancies and aborting the 17,000,000 unaffected fetuses that will carry a single gene for this disease. Such zealots need to be reminded of the consequences should each geneticist be allowed an equal assault on his favorite genetic disorder, given that each human being is a carrier for some four to eight such recessive, lethal genetic diseases.

3. *Control of Human Achievement.* Although human achievement depends at least in part upon genetic endowment, heredity determines only the material upon which experience and education impose the form. The limits of many capacities and powers of an individual are indeed genetically determined, but the nurturing and perfection of these capacities depend upon other influences. Neurological and psychological manipulation hold forth the promise of controlling the development of human capacities, particularly those long considered most distinctively human: speech, thought, choice, emotion, memory, and imagination.

These techniques are now in a rather primitive state because we understand so little about the brain and mind. Nevertheless, we have already seen the use of electrical stimulation of the human brain to produce sensations of intense pleasure and to control rage, the use of brain surgery (for example, frontal lobotomy) for the relief of severe anxiety, and the use of aversive conditioning with electric shock to treat sexual perversion. Operant-conditioning techniques are widely used, apparently with success, in schools and mental hospitals. The use of so-called consciousness-expanding and hallucinogenic drugs is widespread, to say nothing of tranquilizers and stimulants. We are promised drugs to modify memory, intelligence, libido, and aggressiveness.

The following passages from a recent book by Yale neurophysiologist José Delgado—a book instructively entitled *Physical Control of the Mind: Toward a Psychocivilized Society*—should serve to make this discussion more concrete. In the early 1950's, it was discovered that, with electrodes placed in certain discrete regions of their brains, animals would repeatedly and indefatigably press levers to stimulate their own brains, with obvious resultant enjoyment. Even starving animals preferred stimulating these so-called pleasure centers to eating. Delgado comments on the electrical stimulation of a similar center in a human subject (*4*, p. 185).

"[T]he patient reported a pleasant tingling sensation in the left side of her body 'from my face down to the bottom of my legs.' She started giggling and making funny comments, stating that she enjoyed the sensation 'very much.' Repetition of these stimulations made the patient more communicative and flirtatious, and she ended by openly expressing her desire to marry the therapist."

And one further quotation from Delgado (*4*, p. 88).

"Leaving wires inside of a thinking brain may appear unpleasant or dangerous, but actually the many patients who have undergone this experience have not been concerned about the fact of being wired, nor have they felt any discomfort due to the presence of conductors in their heads. Some women have shown their feminine adaptability to circumstances by wearing attractive hats or wigs to conceal their electrical headgear, and many people have been able to enjoy a normal life as outpatients, returning to the clinic periodically for examination and stimulation. In a few cases in which contacts were located in pleasurable areas, patients have had the opportunity to stimulate their own brains by pressing the button or a portable instrument, and this procedure is reported to have therapeutic benefits."

It bears repeating that the sciences of neurophysiology and psychopharmacology are in their infancy. The techniques that are now available are crude, imprecise, weak, and unpredictable, compared to those that may flow from a more mature neurobiology.

Basic Ethical and Social Problems in the Use of Biomedical Technology

After this cursory review of the powers now and soon to be at our disposal, I turn to the questions concerning the use of these powers. First, we must recognize that questions of use of science and technology are always moral and political questions, never simply technical ones. All private or public decisions to develop or to use biomedical technology—and decisions *not* to do so—inevitably contain judgments about value. This is true even if the values guiding those decisions are not articulated or made clear, as indeed they often are not. Secondly, the value judgments cannot be derived from biomedical science. This is true even if scientists themselves make the decisions.

These important points are often overlooked for at least three reasons.

1. They are obscured by those who like to speak of "the control of nature by science." It is men who control, not that abstraction "science." Science may provide the means, but men choose the ends; the choice of ends comes from beyond science.

2. Introduction of new technologies often appears to be the result of no decision whatsoever, or of the culmination of decisions too small or unconscious to be recognized as such. What can be done is done. However, someone is deciding on the basis of some notions of desirability, no matter how self-serving or altruistic.

3. Desires to gain or keep money and power no doubt influence much of what happens, but these desires can also be formulated as reasons and then discussed and debated.

Insofar as our society has tried to deliberate about questions of use, how has it done so? Pragmatists that we are, we prefer a utilitarian calculus: we weigh "benefits" against "risks," and we weigh them for both the individual and "society." We often ignore the fact that the very definitions of "a benefit" and "a risk" are themselves based upon judgments about value. In the biomedical areas just reviewed, the benefits are considered to be self-evident: prolongation of life, control of fertility and of population size, treatment and prevention of genetic disease, the reduction of anxiety and aggressiveness, and the enhancement of memory, intelligence, and pleasure. The assessment of risk is, in general, simply pragmatic—will the technique work effectively and reliably, now much will it cost, will it do detectable bodily harm, and who will complain if we proceed with development? As these questions are familiar and congenial, there is no need to belabor them.

The very pragmatism that makes us sensitive to considerations of economic cost often blinds us to the larger social costs exacted by biomedical advances. For one thing, we seem to be unaware that we may not be able to maximize all the benefits, that several of the goals we are promoting conflict with each other. On the one hand, we seek to control population growth by lowering fertility; on the other hand, we develop techniques to enable every infertile woman to bear a child. On the one hand, we try to extend the lives of individuals with genetic disease; on the other, we wish to eliminate deleterious genes from the human population. I am not urging that we resolve these conflicts in favor of one side or the other, but simply that we recognize that such conflicts exist. Once we do, we are more likely to appreciate that most "progress" is heavily paid for in terms not generally included in the simple utilitarian calculus.

To become sensitive to the larger costs of biomedical progress, we must attend to several serious ethical and social questions. I will briefly discuss three of them: (i) questions of distributive justice, (ii) questions of the use and abuse of power, and (iii) questions of self-degradation and dehumanization.

Distributive Justice

The introduction of any biomedical technology presents a new instance of an old problem—how to distribute scarce resources justly. We should assume that demand will usually exceed supply. Which people should receive a kidney transplant or an artificial heart? Who should get the benefits of genetic therapy or of brain stimulation? Is "first-come, first served" the fairest principle? Or are certain people "more worthy," and if so, on what grounds?

It is unlikely that we will arrive at answers to these questions in the form of deliberate decisions. More likely, the problem of distribution will continue to be decided ad hoc and locally. If so, the consequence will probably be a sharp increase in the already far too great inequality of medical care. The extreme case will be longevity, which will probably be, at first, obtainable only at great expense. Who is likely to be able to buy it? Do conscience and prudence permit us to enlarge the gap between rich and poor, especially with respect to something as fundamental as life itself?

Questions of distributive justice also arise in the earlier decisions to acquire new knowledge and to develop new techniques. Personnel and facilities for medical research and treatment are scarce resources. Is the development of a new technology the best use of the limited resources, given current circumstances? How should we balance efforts aimed at prevention against those aimed at cure, or either of these against efforts to redesign the species? How should we balance the delivery of available levels of care against further basic research? More fundamentally, how should we balance efforts in biology and medicine against efforts to eliminate poverty, pollution, urban decay, discrimination, and poor education? This last question about distribution is perhaps the most profound. We should reflect upon the social consequences of seducing many of our brightest young people to spend their lives locating the biochemical defects in rare genetic diseases, while our more serious problems go begging. The current squeeze on money for research provides us with an opportunity to rethink and reorder our priorities.

Problems of distributive justice are frequently mentioned and discussed, but they are hard to resolve in a rational manner. We find them especially difficult because of the enormous range of conflicting values and interests that characterizes our pluralistic society. We cannot agree—unfortunately, we often do not even try to agree—on standards for just distribution. Rather, decisions tend to be made largely out of a clash of competing interests. Thus, regrettably, the question of how to distribute justly often gets reduced to who shall decide how to distribute. The question about justice has led us to the question about power.

Use and Abuse of Power

We have difficulty recognizing the problems of the exercise of power in the biomedical enterprise because of our delight with the wondrous fruits it has yielded. This is ironic because the notion of power is absolutely central to the modern conception of science. The ancients conveived of science as the *understanding* of nature, pursued for its own sake. We moderns view science as power, as *control* over nature; the conquest of nature "for the relief of man's estate" was the charge issued by Francis Bacon, one of the leading architects of the modern scientific project (5).

Another source of difficulty is our fondness for speaking of the abstraction "Man." I suspect that we prefer to speak figuratively about "Man's power over Nature" because it obscures an unpleasant reality about human affairs. It is in fact particular men who wield power, not Man. What we really mean by "Man's power over Nature" is a power exercised by some men over other men, with a knowledge of nature as their instrument.

While applicable to technology in general, these reflections are especially pertinent to the technologies of human engineering, with which men deliberately exercise power over future generations. An excellent discussion of this question is found in *The Abolition of Man,* by C. S. Lewis (6).

"It is, of course, a commonplace to complain that men have hitherto used badly, and against their fellows, the powers that science has given them. But that is not the point I am trying to make. I am not speaking of particular corruptions and abuses which an increase of moral virtue would cure: I am considering what the thing called "Man's power over Nature" must always and essentially be. . . .

"In reality, of course, if any one age really attains, by eugenics and scientific education, the power to make its descendants what it pleases, all men who live after it are the patients of that power. They are weaker, not stronger: for though we may have put wonderful machines in their hands, we have pre-ordained how they are to use them. . . . The real picture is that of one dominant age . . . which resists all previous ages most successfully and dominates all subsequent ages most irresistibly, and thus is the real master of the human species. But even within this master generation (itself an infinitesimal minority of the species) the power will be exercised by a minority smaller still. Man's conquest of Nature, if the dreams of some scientific planners are realized, means the rule of a few hundreds of men over billions upon billions of men. There neither is nor can be any simple increase of power on Man's side. Each new power won *by* man is a power *over* man as well. Each advance leaves him weaker as well as stronger. In every victory, besides being the general who triumphs, he is also the prisoner who follows the triumphal car."

Please note that I am not yet speaking about the problem of the misuse or abuse of power. The point is rather that the power which grows is unavoidably the power of only some men, and that the number of powerful men decreases as power increases.

Specific problems of abuse and misuse of specific powers must not, however, be overlooked. Some have voiced the fear that the technologies of genetic engineering and behavior control, though developed for good purposes, will be put to evil uses. These fears are perhaps somewhat exaggerated, if only because biomedical technologies would add very little to our highly developed arsenal for mischief, destruction, and stultification. Nevertheless, any proposal for large-scale human engineering should make us wary. Consider a program of positive eugenics based upon the widespread practice of asexual reproduction. Who shall decide what constitutes a superior individual worthy of replication? Who shall decide which individuals may or must reproduce, and by which method? These are questions easily answered only for a tyrannical regime.

Concern about the use of power is equally necessary in the selection of means for desirable or agreed-upon ends. Consider the desired end of limiting population growth.

An effective program of fertility control is likely to be coercive. Who should decide the choice of means? Will the program penalize "conscientious objectors"?

Serious problems arise simply from obtaining and disseminating information, as in the mass screening programs now being proposed for detection of genetic disease. For what kinds of disorders is compulsory screening justified? Who shall have access to the data obtained, and for what purposes? To whom does information about a person's genotype belong? In ordinary medical practice, the patient's privacy is protected by the doctor's adherence to the principle of confidentiality. What will protect his privacy under conditions of mass screening?

More than privacy is at stake if screening is undertaken to detect psychological or behavioral abnormalities. A recent proposal, tendered and supported high in government, called for the psychological testing of all 6-year-olds to detect future criminals and misfits. The proposal was rejected; current tests lack the requisite predictive powers. But will such a proposal be rejected if reliable tests become available? What if certain genetic disorders, diagnosable in childhood, can be shown to correlate with subsequent antisocial behavior? For what degree of correlation and for what kinds of behavior can mandatory screening be justified? What use should be made of the data? Might not the dissemination of the information itself undermine the individual's chance for a worthy life and contribute to his so-called antisocial tendencies?

Consider the seemingly harmless effort to redefine clinical death. If the need for organs for transplantation is the stimulus for redefining death, might not this concern influence the definition at the expense of the dying? One physician, in fact, refers in writing to the revised criteria for declaring a patient dead as a "new definition of heart donor eligibility" (7, p. 526).

Problems of abuse of power arise even in the acquisition of basic knowledge. The securing of a voluntary and informed consent is an abiding problem in the use of human subjects in experimentation. Gross coercion and deception are now rarely a problem; the pressures are generally subtle, often related to an intrinsic power imbalance in favor of the experimentalist.

A special problem arises in experiments on or manipulations of the unborn. Here it is impossible to obtain the consent of the human subject. If the purpose of the intervention is therapeutic—to correct a known genetic abnormality, for example—consent can reasonably be implied. But can anyone ethically consent to nontherapeutic interventions in which parents or scientists work their wills or their eugenic visions on the child-to-be? Would not such manipulation represent in itself an abuse of power, independent of consequences?

There are many clinical situations which already permit, if not invite, the manipulative or arbitrary use of powers provided by biomedical technology: obtaining organs for transplantation, refusing to let a person die with dignity, giving genetic counselling to a frightened couple, recommending eugenic sterilization for a mental retardate, ordering electric shock for a homosexual. In each situation, there is an opportunity to violate the will of the patient or subject. Such opportunities have generally existed in medical practice, but the dangers are becoming increasingly serious. With the growing complexity of the technologies, the technician gains in authority, since he alone can understand what he is doing. The patient's lack of knowledge makes him deferential and often inhibits him from speaking up when he feels threatened. Physicians *are* sometimes troubled by their increasing power, yet they feel they cannot avoid its exercise. "Reluctantly," one commented to me, "we shall have to play God." With what guidance and to what ends I shall consider later. For the moment, I merely ask: "By whose authority?"

While these questions about power are pertinent and important, they are in one sense misleading. They imply an inherent conflict of purpose between physician and patient, between scientist and citizen. The discussion conjures up images of master and slave, of oppressor and oppressed. Yet it must be remembered that conflict of purpose is largely absent, especially with regard to general goals. To be sure, the purposes of medical scientists are not always the same as those of the subjects experimented on. Nevertheless, basic sponsors and partisans of biomedical technology are precisely those upon whom the technology will operate. The will of the scientist and physician is happily married to (rather, is the offspring of) the desire of all of us for better health, longer life, and peace of mind.

Most future biomedical technologies will probably be welcomed, as have those of the past. Their use will require little or no coercion. Some developments, such as pills to improve memory, control mood, or induce pleasure, are likely to need no promotion. Thus, even if we should escape from the dangers of coercive manipulation, we shall still face large problems posed by the voluntary use of biomedical technology, problems to which I now turn.

Voluntary Self-Degradation and Dehumanization

Modern opinion is sensitive to problems of restriction of freedom and abuse of power. Indeed, many hold that a man can be injured only by violating his will. But this view is much too narrow. It fails to recognize the great dangers we shall face in the use of biomedical technology, dangers that stem from an excess of freedom, from the uninhibited exercises of will. In my view, our greatest problem will increasingly be one of voluntary self-degradation, or willing dehumanization.

Certain desired and perfected medical technologies have already had some dehumanizing consequences. Improved methods of resuscitation have made possible heroic efforts to "save" the severely ill and injured. Yet these efforts are sometimes only partly successful; they may succeed in salvaging individuals with severe brain damage, capable of only a less-than-human, vegetating existence. Such patients, increasingly found in the intensive care units of university hospitals, have been denied a death with dignity. Families are forced to suffer seeing their loved ones so reduced, and are made to bear the burdens of a protracted death watch.

Even the ordinary methods of treating disease and prolonging life have impoverished the context in which men die. Fewer and fewer people die in the familiar surroundings of home or in the company of family and friends. At that time of life when there is perhaps the greatest need for human warmth and comfort, the dying patient is kept company by cardiac pacemakers and defibrillators, respirators, aspirators, oxygenators, catheters, and his intravenous drip.

But the loneliness is not confined to the dying patient in the hospital bed. Consider the increasing number of old people who are still alive, thanks to medical progress. As a group, the elderly are the most alienated members of our society. Not yet ready for the world of the dead, not deemed fit for the world of the living, they are shunted aside. More and more of them spend the extra years medicine has given them in "homes for senior citizens," in chronic hospitals, in nursing homes—waiting for the end. We have learned how to increase their years, but we have not learned how to help them enjoy their days. And yet, we bravely and relentlessly push back the frontiers against death.

Paradoxically, even the young and vigorous may be suffering because of medicine's success in removing death from their personal experience. Those born since penicillin represent the first generation ever to grow up without the experience or fear of probable unexpected death at an early age. They look around and see that virtually all of their friends are alive. A thoughtful physician, Eric Cassell, has remarked on this in "Death and the physician" (*8*, p. 76):

"[W]hile the gift of time must surely be marked as a great blessing, the *perception* of time, as stretching out endlessly before us, is somewhat threatening. Many of us function best under deadlines, and tend to procrastinate when time limits are not set. . . . Thus, this unquestioned boon, the extension of life, and the removal of the threat of premature death, carries with it an unexpected anxiety: the anxiety of an unlimited future.

"In the young, the sense of limitless time has apparently imparted not a feeling of limitless opportunity, but increased stress and anxiety, in addition to the anxiety which results from other modern freedoms: personal mobility, a wide range of occupational choice, and independence from the limitations of class and familial patterns of work. . . . A certain aimlessness (often ringed around with great social consciousness) characterizes discussions about their own aspirations. The future is endless, and their inner demands seem minimal. Although it may appear uncharitable to say so, they seem to be acting in a way best described as "childish"—particulary in their lack of a time sense. They behave as though there were no tomorrow, or as though the time limits imposed by the biological facts of life had become so vague for them as to be nonexistent."

Consider next the coming power over reproduction and genotype. We endorse the project that will enable us to control numbers and to treat individuals with genetic disease. But our desires outrun these defensible goals. Many would welcome the chance to become parents without the inconvenience of pregnancy; others would wish to know in advance the characteristics of their offspring (sex, height, eye color, intelligence); still others would wish to design these characteristics to suit their tastes.

Some scientists have called for the use of the new technologies to assure the "quality" of all new babies (9). As one obstetrician put it: "The business of obstetrics is to produce *optimum* babies." But the price to be paid for the "optimum baby" is the transfer of procreation from the home to the laboratory and its coincident transformation into manufacture. Increasing control over the product is purchased by the increasing depersonalization of the process. The complete depersonalization of procreation (possible with the development of an artificial placenta) shall be in itself, seriously dehumanizing, no matter how optimum the product. It should not be forgotten that human procreation not only issues new human beings, but is itself a human activity.

Procreation is not simply an activity of the rational will. It is a more complete human activity precisely because it engages us bodily and spiritually, as well as rationally. Is there perhaps some wisdom in the mystery of nature which joins the pleasure of sex, the communication of love, and the desire for children in the very activity by which we continue the chain of human existence? Is not biological parenthood a built-in "mechanism," selected because it fosters and supports in parents an adequate concern for and commitment to their children? Would not the laboratory production of human beings no longer be *human* procreation? Could it keep human parenthood human?

The dehumanizing consequences of programmed reproduction extend beyond the mere acts and processes of life-giving. Transfer of procreation to the laboratory will no doubt weaken what is presently for many people the best remaining justification and support for the existence of marriage and the family. Sex is now comfortably at home outside of marriage; child-rearing is progressively being given over to the state, the schools, the mass media, and the child-care centers. Some have argued that the family, long the nursery of humanity, has outlived its usefulness. To be sure, laboratory and governmental alternatives might be designed for procreation and child-rearing, but at what cost?

This is not the place to conduct a full evaluation of the biological family. Nevertheless, some of its important virtues are, nowadays, too often overlooked. The family is rapidly becoming the only institution in an increasingly impersonal world where each person is loved not for what he does or makes, but simply because he is. The family is also the institution where most of us, both as children and as parents, acquire a sense of continuity with the past and a sense of commitment to the future. Without the family, we would have little incentive to take an interest in anything after our own deaths. These observations suggest that the elimination of the family would weaken ties to past and future, and would throw us, even more than we are now, to the mercy of an impersonal, lonely present.

Neurobiology and psychobiology probe most directly into the distinctively human. The technological fruit of these sciences is likely to be both more tempting than Eve's apple and more "catastrophic" in its result (10). One need only consider contemporary drug use to see what people are willing to risk or sacrifice for novel experiences, heightened perceptions, or just "kicks." The possibility of drug induced, instant, and effortless gratification will be welcomed. Recall the possibilities of voluntary self-stimulation of the brain to reduce anxiety, to heighten pleasure, or to create visual and auditory sensations unavailable through the peripheral sense organs. Once these techniques are perfected and safe, is there much doubt that they will be desired, demanded, and used?

What ends will these techniques serve? Most likely, only the most elemental, those most tied to the bodily pleasures. What will happen to thought, to love, to friendship, to art, to judgment, to public-spiritedness in a society with a perfected technology of pleasure? What kinds of creatures will we become if we obtain our pleasure by drug or electrical stimulation without the usual kind of human efforts and frustrations? What kind of society will we have?

We need only consult Aldous Huxley's prophetic novel *Brave New World* for a likely answer to these questions. There we encounter a society dedicated to homogeneity and stability, administered by means of instant gratifications and peopled by creatures of human shape but of stunted humanity. They consume, fornicate, take "soma," and operate the machinery that makes it all possible. They do not read, write, think, love, or govern themselves. Creativity and curiosity, reason and passion, exist only in a rudimentary and multilated form. In short, they are not men at all.

True, our techniques, like theirs, may in fact enable us to treat schizophrenia, to alleviate anxiety, to curb aggressiveness. We, like they, may indeed be able to save mankind from itself, but probably only at the cost of its humanness. In the end, the price of relieving man's estate might well be the abolition of man (*11*).

There are, of course, many other routes leading to the abolition of man. There are many other and better known causes of dehumanization. Disease, starvation, mental retardation, slavery, and brutality—to name just a few—have long prevented many, if not most, people from living a fully human life. We should work to reduce and eventually to eliminate these evils. But the existence of these evils should not prevent us from appreciating that the use of the technology of man, uninformed by wisdom concerning proper human ends, and untempered by an appropriate humility and awe, can unwittingly render us all irreversibly less than human. For, unlike the man reduced by disease or slavery, the people dehumanized à la *Brave New World* are not miserable, do not know that they are dehumanized, and, what is worse, would not care if they knew. They are, indeed, happy slaves, with a slavish happiness.

Some Fundamental Questions

The practical problems of distributing scarce resources, of curbing the abuses of power, and of preventing voluntary dehumanization point beyond themselves to some large, enduring, and most difficult questions: the nature of justice and the good community, the nature of man and the good for man. My appreciation of the profundity of these questions and my own ignorance before them makes me hesitant to say any more about them. Nevertheless, previous failures to find a shortcut around them have led me to believe that these questions must be faced if we are to have any hope of understanding where biology is taking us. Therefore, I shall try to show in outline how I think some of the larger questions arise from my discussion of dehumanization and self-degradation.

My remarks on dehumanization can hardly fail to arouse argument. It might be said, correctly, that to speak about dehumanization presupposes a concept of "the distinctively human." It might also be said, correctly, that to speak about wisdom concerning proper human ends presupposes that such ends do in fact exist and that they may be more or less accessible to human understanding, or at least to rational inquiry. It is true that neither presupposition is at home in modern thought.

The notion of the "distinctively human" has been seriously challenged by modern scientists. Darwinists hold that man is, at least in origin, tied to the subhuman; his seeming distinctiveness is an illusion or, at most, not very important. Biochemists and molecular biologists extend the challenge by blurring the distinction between the living and the nonliving. The laws of physics and chemistry are found to be valid and are held to be sufficient for explaining biological systems. Man is a collection of molecules, an accident on the stage of evolution, endowed by chance with the power to change himself, but only along determined lines.

Psychoanalysts have also debunked the "distinctly human." The essence of man is seen to be located in those drives he shares with other animals—pursuit of pleasure and avoidance of pain. The so-called "higher functions" are understood to be servants of the more elementary, the more base. Any distinctiveness or "dignity" that man has consists of his superior capacity for gratifying his animal needs.

The idea of "human good" fares no better. In the social sciences, historicists and existentialists have helped drive this question underground. The former hold all notions of human good to be culturally and historically bound, and hence mutable. The latter hold that values are subjective: each man makes his own, and ethics becomes simply the cataloging of personal tastes.

Such appear to be the prevailing opinions. Yet there is nothing novel about reductionism, hedonism, and relativism; these are doctrines with which Socrates contended. What is new is that these doctrines seem to be vindicated by scientific advance. Not only do the scientific notions of nature and of man flower into verifiable predictions, but they yield marvelous fruit. The technological triumphs are held to validate their scientific foundations. Here, perhaps, is the most pernicious result of technological progress—more dehumanizing than any actual manipulation or technique, present or future. We are witnessing the erosion, perhaps the final erosion, of the idea of man as something splendid or divine, and its replacement with a view that sees man, no less than nature, as simple more raw material for manipulation and homogenization. Hence, our peculiar moral crisis. We are in turbulent seas without a landmark precisely because we adhere more and more to a view of nature and of man which both gives us enormous power and, at the same time, denies all possibility of standards to guide its use. Though well-equipped, we know not who we are nor where we are going. We are left to the accidents of our hasty, biased, and ephemeral judgments.

Let us not fail to note a painful irony: our conquest of nature has made us the slaves of blind chance. We triumph over nature's unpredictabilities only to subject ourselves to the still greater unpredictability of our capricious wills and our fickle opinions. That we have a method is no proof against our madness. Thus, engineering the engineer as well as the engine, we race our train we know not where (*12*).

While the disastrous consequences of ethical nihilism are insufficient to refute it, they invite and make urgent a reinvestigation of the ancient and enduring questions of what is a proper life for a human being, what is a good community, and how are they achieved (*13*). We must not be deterred from these questions simply because the best minds in human history have failed to settle them. Should we not rather be encouraged by the fact that they considered them to be the most important questions?

As I have hinted before, our ethical dilemma is caused by the victory of modern natural science with its nonteleological view of man. We ought therefore to reexamine with great care the modern notions of nature and of man, which undermine those earlier notions that provide a basis for ethics. If we consult our common experience, we are likely to discover some grounds for believing that the questions about man and human good are far from closed. Our common experience suggests many difficulties for the modern "scientific view of man." For example, this view fails to account for the concern for justice and freedom that appears to be characteristic of all human societies (*14*). It also fails to account for or to explain the fact that men have speech and not merely voice, that men can choose and act and not merely move or react. It fails to explain why men engage in moral discourse, or, for that matter, why they speak at all. Finally, the "scientific view of man" cannot account for scientific inquiry itself, for why men seek to know. Might there not be something the matter with a knowledge of man that does not explain or take account of his most distinctive activities, aspirations, and concerns (*15*)?

Having gone this far, let me offer one suggestion as to where the difficulty might lie: in the modern understanding of knowledge. Since Bacon, as I have mentioned earlier, technology has increasingly come to the basic justification for scientific inquiry. The end is power, not knowledge for its own sake. But power is not only the end. It is also an important *validation* of knowledge. One definitely knows that one knows only if one can make. Synthesis is held to be the ultimate proof of understanding (*16*). A more radical formulation holds that one knows only what one makes: knowing *equals* making.

Yet therein lies a difficulty. If truth be the power to change or to make the object studied, then of what do we have knowledge? If there are no fixed realities, but only material upon which we may work our wills, will not "science" be merely the "knowledge" of the transient and the manipulatable? We might indeed have knowledge of the laws by which things change and the rules for their manipulation, but no knowledge of the things themselves. Can such a view of "science" yield any knowledge about the nature of man, or indeed, about the nature of anything? Our questions appear to lead back to the most basic of questions: What does it mean to know? What is it that is knowable (*17*)?

We have seen that the practical problems point toward and make urgent certain enduring, fundamental questions. Yet while pursuing these questions, we cannot afford to neglect the practical problems as such. Let us not forget Delgado and the "psychocivilized society." The philosophical inquiry could be rendered moot by our blind, confident efforts to dissect and redesign ourselves. While awaiting a reconstruction of theory, we must act as best we can.

What Is To Be Done?

First, we sorely need to recover some humility in the face of our awesome powers. The arguments I have presented should make apparent the folly of arrogance, of the presumption that we are wise enough to remake ourselves. Because we lack wisdom, caution is our urgent need. Or to put it another way, in the absence of that "ultimate wisdom," we can be wise enough to know that we are not wise enough. When we lack sufficient wisdom to do, wisdom consists in not doing. Caution, restraint, delay, abstention are what this second-best (and, perhaps, only) wisdom dictates with respect to the technology for human engineering.

If we can recognize that biomedical advances carry significant social costs, we may be willing to adopt a less permissive, more critical stance toward new developments. We need to reexamine our prejudice not only that all biomedical innovation is progress, but also that it is inevitable. Precedent certainly favors the view that what can be done will be done, but is this necessarily so? Ought we not to be suspicious when technologists speak of coming developments as automatic, not subject to human control? Is there not something contradictory in the notion that we have the power to control all the untoward consequences of a technology, but lack the power to determine whether it should be developed in the first place?

What will be the likely consequences of the perpetuation of our permissive and fatalistic attitude toward human engineering? How will the large decisions be made? Technocratically and self-servingly, if our experience with previous technologies is any guide. Under conditions of laissez-faire, most technologists will pursue techniques, and most private industries will pursue profits. We are fortunate that, apart from the drug manufacturers, there are at present in the biomedical area few large industries that influence public policy. Once these appear, the voice of "the public interest" will have to shout very loudly to be heard above their whisperings in the halls of Congress. These reflections point to the need for institutional controls.

Scientists understandably balk at the notion of the regulation of science and technology. Censorship is ugly and often based upon ignorant fear; bureaucratic regulation is often stupid and inefficient. Yet there is something disingenuous about a scientist who professes concern about the social consequences of science, but who responds to every suggestion of regulation with one or both of the following: "No restrictions on scientific research," and "Technological progress should not be curtailed." Surely, to suggest that *certain* technologies ought to be regulated or forestalled is not to call for the halt of *all* technological progress (and says nothing at all about basic research). Each development should be considered on its own merits.

Although the dangers of regulation cannot be dismissed, who, for example, would still object to efforts to obtain an effective, complete, global prohibition on the development, testing, and use of biological and nuclear weapons?

The proponents of laissez-faire ignore two fundamental points. They ignore the fact that not to regulate is as much a policy decision as the opposite, and that it merely postpones the time of regulation. Controls will eventually be called for—as they are now being demanded to end environmental pollution. If attempts are not made early to detect and diminish the social costs of biomedical advances by intelligent institutional regulation, the society is likely to react later with more sweeping, immoderate, and throttling controls.

The proponents of laissez-faire also ignore the fact that much of technology is already regulated. The federal government is already deep in research and development (for example, space, electronics, and weapons) and is the principal sponsor of biomedical research. One may well question the wisdom of the direction given, but one would be wrong in arguing that technology cannot survive social control. Clearly, the question is not control versus no control, but rather what kind of control, when, by whom, and for what purpose.

Means for achieving international regulation and control need to be devised. Biomedical technology can be no nation's monopoly. The need for international agreements and supervision can readily be understood if we consider the likely American response to the successful asexual reproduction of 10,000 Mao Tse-tungs.

To repeat, the basic short-term need is caution. Practically, this means that we should shift the burden of proof to the *proponents* of a new biomedical technology. Concepts of "risk" and "cost" need to be broadened to include some of the social and ethical consequences discussed earlier. The probable or possible harmful effects of the widespread use of a new technique should be anticipated and introduced as "costs" to be weighed in deciding about the *first* use. The regulatory institutions should be encouraged to exercise restraint and to formulate the grounds for saying "no." We must all get used to the idea that biomedical technology makes possible many things we should never do.

But caution is not enough. Nor are clever institutional arrangements. Institutions can be little better than the people who make them work. However worthy our intentions, we are deficient in understanding. In the *long* run, our hope can only lie in education: in a public educated about the meanings and limits of science and enlightened in its use of technology; in scientists better educated to understand the relationships between science and technology on the one hand, and ethics and politics on the other; in human beings who are as wise in the latter as they are clever in the former.

REFERENCES AND NOTES

1. R.G. Edwards, B.D. Bavister, P.C. Steptoe, *Nature 221*,632 (1969).
2. R.G. Edwards, P.C. Steptoe, J.M. Purdy, *ibid. 227*, 1307 (1970).
3. H.J. Muller, *Science 134*, 643 (1961).
4. J.M.R. Delgado, *Physical Control of the Mind: Toward a Psychocivilized Society* (Harper & Row, New York, 1969).
5. F. Bacon, *The Advancement of Learning, Book I*, H.G. Dick, Ed. (Random House, New York, 1955), p. 193.
6. C.S. Lewis, *The Abolition of Man* (Macmillan, New York, 1965), pp. 69-71.
7. D.D. Rutstein, *Daedalus* (Spring 1969), p. 523.
8. E.J. Cassell, *Commentary* (June 1969), p. 73.
9. B. Glass, *Science 171*, 23 (1971).
10. It is, of course, a long-debated question as to whether the fall of Adam and Eve ought to be considered "catastrophic," or more precisely, whether the Hebrew tradition considered it so. I do not mean here to be taking sides in this quarrel by my use of the term "catastrophic," and, in fact, tend to line up on the negative side of the questions, as put above. Curiously, as Aldous Huxley's *Brave New World* [(Harper & Row, New York, 1969)] suggests, the implicit goal of the biomedical technology could well be said to be the reversal of the Fall and a return of man to the hedonic and immortal existence of the Garden of Eden. Yet I can point to at least two problems. First, the new Garden of Eden will probably have no gardens; the received, splendid world of nature will be buried beneath asphalt, concrete, and other human fabrications, a transformation that is already far along. (Recall that in *Brave New World* elaborate consumption-oriented, mechanical amusement parks—featuring, for example, centrifugal bumble-puppy—had supplanted wilderness and even ordinary gardens.) Second, the new inhabitant of the new "Garden" will have to be a creature for whom we have no precedent, a creature as difficult to imagine as to bring into existence. He will have to be simultaneously an innocent like Adam and a technological wizard who keeps the "Garden" running. (I am indebted to Dean Robert Goldwin, St. John's College, for this last insight.)
11. Some scientists naively believe that an engineered increase in human intelligence will steer us in the right direction. Surely we have learned by now that intelligence, whatever it is and however measured, is not synonymous with wisdom and that, if harnessed to the wrong ends, it can cleverly perpetrate great folly and evil. Given the activities in which many, if not most, of our best minds are now engaged, we should not simply rejoice in the prospect of enhancing IQ. On what would this increased intelligence operate? At best, the programming of further increases in IQ. It would design and operate techniques for prolonging life, for engineering reproduction, for delivering gratifications. With no gain in wisdom, our gain in intelligence can only enhance the rate of our dehumanization.
12. The philosopher Hans Jonas has made the identical point: "Thus the slow-working accidents of nature, which by the very patience of their small increments, large numbers, and gradual decisions, may well cease to be 'accident' in

outcome, are to be replaced by the fast-working accidents of man's hasty and biased decisions, not exposed to the long test of the ages. His uncertain ideas are to set the goals of generations, with a certainty borrowed from the presumptive certainty of the means. The latter presumption is doubtful enough, but this doubtfulness becomes secondary to the prime question that arises when man indeed undertakes to 'make himself': in what image of his own devising shall he do so, even granted that he can be sure of the means? In fact, of course, he can be sure of neither, not of the end, nor of the means, once he enters the realm where he plays with the roots of life. Of one thing only can he be sure: of his power to move the foundations and to cause incalculable and irreversible consequences. Never was so much power coupled with so little guidance for its use." [*J. Cent. Conf. Amer. Rabbis* (January 1968), p. 27.] These remarks demonstrate that, contrary to popular belief, we are not even on the right road toward a rational understanding of and rational control over human nature and human life. It is indeed the height of irrationality triumphantly to pursue rationalized technique, while at the same time insisting that questions of ends, values, and purposes lie beyond rational discourse.

13. It is encouraging to note that these questions are seriously being raised in other quarters—for example, by persons concerned with the decay of cities or the pollution of nature. There is a growing dissatisfaction with ethical nihilism. In fact, its tenets are unwittingly abandoned, by even its staunchest adherents, in any discussion of "what to do." For example, in the biomedical area, everyone, including the most unreconstructed and technocratic reductionist, finds himself speaking about the use of powers for "human betterment." He was wandered unawares onto ethical ground. One cannot speak of "human betterment" without considering what is meant by *the human* and by the related notion of *the good for man*. These questions can be avoided only by asserting that practical matters reduce to tastes and power, and by confessing that the use of the phrase "human betterment" is a deception to cloak one's own will to power. In other words, these questions can be avoided only by ceasing to discuss.

14. Consider, for example, the widespread acceptance, in the legal systems of very different societies and cultures, of the principle and the practice of third-party adjudication of disputes. And consider why, although many societies have practiced slavery, no slave-holder has preferred his own enslavement to his own freedom. It would seem that some notions of justice and freedom, as well as right and truthfulness, are constitutive for any society, and that a concern for these values may be a fundamental characteristic of "human nature."

15. Scientists may, of course, continue to believe in righteousness or justice or truth, but these beliefs are not grounded in their "scientific knowledge" of man. They rest instead upon the receding wisdom of an earlier age.

16. This belief, silently shared by many contemporary biologists, has recently been given the following clear expression: "One of the acid tests of understanding an object is the ability to put it together from its component parts. Ultimately, molecular biologists will attempt to subject their understanding of all

structure and function to this sort of test by trying to synthesize a cell. It is of some interest to see how close we are to this goal." [P. Handler, Ed, *Biology and the Future of Man* (Oxford Univ. Press, New York, 1970), p. 55.]

17. When an earlier version of this article was presented publicly, it was criticized by one questioner as being "antiscientific." He suggested that my remarks "were the kind that gave science a bad name." He went on to argue that, far from being the enemy of morality, the pursuit of truth was itself a highly moral activity, perhaps the highest. The relation of science and morals is a long and difficult question with an illustrious history, and it deserves a more extensive discussion than space permits. However, because some readers may share the questioner's response, I offer a brief reply. First, on the matter of reputation, we should recall that the pursuit of truth may be in tension with keeping a good name (witness Oedipus, Socrates, Galileo, Spinoza, Solzhenitsyn). For most of human history, the pursuit of truth (including "science") was not a reputable activity among the many, and was, in fact, highly suspect. Even today, it is doubtful whether more than a few appreciate knowledge as an end in itself. Science has acquired a "good name" in recent times largely because of its technological fruit; it is therefore to be expected that a disenchantment with technology will reflect badly upon science. Second, my own attack has not been directed against science, but against the use of *some* technologies and, even more, against the unexamined belief—indeed, I would say, superstition—that all biomedical technology is an unmixed blessing. I share the questioner's belief that the pursuit of truth is a highly moral activity. In fact, I am inviting him and others to join in a pursuit of the truth about whether all these new technologies are really good for us. This is a question that merits and is susceptible of serious intellectual inquiry. Finally, we must ask whether what we call "science" has a monopoly on the pursuit of truth. What is "truth"? What is knowable, and what does it mean to know? Surely, these are also questions that can be examined. Unless we do so, we shall remain ignorant about what "science" is, and about what it discovers. Yet "science"—that is, modern natural science—cannot begin to answer them; they are philosophical questions, the very ones I am trying to raise at this point in the text.